Reinforced Polymer Composites

Reinforced Polymer Composites

Processing, Characterization and Post Life Cycle Assessment

Edited by

Pramendra K. Bajpai
Inderdeep Singh

Editors

Prof. Pramendra K. Bajpai
Netaji Subhas University of Technology
Division of Manufacturing Processes and
Automation Engineering
Azad Hind Fauz Marg
Sector 3
Dwarka
110078 New Delhi
India

Prof. Inderdeep Singh
Indian Institute of Technology Roorkee
Department of Mechanical and
Industrial Engineering
Haridwar Highway
Roorkee
247667 Uttarakhand
India

■ All books published by **Wiley-VCH**
are carefully produced. Nevertheless,
authors, editors, and publisher do not
warrant the information contained in
these books, including this book, to
be free of errors. Readers are advised
to keep in mind that statements, data,
illustrations, procedural details or other
items may inadvertently be inaccurate.

Library of Congress Card No.:
applied for

British Library Cataloguing-in-Publication Data
A catalogue record for this book is
available from the British Library.

**Bibliographic information published by
the Deutsche Nationalbibliothek** The
Deutsche Nationalbibliothek lists this
publication in the Deutsche
Nationalbibliografie; detailed
bibliographic data are available on the
Internet at <http://dnb.d-nb.de>.

© 2020 Wiley-VCH Verlag GmbH &
Co. KGaA, Boschstr. 12, 69469
Weinheim, Germany

Print ISBN: 978-3-527-34599-1
ePDF ISBN: 978-3-527-82096-2
ePub ISBN: 978-3-527-82099-3
oBook ISBN: 978-3-527-82097-9

Cover Design   SCHULZ Grafik-Design,
Fußgönheim, Germany
Typesetting   SPi Global, Chennai, India
Printing and Binding   Markono Print Media
Pte Ltd, Singapore

Printed on acid-free paper

10 9 8 7 6 5 4 3 2 1

Contents

1

Overview and Present Status of Reinforced Polymer Composites

Furkan Ahmad[1], Inderdeep Singh[2], and Pramendra K. Bajpai[1]

[1] Netaji Subhas University of Technology, MPAE Division, Sector-3, Dwarka, New Delhi, 110078, India
[2] Indian Institute of Technology Roorkee, Department of Mechanical and Industrial Engineering, Roorkee, Uttarakhand, 247667, India

1.1 Introduction

Humans have been using a number of materials to improve their living standards since ages. In fact, the progress of human civilization has been classified into three categories, popularly known as the Stone Age, the Bronze Age, and the Iron Age, on the basis of materials only. Looking at the current rate of demand and consumption of plastics, it would not be wrong if somebody categorizes the present age as "The Age of Plastics" or "Plastic Age". New materials form the foundation for new technologies and help in understanding nature. The most complex designs in the world can be of no use if suitable material is not used during the fabrication of products with that design. In actual realization of a design, the role of materials is quite indispensable. The limited availability of natural resources has forced material engineers to use materials in a more conscious manner. Therefore, material scientists and engineers are trying to optimize the use of materials in every possible field of application. In the present age, transportation industry is the biggest contributor of carbon footprints in the environment. Lower fuel consumption of automotive vehicles can lower the carbon footprints. In the quest for achieving low fuel consumption, transportation industry is leaning toward materials having high strength to weight ratio. Reinforced polymer composites (RPCs), also known as fiber reinforced polymer composites (FRPCs) are such promising materials for almost every industry looking for low weight and high strength materials [1]. The application spectrum of FRPCs has spread in almost every sector starting from engineered domestic products to the highly sensitive biomedical industry. FRPCs are not only just a replacement for conventional alloys but they also provide engineered properties. Rahmani et al. [2] fabricated the carbon/epoxy-based FRPCs with 40 wt% of fiber. This system of FRPCs was able to achieve a tensile strength of 2500 MPa, which is quite close to the tensile strength of steel. The authors concluded that fiber orientation was the most influencing factor among other factors, namely number of laminates and resin type. The authors suggested the use of $\pm 35°$ angle of plies to obtain better tensile properties along with good flexural properties. Modification in the matrix material can enhance the overall properties of FRPCs. Islam et al. [3] modified the epoxy matrix by incorporating nanoclay and multiwalled carbon nanotubes (MWCNTs). The authors found significant improvement in the static and dynamic mechanical properties of the developed carbon fiber-based FRPCs. Cho et al. [4] enhanced the in-plane shear strength and shear modulus of carbon fiber reinforced epoxy composites by incorporating graphite nanoplatelets in the epoxy matrix using the sonication method. Increasing the volume fraction of reinforcement can increase the mechanical properties of the developed FRPCs. Aramide et al. [5] fabricated glass fiber/epoxy-based FRPCs with varying volume fraction of fibers from 5% to 30%. The authors found that the mechanical strength increased as the fiber volume fraction increased up to 30%. Treinyte et al. [6] fabricated poly (vinyl alcohol)-based pots. Forestry and wood processing waste was used as filler in the matrix. The authors claimed that the manufactured pots showed 45% lower water evaporation rate in comparison to regular peat pots.

Architecture of the reinforcement also affects the mechanical performance of developed FRPCs products. A range of reinforcement architectures is available in the market such as short fiber, unidirectional prepregs, 2D and 3D woven mats, braided mats, and knitted mats. Every architecture has its own merits and demerits – 2D woven mats show better in-plane mechanical properties but they lack in out-of plane properties while 3D woven mats offer better out-of-plane properties in comparison to others [7]. Erol et al. [8] investigated the effect of yarn material and weaving pattern on the macroscopic properties of FRPCs and concluded that weave pattern greatly influenced the tensile and shear properties of the developed composites. Some authors [9] have even used 3D and 5D braided reinforcement for the development of FRPCs. The authors concluded that braided architecture affects the fracture mechanism in a significant way. Kostar et al. [10] used two-sided co-braided carbon and Kevlar hybrid reinforcement for the development of FRPCs and concluded that the tensile strength and modulus of hybrid reinforcement-based FRPCs were 13% and 80% higher than those with simple reinforcement. FRPs have evolved over a long time period as shown in Figure 1.1.

Environmental problems and difficulty in the recycling associated with synthetic composites have led to the development of biocomposites/green composites. Biocomposites are eco-friendly materials with adequate mechanical properties. Fombuena et al. [11] fabricated biocomposites using bio-fillers derived from sea-shell waste as reinforcement in bio-based epoxy matrix. The authors found impressive improvement in mechanical properties of bio-based epoxy when reinforced with bio-fillers. End of life (EOL) impact of synthetic fibers and polymers is negative to the environment. Duflou et al. [12] showed that low mechanical strength of flax fiber is an obstruction in the replacement of glass fiber but it can be used in many applications where high mechanical strength is not the primary requirement. Effect of moisture on the mechanical performance of natural fiber-based biocomposites is yet another concern while using biocomposites. Baghaei et al. [13] developed poly lactic acid (PLA)-based biocomposites and analyzed the moisture absorption behavior. The authors found that the moisture absorption characteristic of the developed composite

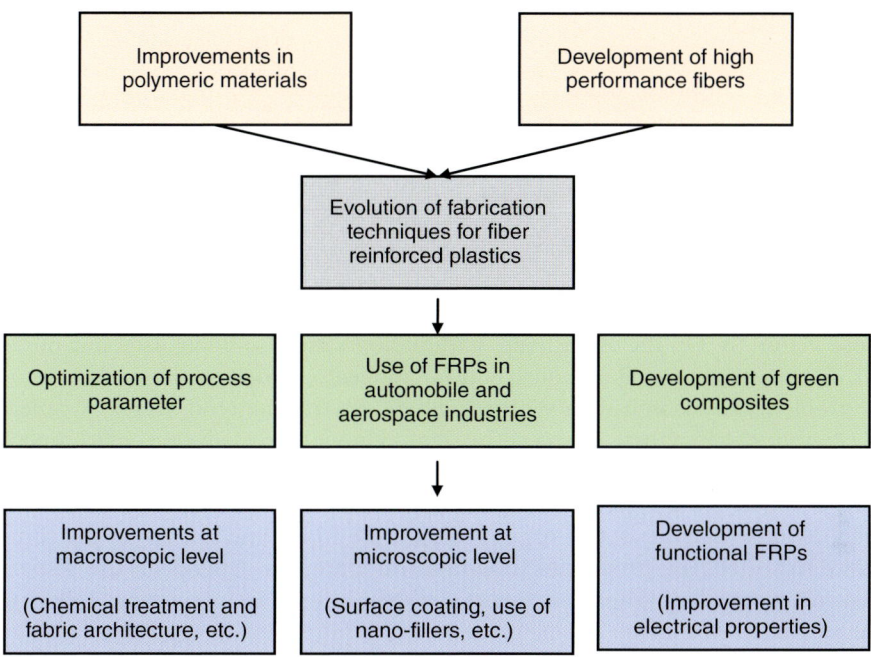

Figure 1.1 Development stages of FRPs.

was reduced when the reinforcement was used in the woven form instead of the nonwoven form.

Hybridization can improve the mechanical strength of green composites. Hassanin et al. [14] developed a biocomposite particle board using a mixture of wood particles and short glass fibers covered with an outer layer of jute fabric. The particle board showed excellent mechanical and physical properties in comparison to commercially available particle boards. Chaudhary et al. [15] hybridized the reinforcement and found improved mechanical and thermal properties of the developed biocomposites. The authors used three types of woven fibers mats, namely jute, hemp, and flax, as reinforcement in epoxy matrix.

Chemical treatment of fibers/surface modification of fibers is also a promising method for improvement in the mechanical properties of FRPCs. Alkali, acryl, benzyl, and silane solutions are commonly used for the treatment of fibers [16]. Asaithambi et al. [17] treated banana fibers before using them as reinforcement in PLA-based FRPCs. Banana fibers were first pretreated with 5% NaOH solution at room temperature for around two hours, and then the chemical treatment of the fiber was completed using benzoyl peroxide. Significant improvement in the mechanical properties developed with treated FRPCs was found in comparison to those developed with untreated FRPCs. Rahman and Khan [18] used ethylene dimethyl acrylate (EMA) for the surface modification of coir fibers along with UV treatment for the aging of fibers. The authors concluded that the mechanical properties of FRPCs developed using treated fiber were better than those of untreated fiber reinforced FRPCs.

1.2 FRPCs

FRPCs are multiphase materials comprised of natural/synthetic fiber as reinforcement and thermoset/thermoplastic polymer as matrix, resulting in synergistic properties that cannot be achieved from a single component alone. In general, reinforcement is in the form of long continuous fibers but they can be used in various other forms such as short fibers, fillers, or whiskers. The fibrous form of reinforcement is used in composite materials because they are stronger and stiffer than any other form [19]. Synthetic fibers (carbon, glass, aramid, etc.) can provide more strength than most of the metals along with being lighter than those materials. On the other hand, natural fibers are also being used in a number of structural as well as nonstructural applications due to the environmental problems associated with synthetic fibers. Matrix material, which is generally continuous in nature, protects the reinforcement from adverse environment and transfers the load to reinforcement from the point of application of load [12]. The matrix material holds the flexible reinforcements together to make it a solid. Matrix material is also responsible for the finish and texture of the composite material. The properties of composite materials depend on the dispersion and properties of the constituents and their interfacial interaction. Tailoring the properties of a material according to the requirement of application can be easily done in composite materials [20]. Table 1.1 shows the commonly used natural and synthetic polymers and fibers used as matrix and reinforcement, respectively.

1.2.1 Fabrication of Fiber Reinforced Composites

Fabrication methods of FRPCs still require a lot of attention in order to produce defect-free high quality products. Some unique features of primary and secondary processing of FRPCs are tabulated in Table 1.2.

Table 1.1 Matrix and reinforcement materials used in reinforced polymer composites.

Matrix	Natural	Synthetic
	Polysaccharides such as homoglycans, cellulose, chitin, chitosan, heteroglycans, such as alginate, agar, and agarose, carrageenan, pectins, gums, and proteoglycans, protein, peptides, and enzymes	Polyolefins, poly(tetrafluoroethylene) (PTFE), poly(vinylchloride)(PVC), silicone, methacrylates, aliphatic polyesters, polyethers, poly(amino acids), polyamides, polyurethanes, epoxy, polycarbonates
Reinforcement	Natural	Synthetic
	Animal-based – silk, wool, hair; Plant-based – bast fibers (jute, flax, ramie, hemp, kenaf, roselle, etc.), leaf fibers (sisal, banana, agava, etc.), seed, fruit, wood, and stalk fibers	Carbon, glass, Aramid/Kevlar, graphite, aromatic polyester fibers, boron, silica carbide

1.2.2 Present Status of FRPCs

RPC products such as pipes are being used in various adverse conditions such as in offshore and marine applications. These pipes are exposed to severe climatic conditions ranging from −40 to 80 °C [21]. Benyahia et al. [21] tested the mechanical properties of a filament wound glass/epoxy pipe of 86 mm diameter and 6.2 mm thickness. The authors estimated that there was degradation of mechanical properties at higher temperatures. Ellyin and Maser [22] investigated the effect of moisture at elevated temperature on the mechanical properties of glass fiber reinforced polymer (GFRP) composite tubes. At lower temperature, the ductility of the specimen was found to be decreased drastically and the stiffness was increased. Above the glass transition temperature, there was sudden degradation in the mechanical properties of composite pipes. In recent progress, shape memory alloy (SMA) wires are being incorporated into the FRPCs as reinforcement to increase the functionality of the developed composites such as shape recovery, high damping capacity, generation of high recovery stresses, and controlled overall thermal expansion. SMA wires not only improve the functionality of the FRPCs but also offer improved mechanical properties [23]. Paine and Rogers [24] concluded that the low velocity impact properties of FRPCs can be improved by incorporating SMA wires. Incorporation of just 2.8% volume fraction of SMA wires as reinforcement was able to increase the impact delamination resistance by 25% in comparison to the FRPCs without the SMA wire reinforcement. Pappada et al. [25, 26] fabricated hybrid glass fiber reinforced vinyl ester-based FRPC material and incorporated SMA wires in two forms, namely unidirectional SMA wires and knitted SMA wires. The authors assessed impact properties and found that FRPCs reinforced with SMA wires achieved higher impact properties than FRPCs with unidirectional SMA wires.

Polymer nanocomposites are also a relatively new class of materials. Nanocomposites are generally fabricated by incorporating one or more constituents of the size of the order of nanometers. These constituents are generally inorganic in nature and known as fillers, and not as reinforcement, due to their small size. Various researchers have reported impressive properties of nanocomposites such as high modulus and strength, high resistance to heat, and reduced flammability. However, effective dispersion of the nano-sized fillers throughout the polymer matrix is still a challenge, and moreover this dispersion controls and determines the physical, chemical, and mechanical properties of the developed FRPC products [27–29]. The authors have used an in situ approach to homogenize the dispersion of nano-sized fillers. In this approach, nano-fillers are directly synthesized with the polymer using some suitable precursor [30, 31]. Although the in situ approach provides controlled dispersion of nano-fillers, it involves complex procedures and processing steps along with expensive reactants [32, 33]. Various researchers used the ball milling method to fabricate nanocomposites. In this method, first both the constituents, polymer and nano-fillers, are mixed with each other in solid state using ball mills and then the mixture is melted to polymerize. Although the morphology of the fillers changes in the ball mill, this change positively affects the composites by enriching the filler compatibility with the polymer. The ball milling method is not just an alternative to ex situ

Table 1.2 Primary and secondary processing methods for FRPCs.

Processing	Fabrication technique	Features
Primary processing methods	Hand lay-up	Minimum infrastructural requirement; low initial capital requirement; only for thermosetting resins; lower production rate; and low volume fraction of the reinforcement
	Spray lay-up	Extension of hand lay-up technique; reinforcement in the form of chopped fibers only
	Compression molding	Use of heat and pressure both simultaneously; dimensionally accurate and finished products; process parameters need to be optimized; both thermosetting and thermoplastic polymers can be used; higher initial capital requirement compared to hand lay-up
	Injection molding	Reinforcement only in the form of short fibers; damage of fibers in barrel due to shearing action of screw. Highly accurate dimensions of the product; used for mass production
	Pultrusion process	Resin impregnated continuous fibers are passed through a heating die for curing; automated process used for continuous production; only products with constant cross-sectional area depending on the die can be manufactured
	Resin transfer molding	Liquid resin system is forced into the mold; high fiber volume fraction can be achieved. Good surface finish with minimum material wastage
	Filament winding	Continuous fiber strands as reinforcement; controlled fiber orientation; high production rate; high capital investment; not possible to produce female features of products and expensive mandrel
	Vacuum assisted resin transfer molding	Uses vacuum to ensure zero voids; superior quality composites using autoclave (a strong heating container that is used for applying heat and pressure at the time of curing of the composite laminates)
Secondary processing methods	Conventional machining	Drilling with twist drill is the most used conventional method to produce holes in laminates. Requires milling machine or drilling machine. Spindle speed, feed rate, and drill geometry are influential parameters. Delamination, fiber linting, and fiber pull-out are the most common defects
	Unconventional machining	Abrasive water jet (AWJ) reduces the thermal damage that could be generated in conventional machining.
		Laser beam (LB) cutting is also being used for holes generation in composite laminates. High energy input is required.
		Ultrasonic machining (USM) can also be used for hole making in the composite laminates

fabrication of FRPCs but is also an environment friendly and economical method to produce nano-filler reinforced FRPCs [34]. Some authors [35] have also used reinforcing metallic powders such as copper powder of 29.5 and 260 μm size in the polyvinyl butyral (PVB) polymer matrix to fabricate polymer composites. Fan and Wang [36] developed a transparent protective polymer composite material with lightweight property, which could be used against high speed impact loading.

The behavior and performance of FRPCs changes from application to application. FRPCs exposed to various tribological environments lead to the necessity to evaluate the tribological performance. Tribology of FRPCs is quite complex than metal tribology due to the fact that polymers do not obey the well-established laws of friction at high temperature [37]. Xue and Wang [38] studied the effect of filler particle size on the wear and frictional properties of polymer composites. The authors concluded that addition of nano-sized SiC particles into the polymer matrix effectively reduced the friction and wear of the neat polymer. The nano-sized particles form a continuous and thin layer between the interface, which results in reduction of friction and wear. Xing and Li [39] also confirmed a similar behavior of FRPCs with the incorporation of nano-sized fillers. Gears, bearings, shoe soles, and brake pads for automobile applications are some of the mostly used tribological applications of FRPCs [40–42]. Researchers have suggested a number of methods to reduce the friction and wear at the interface between the FRPC product and the metal/nonmetal surface. Microencapsulation of liquid lubricant was found to be an effective method to improve the tribological properties of polymers [43]. Guo et al. [44, 45] demonstrated that the friction coefficient of epoxy-based FRPCs can be reduced up to 75% by incorporating just 10 wt% oil-loaded microcapsules. The authors have claimed to develop self-lubricating polymer-based materials with the help of encapsulation method. Khun et al. [46] and Imani et al. [47] added wax-loaded microcapsules in epoxy matrix composites and found that friction and wear were very much reduced in comparison to that in the neat epoxy polymer composite. In another study, Khun et al. [48] used the two types of microcapsules in the polymer composite. One type of microcapsules were loaded with wax and another type of capsules were loaded with MWCNTs. The authors concluded that tribological and mechanical properties were enhanced simultaneously. Wax-loaded capsules were found to be responsible for improved tribological properties while MWCNTs loaded capsules result in improved mechanical properties, which was achievable with only wax-loaded capsules. Encapsulation may help in the development of self-healing materials as explained by some authors [49].

Self-reinforced composites (SRCs) are yet another category of FRPCs in which only a single polymer is used. Hard/processed form of the same polymer is used as reinforcement that is being used as matrix material [50]. Huang [51] developed a polypropylene (PP)-based SRC using melt-flow induced crystallization. Li and Yao [52] and Makela et al. [53] developed PLA polymer-based fibers that could be used as reinforcement in the SRCs. Similarly, Tormala [54] developed PLA-based SRCs for medical applications. In the same series, Hine and Ward [55] developed PET-based SRCs, Gilbert et al. [56] developed polymethyl methacrylate (PMMA)-based SRCs, and Gindl and Keckes [57] manufactured cellulose-based SRCs.

Gemi [58] developed glass and carbon-based hybrid composite pipes and studied the effect of stacking sequence. The authors concluded that glass–carbon–glass sequence of reinforcement during the winding of fibers leads to no leakage property of pipes.

The superior electrical, mechanical, and thermal properties of graphene make it very useful in the field of FRPCs [59]. Graphene, in the form of 3D foam and gel is being used in FRPCs products in biomedical and electronics applications [60, 61]. Various authors [62, 63] reported impressive improvement in the mechanical properties of epoxy composites with the incorporation of 3D foam. Sun et al. [64] reported that the incorporation of 3D graphene foam in polymer significantly improved electrical properties. Jusza et al. [65] developed luminescent composite materials for possible applications in opto-electronics, sensor networks, and imaging field. Complex technology and expensive manufacturing methods of optically active two-phase composite materials have made them commercially unavailable.

Carbon nanotubes, popularly known as CNTs, are filler/reinforcement that are being used in polymers to fabricate composites with improved physical, mechanical, and electrical properties [66–69]. Nanomaterials are those materials that have dimensions below 100 nm [70]. Several authors [71, 72] reported an increment of over 300% in the tensile strength of FRPCs reinforced with CNT-based nano-fillers. Incorporation of nanocarbons resulted in the increment of electrical properties up to over 14 orders of magnitude [73]. Carbon quantum dots (CQD), a form of nanocarbon material, is also being used as reinforcement due to their tunable optical and photochemical properties. Another emerging class of FRPCs is thermally conductive polymer composites and nanocomposites [74]. Studies [75, 76] have reported that polymers reinforced with aligned molecular chain can obtain higher thermal conductivity than that of many metals. Rajapakse et al. [77] prepared electronically conductive nanocomposites for potential application as a cathode material.

Another advancement in the field of FRPCs is the production of shape memory polymer composites along with self-healing properties [78]. FRPCs are being widely used in the field of electronics and biomedical and energy applications from the last decades. However, low thermal conductivity and insufficient thermal stability have restricted FRPCs usage to a limited number of applications [79].

Along with their advantages, there are some disadvantages associated with FRPCs as well. The disadvantage with FRPCs is the need for recycling and disposal methods after the finite life of the FRPC product. Li et al. [80] investigated the environmental and financial problems associated with the manufacturing of carbon FRPCs. The authors suggested the use of mechanical recycling of FRPCs instead of landfilling and incineration. Landfilling method for disposal of FRPCs was found to be modest with moderate landfilling tax. However, incineration method results in the production of greenhouse gases causing severe damage to environment. Longana et al. [81] suggested another method of recycling known as multiple closed loop recycling of carbon FRPCs. In this method, reclaimed carbon fibers (rCF) are again used to remanufacture a number of products once a virgin carbon fiber (vCF) product has completed its defined life.

1.3   FRPCs Applications and Future Prospects

The high strength to weight ratio of FRPCs makes them irreplaceable in a number of applications in the automobile and aerospace industries [82–86]. Dhruv, the advanced light helicopter (ALH) manufactured by Hindustan Aeronautics Limited for the service of the Indian army, has around 60% of structural area made up of FRPC components and sandwich structures [87]. A number of products are being successfully used in various automotive and other applications as reported in Table 1.3. A number of medical devices have been developed using biodegradable polymers alone. Drug-eluting stents, orthotropic devices, disposable medical devices, drug delivery devices, and stents for urological applications are some biomedical applications of polymers [101]. Tian et al. [101] stated that along with the nontoxic nature and low biodegradability of polymers, mechanical strength is also required in a number of medical applications. To strengthen these biodegradable polymers, fibers are being incorporated in the polymers according to the requirement of application. Carbon fiber reinforced epoxy composite materials are being used to fabricate external fixation equipment used for fractured bones. Bone plates are being used for the development of internal fixation equipment. The authors [102] have reported carbon fiber reinforced polyether ether ketone (PEEK)-based composite as a biocompatible material for bone plate. Lin et al. [103] proposed short glass fiber reinforced PEEK composite material for the fabrication of intramedullary nails, which are generally used to fix fractures of long bones. These nails are fixed in the intramedullary cavity using a screw mechanism. Kettunen et al. [104] used carbon fiber to fabricate composite material for these nails. Some authors [105, 106] have successfully used FRPCs as bone grafting materials. Carbon fiber-based FRPCs are intensively being used to fabricate stems for total hip replacement [107, 108]. Deng and shalaby [109] used ultrahigh molecular weight polyethylene (UHMWPE) to fabricate self-reinforced composite materials for possible application in knee replacement. In dental applications, CF/epoxy-based FRPCs are being used to fabricate dental post [110]. Usually, gold bridges were used to replace one or more teeth but their high cost and time-consuming fabrication process have led to the development of FRPCs-based bridges [111]. FRPCs are also being used to fabricate orthodontic arch-wires. These wires are generally fitted over the teeth in order to align them [112, 113]. Artificial legs, used to support amputees during walk, were generally made of metallic materials. Owing to the high weight of metals and low corrosion resistance, FRPCs have replaced these metallic prosthetic limbs. As of now, all the three components of prosthetic leg, namely shaft, socket, and foot, are being manufactured using FRPCs [114–116]. Moving tables, used in CT and MRI scanners, are being manufactured using FRPCs due to the requirement of lightweight and high strength material [117]. Calcium phosphate (CaP)/polymer composite materials are highly recommended materials in bone replacement due to high compressive and flexural strength [118].

Table 1.3 Applications of reinforced polymer composites.

S. No.	Composite	Processing technique	Application field	Component	References
1.	Glass fiber/unsaturated polyester	Hand lay-up method	Automobile	Front bumper	[88]
2.	Sisal, roselle fiber, banana/epoxy grade 3554 A	Hand lay-up method	Automobile	Visor in two-wheeler Indicator cover Pillion seat cover Rear view mirror cover	[89]
3.	Glass, carbon fiber/epoxy	—	Aerospace	Vertical stabilizer	[89]
4.	GFRPC/epoxy/polyester/pp	—	Electronic	Computer, electric motor covers cell phones	[90]
			Home and furniture	Roof sheet, sun shade, book racks, etc.	
			Aerospace	Luggage rack, bulkheads, ducting, etc.	
			Boats and marine	Boat frame	
			Medical	X-ray beds	
			Automobile	Body panel, seat cover, bumper, and engine cover	
5.	CFRP laminates	Vacuum bagging	Aerospace	Upper deck floor berns Pressure bulkhead Centre wing box, fin box, rudder HTP box	[91]
6.	Glass, carbon, aramid/polyester, vinyl ester, epoxy	Filament winding, resin infusion, prepreg, etc.	Energy industries	Wind turbine blades	[92]

#	Material	Process	Sector	Application	Ref.
7.	CFRP	—	Automobile	Citroen car body	[93]
8.	CF-GF/epoxy (hybrid)	—	Aerospace	Pilot's cabin door	[94]
	Boron–graphite (hybrid)	—		Fighter aircraft components	
	CF–aramid/thermoplastic hybrid	—	Safety	Helmet	
	GFPR, CFPR (hybrid)	—	Civil	Bridge girder	
9.	CF/epoxy	Extrusion, compression molding	Automobile	Stiffener, floor panel, side sill inner	[95]
10.	CFRP	—	Automobile	Door sill stiffeners	[96]
				Engine bay subframe	
11.	CFRP/vinyl ester	Compression molding	Automobile	Fender support, headlamp supports, door components	[97]
	GFRP/vinyl ester			Door inner panel, windshield surround outer and inner panel, door components	
12.	CF/Epoxy	—	Biomedical	Prosthetic limbs (foot)	[98]
13.	Kevlar/CF/PMMA	—	Biomedical	Bone cement (used for fixing the bones)	[99]
14.	E-glass/epoxy	Pultrusion	Electrical applications	Insulating material for high voltage line	[100]

1.4 Conclusion

RPCs are engineered materials used in a wide spectrum of applications ranging from domestic products to biomedical devices. Natural and synthetic fibers are both being reinforced in FRPCs according to the application. FRPCs offer a number of advantages over conventional monolithic materials such as corrosion resistance, light weight and high strength to weight ratio. Automobiles, aircrafts, boats, ships, recreational goods, chemical equipment, and civil building and bridges are some common applications of FRPCs. Biomedical applications such as prosthetic legs and bone cement are relatively new applications of FRPCs-based materials. The consumption of FRPCs in the near future is expected to increase but a lot research is needed in the recycling and disposal methods of synthetic FRPCs.

References

1 Bajpai, P.K., Ahmad, F., and Chaudhary, V. (2017). Processing and characterization of bio-composites. In: *Handbook of Ecomaterials* (ed. L.M. Torres-Martinez, O.V. Kharissova and B.I. Kharisov). Springer International Publishing AG, https://doi.org/10.1007/978-3-319-48281-1_98-1.

2 Rahmani, H., Najafi, S.H.M., and Ashori, A. (2014). Mechanical performance of epoxy/carbon fiber laminated composites. *Journal of Reinforced Plastics and Composites* 33 (8): 733–740.

3 Islam, M.E., Mahdi, T.H., Hosur, M.V., and Jeelani, S. (2015). Characterization of carbon fiber reinforced epoxy composites modified with nanoclay and carbon nanotubes. *Procedia Engineering* 105: 821–828.

4 Cho, J., Chen, J.Y., and Daniel, I.M. (2007). Mechanical enhancement of carbon fiber/epoxy composites by graphite nanoplatelet reinforcement. *Scripta Materialia* 56: 685–688.

5 Aramide, F.O., Atanda, P.O., and Olorunniwo, O.O. (2012). Mechanical properties of a polyester fiber glass composite. *International Journal of Composite Materials* 2: 147–151.

6 Treinyte, J., Bridziuviene, D., Fataraite-Urboniene, E. et al. (2018). Forestry wastes filled polymer composites for agricultural use. *Journal of Cleaner Production* 205: 388–406.

7 Liu, Q. and Hughes, M. (2008). The fracture behaviour and toughness of woven flax fibre reinforced epoxy composites. *Composites: Part A* 39: 1644–1652.

8 Erol, O., Powers, B.M., and Keefe, M. (2017). Effects of weave architecture and mesoscale material properties on the macroscale mechanical response of advanced woven fabrics. *Composites: Part A* 101: 554–566.

9 Yuanyuan, Z., Ying, S., Jialu, L. et al. (2016). Tensile response of carbon-aramid hybrid 3D braided composites. *Materials and Design* 116: 246–252.

10 Kostar, T.D., Chou, T.W., and Popper, P. (2000). Characterization and comparative study of three-dimensional braided hybrid composites. *Journal of Material Science* 35: 2175–2183.

11 Fombuena, V., Bernardi, L., Fenollar, O. et al. (2014). Characterization of green composites from biobased epoxy matrices and bio-fillers derived from seashell wastes. *Materials and Design* 57: 168–174.

12 Duflou, J.R., Yelin, D., Acker, K.V., and Dewulf, W. (2014). Comparative impact assessment for flax fibre versus conventional glass fibre reinforced composites: are bio-based reinforcement materials the way to go? *CIRP Annals – Manufacturing Technology* https://doi.org/10.1016/j.cirp.2014.03.061.

13 Baghaei, B. and Skrifvars, M. (2016). Characterisation of polylactic acid biocomposites made from prepregs composed of woven polylactic acid/hemp–Lyocell hybrid yarn fabrics. *Composites: Part A* 81: 139–144.

14 Hassanin, A.H., Hamouda, T., Candan, Z. et al. (2016). Developing high-performance hybrid green composites. *Composites Part B* https://doi.org/10.1016/j.compositesb.2016.02.051.

15 Chaudhary, V., Bajpai, P.K., and Maheshwari, S. (2018). Studies on mechanical and morphological characterization of developed jute/hemp/flax reinforced hybrid composites for structural applications. *Journal of Natural Fibers* 15 (1): 80–97.

16 Ahmad, F., Chaudhary, V., Ahmad, Z., and Bajpai, P.K. (2017). Effect of fiber selection and fiber treatment on the composite performance. *International Journal of Scientific and Engineering Research* 8: 55–57.

17 Asaithambi, B., Ganesan, G., and Kumar, S.A. (2014). Bio-composites: development and mechanical characterization of banana/sisal fibre reinforced poly lactic acid (PLA) hybrid composites. *Fibers and Polymers* 15: 847–854.

18 Rahman, M.M. and Khan, M.A. (2007). Surface treatment of coir (Cocosnucifera) fibers and its influence on the fibers' physico-mechanical properties. *Composites Science and Technology* 67: 2369–2376.

19 Shah, D.U., Schubel, P.J., and Clifford, M.J. (2013). Can flax replace E-glass in structural composites? A small wind turbine blade case study. *Composites: Part B* 52: 172–181.

20 Ahmad, F. and Bajpai, P.K. (2017). Finite element analysis and simulation of flax/epoxy composites under tensile loading. *International Journal of Advanced Research Science and Engineering* 6 (2): 436–441.

21 Benyahia, H., Tarfaoui, M., El Moumen, A. et al. (2018). Mechanical properties of offshoring polymer composite pipes at various temperatures. *Composites Part B: Engineering* 152: 231–240.

22 Ellyin, F. and Maser, R. (2004). Environmental effects on the mechanical properties of glass-fiber epoxy composite tubular specimens. *Composite Science and Technology* 64 (12): 1863–1874.

23 Cohades, A. and Michaud, V. (2018). Shape memory alloys in fibre-reinforced polymer composites. *Advanced Industrial and Engineering Polymer Research* https://doi.org/10.1016/j.aiepr.2018.07.001.

24 Paine, J.S.N. and Rogers, C.A. (1994). The response of SMA hybrid composite materials to low velocity impact. *Journal of Intelligent Material Systems and Structures* 5: 530–535.

25 Pappada, S., Gren, P., Tatar, K. et al. (2009). Mechanical and vibration characteristics of laminated composite plates embedding shape memory alloy superelastic wires. *Journal of Material Engineering and Performances* 18: 531–537.

26 Pappada, S., Rametta, R., Largo, A., and Maffezzoli, A. (2012). Low-velocity impact response in composite plates embedding shape memory alloy wires. *Polymer Composites* 33: 655–664.

27 Alexandre, M. and Dubois, P. (2000). Polymer-layered silicate nanocomposites: preparation, properties and uses of a new class of materials. *Material Science and Engineering: R: Reports* 28: 1–63.

28 Manias, E., Touny, A., Wu, L. et al. (2001). Polypropylene/montmorillonite nanocomposites. Review of the synthetic routes and materials properties. *Chemistry of Materials* 13: 3516–3523.

29 Ma, P.-C., Siddiqui, N., Marom, G., and Kim, J.-K. (2010). Dispersion and functionalization of carbon nanotubes for polymer-based nanocomposites: a review. *Composites Part A: Applied Science and Manufacturing* 41: 1345–1367.

30 Liu, T., Burger, C., and Chu, B. (2003). Nanofabrication in polymer matrices. *Progress in Polymer Science* 28: 5–26.

31 Ajayan, P.M., Schadler, L.S., and Braun, P.V. (eds.) (2003). *Nanocomposite Science and Technology*. Weinheim: FRG: Wiley-VCH Verlag GmbH & Co. KGaA.

32 Wilberforce, S.I.J., Finlayson, C.E., Best, S.M., and Cameron, R.E. (2011). The influence of the compounding process and testing conditions on the compressive mechanical properties of poly (d, l-lactide-co-glycolide)/a-tricalcium phosphate nanocomposites. *Journal of Mechanical Behaviour of Biomedical Materials* 4: 1081–1089.

33 Sorrentino, A., Gorrasi, G., and Vittoria, V. (2007). Potential perspectives of bio-nanocomposites for food packaging applications. *Trends in Food Science and Technology* 18: 84–95.

34 Delogu, F., Gorrasi, G., and Sorrentino, A. (2017). Fabrication of polymer nanocomposites via ball milling: present status and future perspectives. *Progress in Materials Science* 86: 75–126.

35 Eichner, E., Heinrich, S., and Schneider, G.A. (2018). Influence of particle shape and size on mechanical properties in copper-polymer composites. *Powder Technology* https://doi.org/10.1016/j.powtec.2018.07.100.

36 Fan, J. and Wang, C. (2018). Dynamic compressive response of a developed polymer composite at different strain rates. *Composites Part B* https://doi.org/10.1016/j.compositesb.2018.06.025.

37 Friedrich, K. (2018). Polymer composites for tribological applications. *Advanced Industrial and Engineering Polymer Research* https://doi.org/10.1016/j.aiepr.2018.05.001.

38 Xue, Q. and Wang, Q. (1997). Wear mechanisms of polyetheretherketone composites filled with various kinds of SiC. *Wear* 213: 54–58.

39 Xing, X.S. and Li, R.K.Y. (2004). Wear behavior of epoxy matrix composites filled with uniform sized submicron-spherical silica particles. *Wear* 256: 21–26.

40 Yousef, S. (2016). Polymer nanocomposite components: a case study on gears. In: *Light Weight Composite Structures in Transport Design, Manufacturing, Analysis and Performance* (ed. J. Njuguna), 385–420. Woodhead Publishing.

41 Koike, H., Kida, K., Santos, E.C. et al. (2012). Selflubrication of PEEK polymer bearings in rolling contact fatigue under radial loads. *Tribology International* 49: 30–38.

42 Friedrich, K. and Schlarb, A.K. (2008). *Tribology of Polymeric Nanocomposites*. Amsterdam: Elsevier.

43 Su, F.H., Zhang, Z.Z., Wang, K. et al. (2006). Friction and wear properties of carbon fabric composites filled with nano-Al_2O_3 and nano-Si_3N_4. *Composites Part A: Applied Science and Manufacturing* 37: 1351–1357.

44 Guo, Q.B., Lau, K.T., Rong, M.Z., and Zhang, M.Q. (2010). Optimization of tribological and mechanical properties of epoxy through hybrid filling. *Wear* 269: 13–20.

45 Guo, Q.B., Lau, K.T., Rong, M.Z., and Zhang, M.Q. (2009). Imparting ultra-low friction and wear rate to epoxy by the incorporation of microencapsulated lubricant? *Macromolecular Material Engineering* 294: 20–24.

46 Khun, N.W., Zhang, H., Yue, C.Y., and Yang, J.L. (2014). Self-lubricating and wear resistant epoxy composites incorporated with microencapsulated wax. *Journal of Applied Mechanics* 81: 1–9.

47 Imani, A.H., Zhang, H., Owais, M. et al. (2018). Wear and friction of epoxy based nanocomposites with silica nanoparticles and wax containing microcapsules. *Composites Part A: Applied Science and Manufacturing* 107: 607–615.

48 Khun, N.W., Zhang, H., Yang, J.L., and Liu, E. (2013). Mechanical and tribological properties of epoxy matrix composites modified with microencapsulated mixture of wax lubricant and multi-walled carbon nanotubes. *Friction* 1: 341–349.

49 Oyman, Z.O. (2009). *An Overview of Research on Self Healing Coatings*, 1–3. BoyaTurk, Special Edition.

50 Gao, C., Yu, L., Liu, H., and Chen, L. (2012). Development of self-reinforced polymer composites. *Progress in Polymer Science* 37: 767–780.

51 Huang, H.X. (1998). Self-reinforcement of polypropylene by flow-induced crystallization during continuous extrusion. *Journal of Applied Polymer Science* 67: 2111–2118.

52 Li, R. and Yao, D. (2008). Preparation of single poly(lactic acid) composites. *Journal of Applied Polymer Science* 107: 2909–2916.

53 Makela, P., Pohjonen, T., Tormala, P. et al. (2002). Strength retention properties of self-reinforced poly-l-lactide (SR-PLLA) sutures compared with polyglyconate (Maxon(R)) and polydioxanone (PDS) sutures: an in vitro study. *Biomaterials* 23: 2587–2592.

54 Tormala, P. (1992). Biodegradable self-reinforced composite materials; manufacturing structure and mechanical properties. *Clinical Materials* 10: 29–34.

55 Hine, P.J. and Ward, I.M. (2004). Hot compaction of woven poly(ethylene terephthalate) multifilaments. *Journal of Applied Polymer Science* 91: 2223–2233.

56 Gilbert, J.L., Ney, D.S., and Lautenschlager, E.P. (1995). Self-reinforced composite poly(methyl methacrylate): static and fatigue properties. *Biomaterials* 16: 1043–1055.

57 Gindl, W. and Keckes, J. (2007). Drawing of self-reinforced cellulose films. *Journal of Applied Polymer Science* 103: 2703–2708.

58 Gemi, L. (2018). Investigation of the effect of stacking sequence on low velocity impact response and damage formation in hybrid composite pipes under internal pressure. A comparative study. *Composites Part B: Engineering* 153: 217–232.

59 Idowu, A., Boesl, B., and Agarwal, A. (2018). 3D graphene foam-reinforced polymer composites – a review. *Carbon* https://doi.org/10.1016/j.carbon.2018.04.024.

60 Li N, Zhang Q, Gao S, Song Q, Huang R, Wang L. Biocompatible and conductive scaffold. 2013, doi:https://doi.org/10.1038/srep01604.

61 Bello, A., Fashedemi, O.O., Lekitima, J.N. et al. (2013). High-performance symmetric electrochemical capacitor based on graphene foam and nanostructured manganese oxide. *AIP Advances* 3 (8): 82118.

62 Embrey, L., Nautiyal, P., Loganathan, A. et al. (2017). Three-dimensional graphene foam induces multifunctionality in epoxy nanocomposites by simultaneous improvement in mechanical, thermal, and electrical properties. *Applied Materials and Interfaces* 9 (45): 39717–39727.

63 Qiu, Y., Liu, J., Lu, Y. et al. (2016). Hierarchical assembly of tungsten spheres and epoxy composites in three-dimensional graphene foam and its enhanced acoustic performance as a backing material. *ACS Applied Materials Interfaces* 8: 18496–18504.

64 Sun, X., Liu, X., Shen, X. et al. (2015). Graphene foam/carbon nanotube/poly(dimethylsiloxane) composites for exceptional microwave shielding. *Composites Part A: Applied Science and Manufacturing* 85: 199–206.

65 Jusza, A., Anders, K., Polis, P. et al. (2014). Luminescent properties in the visible of Er3+/Yb3+ activated composite materials. *Optical Materials* https://doi.org/10.1016/j.optmat.2014.03.018.

66 Kroto, H.W., Heath, J.R., Obrien, S.C. et al. (1985). C-60 – Buckminsterfullerene. *Nature* 318: 162–163.

67 Hirsch, A. (2010). The era of carbon allotropes. *Natural Materials* 9: 868–871.

68 Iijima, S. (1991). Helical microtubules of graphitic carbon. *Nature* 354: 56–58.

69 Novoselov, K.S., Geim, A.K., Morozov, S.V. et al. (2004). Electric field effect in atomically thin carbon films. *Science* 306: 666–669.

70 Lee, J.K.Y., Chen, N., Peng, S. et al. (2018). Polymer-based composites by electrospinning: preparation and functionalization with nanocarbons. *Progress in Polymer Science* https://doi.org/10.1016/j.progpolymsci.2018.07.002.

71 Bauhofer, W. and Kovacs, J.Z. (2009). A review and analysis of electrical percolation in carbon nanotube polymer composites. *Composite Science and Technology* 69: 1486–1498.

72 McKeon-Fischer, K., Flagg, D., and Freeman, J. (2011). Coaxial electrospun poly (ε-caprolactone), multiwalled carbon nanotubes, and polyacrylic acid/polyvinyl alcohol scaffold for skeletal muscle tissue engineering. *Journal of Biomedical Materials Research Part A* 99: 493–499.

73 Markowski, J., Magiera, A., Lesiak, M. et al. (2015). Preparation and characterization of nanofibrous polymer scaffolds for cartilage tissue engineering. *Journal of Nanomaterials* 1, 564087–9.

74 Leung, S.N. (2018). Thermally conductive polymer composites and nanocomposites: processing-structure-property relationships. *Composites Part B* https://doi.org/10.1016/j.compositesb.2018.05.056.

75 Loomis, J., Ghasemi, H., Huang, X. et al. (2014). Continous fabrication platform for highly aligned polymer films. *Technology* 2: 1–11.

76 Mehra, N., Mu, L., Ji, T. et al. (2018). Thermal transport in polymeric materials and across composite interfaces. *Applied Material Today* 12: 92–130.

77 Rajapakse, R.M.G., Murakami, K., Bandara, H.M.N. et al. (2010). Preparation and characterization of electronically conducting polypyrrole-montmorillonite nanocomposite and its potential application as a cathode material for oxygen reduction. *Electrochimica Acta* 55: 2490–2497.

78 Mu, T., Liu, L., Lan, X. et al. (2018). Shape memory polymers for composites. *Composites Science and Technology* 160: 169–198.

79 Saba, N. and Jawaid, M. (2018). A review on thermomechanical properties of polymers and fibers reinforced polymer composites. *Journal of Industrial and Engineering Chemistry* https://doi.org/10.1016/j.jiec.2018.06.018.

80 Li, X., Bai, R., and McKechnie, J. (2016). Environmental and financial performance of mechanical recycling of carbon fibre reinforced polymers and comparison with conventional disposal routes. *Journal of Cleaner Production* 127: 451–460.

81 Longana, M.L., Ong, N., Yu, H., and Potter, K.D. (2016). Multiple closed loop recycling of carbon fibre composites with the HiPerDiF (High Performance Discontinuous Fibre) method. *Composite Structures* 152: 271–277.

82 Ahmad, F. and Bajpai, P.K. (2018). Analysis and evaluation of drilling induced damage in fiber reinforced polymer composites: a review. *IOP Conference Series: Materials Science and Engineering* https://doi.org/10.1088/1757-899X/455/1/012105.

83 Bajpai, P.K. and Singh, I. (2013). Drilling behavior of sisal fiber-reinforced polypropylene composite laminates. *Journal of Reinforced Plastics and Composites* 32: 1569–1576.

84 Abrate, S. and Walton, D. (1992). Machining of composite materials. Part II: Non-traditional methods. *Composites Manufacturing* 2: 85–94.

85 Sheikh-Ahmad, J.Y. (2009). *Machining of Polymer Composites*. Springer Science and Business Media. ISBN: 978-0-387-35539-9.

86 Girot, F.A., Lacalle, L.N.L.D., Lamikiz, A., and Iliescu, D. (2009). Machining composite material, Chapter 2. In: *Machinability Aspects of Polymer Matrix Composites* (ed. J.P. Davim). ISTE Publication, Wiley, https://www.researchgate.net/publication/268386427.

87 Udupa, G., Rao, S.S., and Gangadharan, K.V. (2014). A review of carbon nanotube reinforced aluminium composite and functionally graded composites as a future material for aerospace. *International Journal of Modern Engineering Research* 4: 13–22.

88 Achema, F., Yahaya, B.S., Apeh, E.S., and Akinyeye, J.O. (2017). Application of glass fibre reinforced composite in the production of light weight car bumper (a case study of the mechanical properties). *International Journal of Engineering Research and Technology* 6: 575–579.

89 Fox, M.R., Schultheisz, C.R., Reeder, J.R. et al. Materials examination of the vertical stabilizer from American Airlines Flight 587. https://ntrs.nasa.gov/search.jsp?R=20050238475 2019-01-09T09:26:42+00:00Z (accessed 19 January 2019).

90 Sathishkumar, T.P., Satheeshkumar, S., and Naveen, J. (2014). Glass fiber-reinforced polymer composites – a review. *Journal of Reinforced Plastics and Composites* 33: 1258–1275.

91 Vlot, A. and Gunnink, J.W. (2001). *Fibre Metal Laminates*. Dordrecht: Kluwer Academic Publishers.

92 Aymerich, F. (2012). *Composite Materials for Wind Turbine Blades: Issues and Challenges*. University of Patras.

93 Dubravcik, I.M. and Kender, I.S. (2014). Composite materials application in car production. *Transfer Inovacii* 29: 282–285.

94 Gururaja, M.N. and Rao, A.N.H. (2012). A review on recent applications and future prospectus of hybrid composites. *International Journal of Soft Computing and Engineering (IJSCE)* 1: 352–355.

95 Ishikawa, T., Amaoka, K., Masubuchi, Y. et al. (2017). Overview of automotive structural composites technology developments in Japan. *Composites Science and Technology* https://doi.org/10.1016/j.compscitech.2017.09.015.

96 Feraboli, P., Masini, A., and Bonfatti, A. (2007). Advanced composites for the body and chassis of a production high performance car. *International Journal of Vehicle Design* 44: 233–246.

97 Boeman, R.G. and Johnson, N.L. (2002). Development of a cost competitive, composite intensive, body-in-white. *Proceedings of 2002 Future Car Congress*, Arlington, Virginia (3–5 June 2002). SAE technical paper.

98 Toh, S.L., Goh, J.C.H., Tan, P.H., and Tay, T.E. (1993). Fatigue testing of energy storing prosthetic feet. *Prosthetics and Orthotics International* 17: 180–188.

99 Piller, R.M., Blackwell, R., Macneb, I., and Cameron, H.U. (1976). Carbon fiber-reinforced bone cement in orthopedic surgery. *Journal of Biomedical Materials Research* 10: 893–906.

100 Taherian, R. (2008). Application of polymerbased composites: polymer-based composite insulators. In: *Electrical Conductivity in*

Polymer-Based Composites Experiments, Modelling and Applications (ed. R. Taherian and A. Kausar), 131–181. William Andrew.

101 Tian, H., Tang, Z., Zhuang, X. et al. (2012). Biodegradable synthetic polymers: preparation, functionalization and biomedical application. *Progress in Polymer Science* 37: 237–280.

102 Moyen, B.J.-L., Lahey, P.J., Weinberg, E.H., and Harris, W.H. (1978). Effects on intact femora of dogs of the application and removal removal of metal plates. *Journal of Bone and Joint Surgery* 60: 940–947.

103 Lin, T.W., Corvelli, A.A., Frondoza, C.G. et al. (1997). Glass PEEK composite promotes proliferation and osteo-calcin production of human osteoblastic cells. *Journal of Bio-Medical Materials Research* 36: 37–144.

104 Kettunen, J., Makela, A., Miettinen, H. et al. (1999). Fixation of femoral shaft osteotomy with an intramedullary composite rod: an experimental study on dogs with a two year follow-up. *Journal of Biomaterials Science Polymer Edition* 10: 33–45.

105 Marcolongo, M., Ducheyne, P., Garino, J., and Schepers, E. (1998). Bioactive glass fibers/polymeric composites bond to bone tissue. *Journal of Biomedical Materials Research* 39: 161–170.

106 Claes, L., Schultheis, M., Wolf, S. et al. (1999). A new radiolucent system for vertebral body replacement: its stability in comparison to other systems. *Journal of Biomedical Materials Research Applied Biomaterials* 48: 82–89.

107 Simoes, J.A., Marques, A.T., and Jeronimidis, G. (1999). Design of a controlled-stiffness composite proximal femoral prosthesis. *Composites Science and Technology* 60: 559–567.

108 Wintermantel, E., Bruinink, A., Eckert, K. et al. (1998). Tissue engineering supported with structured biocompatible materials: goals and achievements. In: *Materials in Medicine* (ed. M.O. Speidel and P.J. Uggowitzer), 1–136. Zurich: vdf Hochschulverlag AG an der ETH.

109 Deng, M. and Shalaby, S.W. (1988). Properties of self-reinforced ultra-high molecular weight polyethylene composites. *Biomaterials* 70: 1372–1376.

110 Ramakrishna S, Ganesh V.K., Teoh S.H. et al. (1998). Fiber reinforced composite product with graded stiffness. Singapore Patent Application No. 9800874-1.

111 Bjork, N., Ekstrand, K., and Ruyter, I.E. (1986). Implant-fixed dental bridges from carbon/graphite fiber reinforced poly(methyl methacrylate). *Biomaterials* 7: 73–75.

112 Jancar, J. and Dibenedetto, A.T. (1993). Fiber reinforced thermoplastic composites for dentistry part I hydrolytic stability of the interface. *Journal of Materials Science: Materials in Medicine* 4: 555–561.

113 Jancar, J., Dibenedetto, A.T., and Goldberg, A.J. (1993). Fiber reinforced thermoplastic composites for dentistry. Part II: Effect of moisture on flexural properties of unidirectional composites. *Journal of Materials Science: Materials in Medicine* 4: 562–568.

114 Tallent, M.A., Cordova, C.W., Cordova, D.S., and Donnelly, D.S. (1990). Thermoplastic fibers for composite reinforcement. In: *International Encyclopedia of Composites* (ed. S.M. Lee). New York: VCH Publishers.

115 Ganesh, V.K. and Ramakrishan, S. (1998). Prosthesis: another concept. The European periodical for technical textile user. *Quarterly Magazine* 29: 56–60.

116 Goh, J.C.H., Tan, P.H., Toh, S.L., and Tay, T.E. (1994). Gait analysis study of an energy-storing prosthesis foot – a preliminary study. *Gait and Posture* 2: 95–101.

117 Ko, F.K. (1999). Fiber reinforced composites in medical applications. In: *Lectures on Textile Structural Composites* (ed. T.-W. Chou and F.K. Ko). Taipei, Taiwan: China Textile Institute.

118 Johnson, A.J.W. and Herschler, B.A. (2011). A review of the mechanical behavior of CaP and CaP/polymer composites for applications in bone replacement and repair. *Acta Biomaterialia* 7: 16–30.

2

Fabrication of Short Fiber Reinforced Polymer Composites

Ujendra K. Komal[1], Manish K. Lila[1], Saurabh Chaitanya[2], and Inderdeep Singh[1]

[1] *Indian Institute of Technology Roorkee, Department of Mechanical and Industrial Engineering, Roorkee, Uttarakhand, 247667, India*
[2] *Changwon National University, Department of Mechanical Engineering, Changwon, 51140, South Korea*

2.1 Introduction

Composite materials have gained significant attraction toward the replacement of conventional materials such as metals, polymers, ceramics, and wood due to their numerous advantages such as stiffness and specific strength. The use of composite materials is becoming increasingly common in our lives. Composite materials can also be considered as today's material because their application areas have now spread in various industries and range from aircraft body to children's toys. However, they have the potential to be called tomorrow's materials due to the properties they can offer in the future, such as nanocomposites and smart materials. Composite materials can be classified based on different aspects such as matrix and reinforcing materials and the size and shape of materials. The classification of composite materials is shown in Figure 2.1.

Nowadays, polymer matrix composites are gaining importance due to their low weight and affordable cost. They have the potential to replace the conventional material in many specific areas. Their corrosion resistance makes them suitable for use in marine and construction fields. They are light in weight and have good vibration damping properties, which makes them suitable for use in sports equipment, automobiles, and space industry. In polymer matrix composites, the most common fibers are glass fiber, carbon fiber, and aramid fibers. But the growing environmental rules and ecological concerns demand for renewable composites. Therefore, in the present scenario researchers are striving to develop natural fiber reinforced composites. Natural fibers are derived from animals or plants. These fibers are categorized as environment friendly and have comparable properties to synthetic fibers. They are light in weight, biodegradable, cheap, and abundantly available in nature. The classification of natural fibers is shown in shown Figure 2.2.

Natural fiber-based composites (sometimes biocomposites) have revealed comparable or sometimes even better properties than synthetic fiber-based

Reinforced Polymer Composites: Processing, Characterization and Post Life Cycle Assessment,
First Edition. Edited by Pramendra K. Bajpai and Inderdeep Singh.
© 2020 Wiley-VCH Verlag GmbH & Co. KGaA. Published 2020 by Wiley-VCH Verlag GmbH & Co. KGaA.

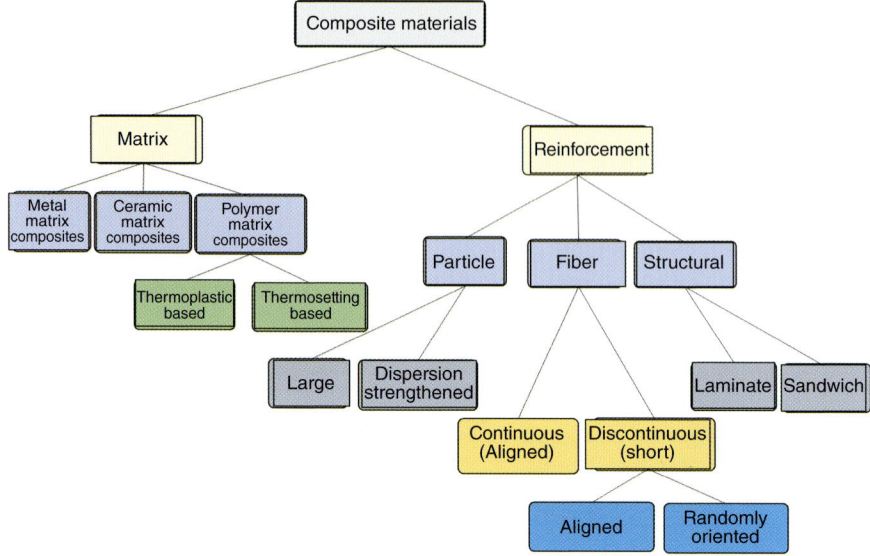

Figure 2.1 Classification of composite materials.

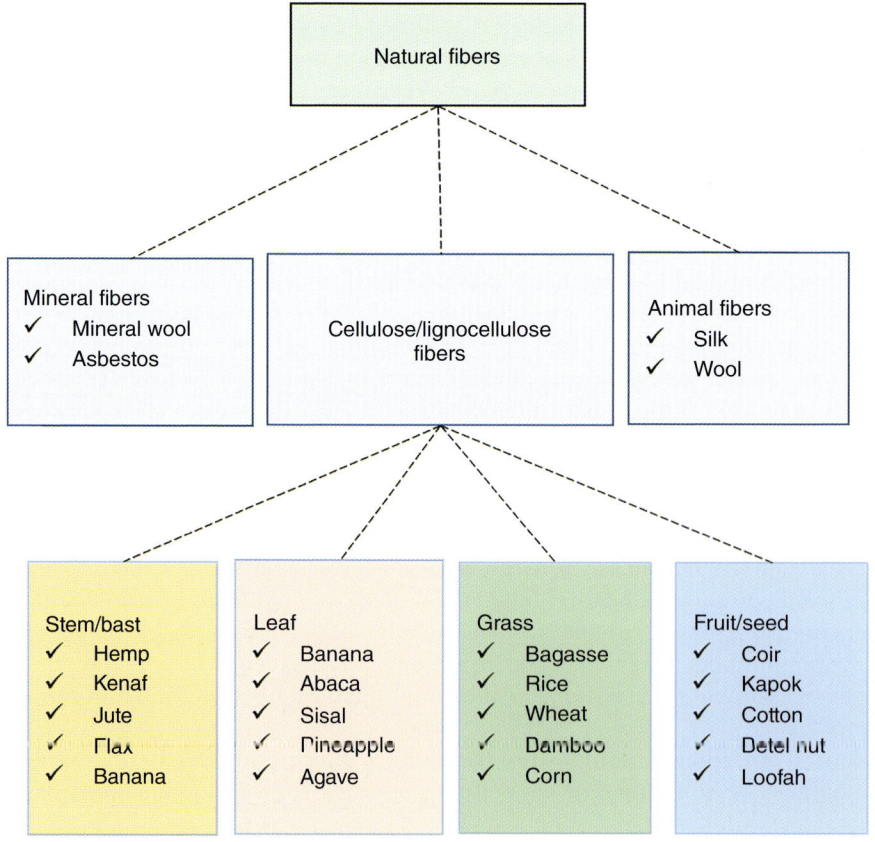

Figure 2.2 Classification of natural fibers.

Table 2.1 Natural fibers vs. synthetic fibers [2–5].

Type	Name	Density (g/cm³)	Tensile strength (MPa)	Young's modulus (GPa)	Elongation at break (%)
Synthetic fibers	Glass	2.5–2.6	2200–3600	65–75	3
	Carbon	1.4–1.8	3000–4000	250–500	1–1.5
	Aramid	1.4	3000–3150	63–67	3.3–3.7
Natural fibers	Hemp	1.4–1.6	550–900	70	1.6
	Sisal	1.3–1.5	600–700	38	2–3
	Jute	1.3–1.5	200–800	10–30	1.8
	Ramie	1.5	500	44	2
	Cotton	1.5–1.6	290–490	12	3–10
	Flax	1.4–1.5	800–1500	60–80	1.2–1.6
	Coir	1.2–1.5	180–220	6	15–25
	Pineapple	0.8–1.6	400–627	1.44	14.5
	Kenaf	1.45	930	53	1.6
	Bagasse	1.25	290	17	–
	Bamboo	0.6–1.1	140–230	11–17	–
	Abaca	1.5	400	12	3–10

composites. Biocomposites can also be tailored easily to meet the specific requirements of products for different engineering applications [1].

A comparative analysis depicting the properties of a few natural fibers and synthetic fibers is shown in Table 2.1. Polymer composites based on the natural fibers are either partially or fully biodegradable in nature depending upon their constituents. Natural fiber reinforced nonbiodegradable polymeric composites are called partially biodegradable composites. When natural fibers are used to reinforce the biodegradable polymers, these can be termed as biodegradable composites or simply biocomposites. Figure 2.3 depicts the classification of polymer composites based on their disposal characteristics.

In order to manufacture composite products with extraordinary characteristics, it becomes necessary to focus on various aspects of processing techniques. The current chapter attempts a brief discussion on processing of short fiber reinforced polymer composites.

2.2 Challenges with Composite Materials

Composites possess several favorable characteristics such as high strength to weight ratio, comparable mechanical properties, and corrosion resistance. However, there are various challenges associated with the composite materials, which

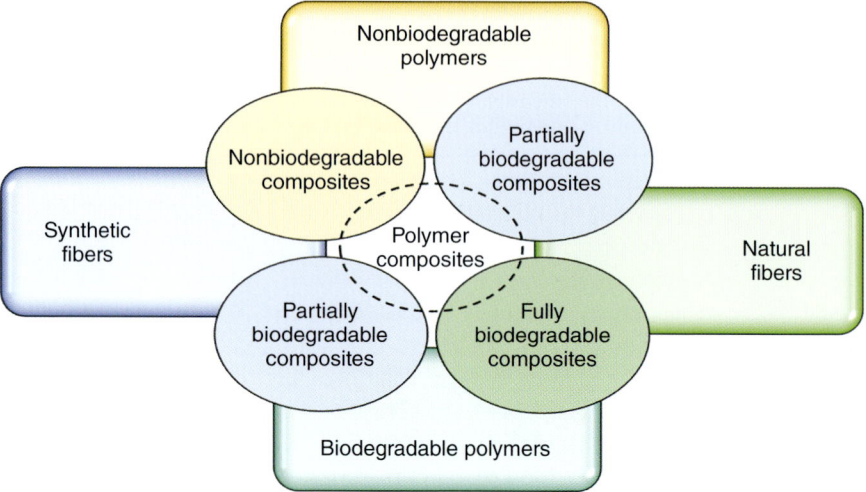

Figure 2.3 Classification of polymer matrix composites.

becomes necessary in order to fabricate composite products with excellent properties. Some of these challenges have been discussed here:

- Composites are tailor-made materials, i.e. they can be engineered according to the requirement. However, it is a challenging task to select the best processing technique with optimized processing parameters in order to meet specific requirements.
- The properties (physical, chemical, thermal, and mechanical) of both the constituents (reinforcement and matrix) are completely different. Therefore, the blending/melt-mixing of these significantly different materials is key to achieving the required properties.
- The interfacial characteristics between the constituents play a crucial role in governing the overall properties of the composite materials. The interfacial zones are prone to failure under loading and it is a challenge to control these characteristics. There are a number of approaches that have been tried worldwide to improve the interfacial characteristics between the constituents such as treatment (physical and chemical) and coating of fibers, and inclusion of fillers, additives, and catalyst during blending.
- As compared to conventional materials, the constituents of composites are completely distinct and their properties are also different. Therefore, the machines and tooling requirements are entirely distinct from those for conventional materials. The tooling and operating parameters for reinforcement may not be suitable for the matrix or vice versa.

Thermoset-based polymer composites are generally manufactured using hand lay-up, spray lay-up, pultrusion, and resin transfer molding. Thermoplastic-based composites are generally developed using injection molding and compression molding. The addition of natural fibers into the polymeric matrix increases the complexity in the process and in order to develop the desired composite products, it becomes necessary to do retrofitting in the existing setup, tooling, and operating parameters. Moreover, several approaches have been tried to improve the interaction between the constituents. A brief discussion on the most popular approaches tried by researchers is covered in the following section.

2.3 Preprocessing of Natural Fibers and Polymeric Matrix

Natural fibers pose compatibility issues with the matrix during development of polymer composites. Natural fibers comprise of cellulosic fibrils, attached together by noncellulosic content such as hemicellulose, lignin, and pectin [6]. The properties of the natural fibers mainly depend upon their lignocellulosic content. Excessive wax and hemicellulose cause the formation of weak interfacial interaction among the fibers and matrix. Moreover, the noncellulosic content is also responsible for the hydrophilic nature and early thermal degradation of the fibers during processing. To overcome these challenges, several preprocessing approaches have been suggested in the literature such as modification of the surface of fibers and compounding of fibers and matrix.

2.3.1 Fiber Surface Modification

The aim of this approach is to improve the interfacial interaction between the fibers and matrix by removing the noncellulosic content from the surface of the fibers using any surface modification technique. This technique can be generally divided into physical and chemical treatments. Physical treatment includes heat treatment, calendaring, and the use of electric discharge [7]. Electric discharge treatments (corona and plasma treatment) of natural fibers have attracted a lot of attention worldwide as these techniques demonstrate the ability to alter the surface energy of the fibers without affecting the bulk properties.

The modification of fiber surface using chemical treatment is the technique most widely explored by researchers. Chemicals such as sodium hydroxide, silane, acetic acid, and permanganate are some of the most widely used chemicals for the chemical treatment of fibers. Chemical treatment removes the excessive noncellulosic constituents from the surface of the fibers, which results in enhancement in interfacial characteristics between the polymers and matrix [4, 8–10]. Brief information on the widely explored surface modification techniques is shown in Table 2.2.

2.3.2 Compounding of Natural Fibers and Polymeric Matrix

Compounding is done to achieve the uniform distribution and alignment of fibers within the polymeric composites. There are various compounding methods prior to final fabrication that have been suggested in the literature. The most widely used methods are extrusion, pultrusion, and melt blending. These methods are generally used for the compounding of thermoplastic-based

Table 2.2 Surface modification techniques.

Reinforcement	Matrix	Processing routes	Fiber treatment/ compatibilizer/filler	References
Banana fiber	PLA	Melt blending followed by compression molding	Alkali (NaOH) treatment and silane coupling	[11]
Hemp	PLA	Hot press	Alkali (NaOH) treatment	[12]
Banana fiber	PP	Extrusion–injection molding	Sodium hydroxide (NaOH) treatment	[13]
Bagasse fiber	Unsaturated polyester	Vacuum bagging	Sodium hydroxide (NaOH) and acrylic acid	[10]
Sisal fibers	PP	Melt-mixing and solution-mixing	Sodium hydroxide, maleic anhydride, and permanganate	[14]
Wood flour	PLA	Extrusion followed by injection molding	Methylenediphenyl-diisocyanate and poly(ethylene–acrylic acid)	[15]
Banana and sisal fibers	PLA	Extrusion–injection molding	Alkali (NaOH) and benzoyl peroxide treatment	[16, 17]

composites. The extrusion method has been suggested as the most widely used compounding method in the literature. The working principle of a typical extrusion machine is shown in Figure 2.4. Depending upon the number of screws used, the extruder can be a single screw extruder or a twin screw extruder.

During compounding, the polymeric pellets and fibers are mixed rigorously and fed into the hopper of the extruder. The mixture is then carried inside the barrel with the help of a rotating screw. As the mixture travels toward the die, the pellets melt and compounding with fibers take place due to heating and the rotating action of the screw. The barrel is generally divided into the feed zone, melt zone, compression zone, and metering zone (hopper to die end). The temperature of these zones can be judiciously selected and controlled to avoid the thermal degradation of the constituents during processing. The composite strand in the form of wire is then passed through the water bath for cooling and then pelletized using a pelletizer. These pellets can be used for the fabrication of composite specimens using injection molding process.

2.4 Processing of Polymeric Matrix Composites

The behavior and performance of the composite products depend upon the properties, size and shape of its constituents, processing techniques, operating

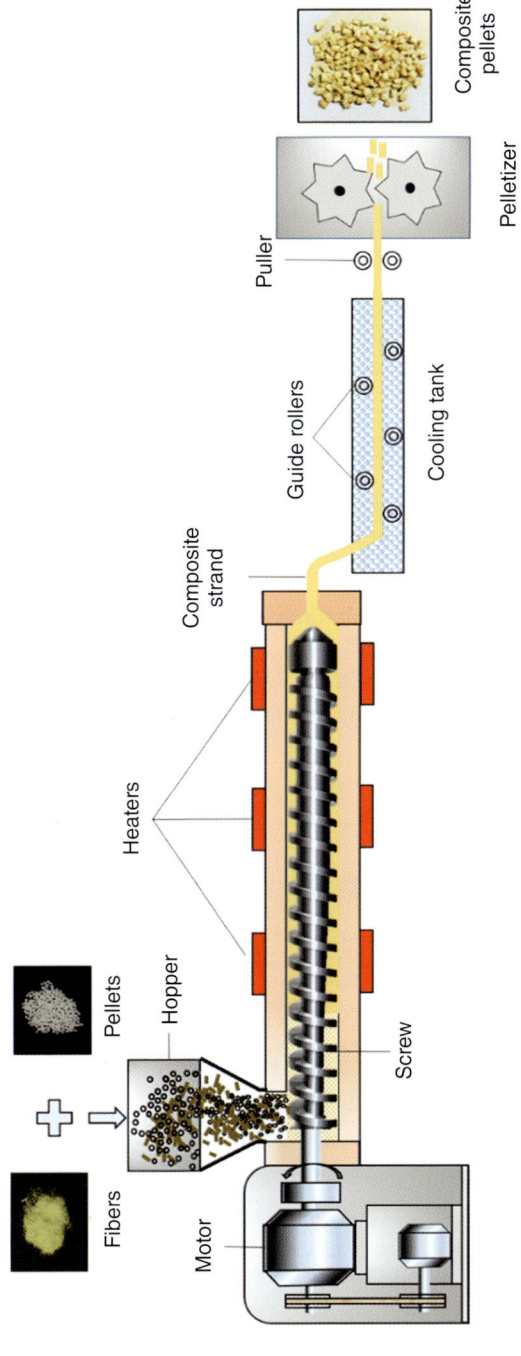

Figure 2.4 Schematic of extrusion setup.

Pellets

Fibers

Hopper

Heaters

Motor

Screw

Composite
strand

Guide rollers

Cooling tank

Puller

Pelletizer

Composite
pellets

parameters, the distribution and orientation of fibers, and the interfacial characteristics between the constituents [9, 14, 18]. The processing of polymer-based composites can be categorized into primary and secondary processing. In polymeric matrix composites, the matrix material can be either thermoplastics or thermosets. Generally, thermosets and thermoplastics are available in the form of resin and granules, respectively. There are a number of processing techniques (hand lay-up, injection molding, resin transfer molding, and compression molding) that can be used for converting these matrix materials into the final products. Secondary processing is the crucial and next stage to manufacture the final composite products. It includes machining, drilling, and joining of the composite products manufactured by primary processing techniques. The primary processing techniques of polymer-based composites can be mainly categorized as open mold and closed mold processes (Figure 2.5). Each process has its own benefits and limitations based on the materials, part accuracy, wastage, and cost.

2.4.1 Selection of Processing Techniques

The selection of appropriate techniques for the fabrication of polymer composites involves several factors to be taken into account. Some of these factors are as follows:

- The size, shape, orientation, and characteristics of the constituents (fibers and matrix)
- The size, shape, and properties of the resultant composites
- Manufacturing limitations of fibers and matrix
- The manufacturing cost.

The choice of processing method plays a significant role during the development of composite materials. The geometry required for the composite parts

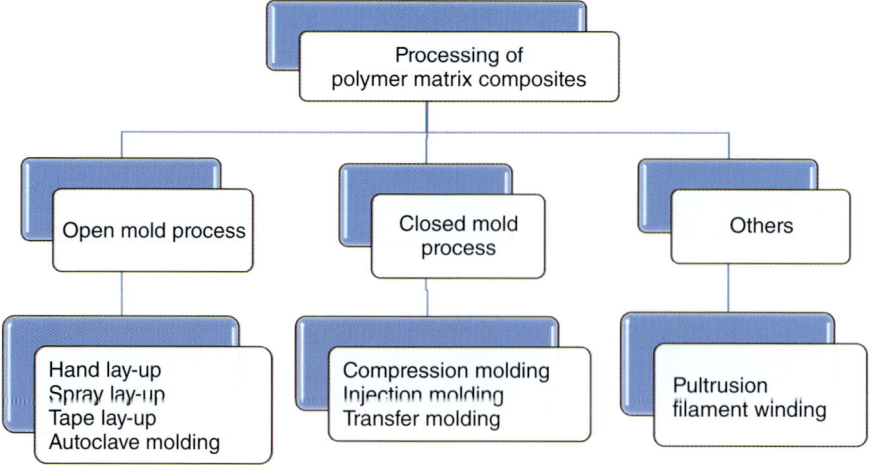

Figure 2.5 Processing techniques of polymer-based composites.

also plays a vital role in selection. Pultrusion and direct extrusion processes are employed for the development of long products having uniform cross-sectional profile. Pultrusion process is generally used for the development of composites reinforced with continuous strands of natural fibers while extrusion process can be employed for the development of products incorporating short fibers as reinforcement. In case of thermoplastic-based composites, these processes have been reportedly used as pre-compounding processes before injection or compression molding.

Compression molding is usually employed to develop products with simple geometries, while the composite parts of complex geometry with high precision and dimensional accuracy are usually made by injection molding process. Moreover, owing to its excellent production rate, this is the most widely used processing route for the manufacturing of plastic parts in the industry. As natural fibers are derived from various parts of the plant, they have the limitation of maximum fiber length that can be extracted from the plant source.

Therefore, their use in the form of continuous fibers is usually fulfilled by spinning the fibers into yarns. These yarns are then used to feed the pultrusion machine or in case of compression molding process, these spun yarns are weaved into the desired orientation and then hot-compressed between the polymer sheets to produce natural fiber-based composites. The use of long or continuous fibers increases the cost of fiber production as well as adds to the process complexity. Short fiber extraction is rather easier and cheaper compared to long/continuous fibers. Short fibers are easier to mix and blend with the polymer matrix and can be processed using extrusion and injection molding processes. A comparative analysis of some of the most widely used processing routes is depicted in Table 2.3.

A brief review on the different processing techniques employed for the fabrication of natural fiber-based composites is shown in Table 2.4.

It has been observed that the injection molding process has been used extensively for the fabrication of natural fiber-based polymeric composites.

2.4.2 Injection Molding

It is one of the most widely used techniques for the fabrication of the plastic and composite products in the industry. The working principle of injection molding machine is the same as the extrusion process. The mixing of the fibers and matrix takes place in the heated barrel, and the metered amount of the mixture is then injected through a nozzle into the mold cavity. After injection, cooling takes place in the mold cavity. A variety of cooling medium (air, water, and oil) can be used as per the requirement. After cooling, the composite part is ejected by giving a slight tap with the help of ejector pins. The main parts of a typical injection molding machine are depicted in Figure 2.6.

2.4.2.1 Operating Parameters
A typical injection molding machine can generally be used for the fabrication of thermoplastics and thermoplastic-based composite products. Natural fibers are hydrophilic in nature and therefore, absorb moisture from the atmosphere. The

Table 2.3 Comparison of processing techniques.

Characteristics	Injection molding	Compression molding	Resin transfer molding
Properties	• Composites based on thermosets and thermoplastics can be easily processed • Small to large size products with complex geometries and high dimensional accuracy can be easily produced • Production rate is high • Excellent dimensional accuracy can be achieved	• Composites based on thermosets and thermoplastics can be easily processed • Good dimensional accuracy can be achieved • Good flexibility in part design is possible • Extra features such as inserts, bosses, and attachment can be molded during the processing	• High injection pressure is not required during processing • Material wastage is minimal • Parts with excellent dimensional accuracy and surface finish can be produced • Any combination of reinforced materials in any orientation can be achieved • Ability to incorporate inserts and other attachments into molding
Limitations	• Mainly suitable for short fiber type of reinforcement • Excessive damage to the fibers and matrix may take place during the processing • High initial tooling and operational cost	• Production rate is slow • Large curing time • Defects such as uneven parting lines can be present in the final product • It is a labor intensive process	• Mold cavity limits the size of the composite products • High tooling cost • Composites based on short fiber reinforcement may not be suitable for processing
Applications	• The process is well suited for high production volume at low cost especially for automotive parts • It can easily be used for making automotive door panels, air spoilers, fenders, and body panels	• This process can be used for producing the variety of products such as toys, kitchen goods, automotive, airplane, and electrical parts	• Complex structures can be produced • Automotive body parts, containers, and bathtubs are generally manufactured

fiber is first dried using different methods (hot air oven, circulation dryer, and dehumidification dryer) as per the requirement. In injection molding process, the short fibers (chopped) and matrix are either pre-compounded or directly fed into the hopper of the injection molding machine. In order to get a good distribution of fibers within the composites, many researchers have recommended the use of extrusion machine (single or twin screw) for compounding of the constituents. During injection molding, the natural fibers have to travel from hopper to nozzle

Table 2.4 Fabrication techniques.

Reinforcement	Matrix	Processing routes	References
Microcrystalline cellulose, cellulose fibers, and wood flour	PLA	Extrusion–injection molding	[19]
Corn husk flour	PLA	Extrusion–injection molding	[20]
Banana fiber	PP	Extrusion followed by injection molding	[13, 21]
Bagasse fiber	Unsaturated polyester	Vacuum bagging	[10]
Sisal fibers	PP	Melt-mixing and solution-mixing	[14]
Hemp	PLA	Micronizer + compression molding	[22]
Banana and sisal fibers	PLA	Extrusion–injection molding	[16]

PP, polypropylene; PLA, poly-lactic acid.

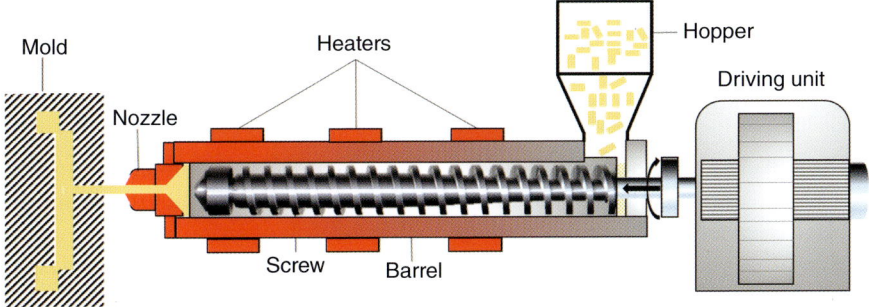

Figure 2.6 Injection molding process.

where excessive fiber damage may take place due to heating and shearing action. In order to encounter such type of challenges, researchers have recommended the use of a separate hopper (for feeding the natural fibers) near the nozzle. The behavior (physical, chemical, and thermal properties) of the fibers and matrix is entirely different from each other. The operating parameters selected for reinforcement may not be suitable for the matrix or vice versa. Therefore, sound understanding of the effect of these processing parameters on the quality and performance of the composites and judicious selection of processing parameters is necessary to produce the composite parts with exceptional properties. The important processing parameters involved during injection molding process are shown in Figure 2.7.

Optimization of these parameters is necessary for the reduction in cycle time and operating cost, which leads to increased productivity.

Barrel Temperature The temperature of barrel is a crucial parameter and has significant effect on the behavior of the final products. The melt-mixing of fibers and

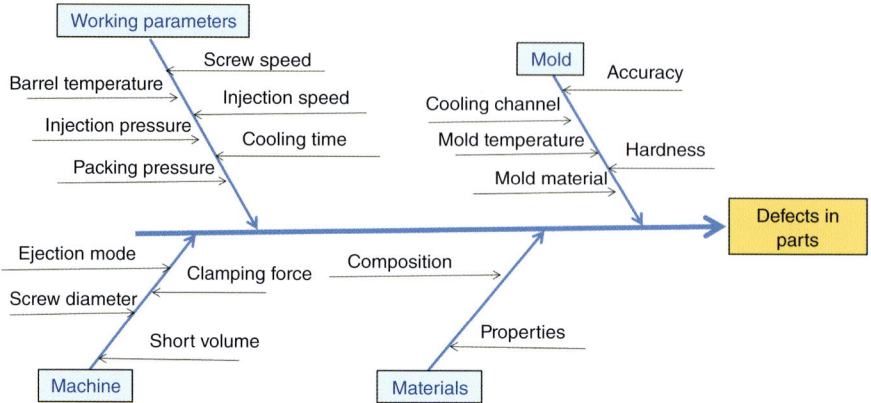

Figure 2.7 Important processing parameters during injection molding process.

matrix takes place in the heated barrel. The heating in barrel can be divided into different zones (from feeding zone to nozzle). The temperature in these zones has to be controlled judiciously. Inside the barrel, the heat is generated in two ways: external heating due to heaters and internal heating due to shearing action of screw during plasticizing. To increase the flow ability (reduction in viscosity) of the mixture, the temperature of the barrel must be higher, but at the same time excessive heating can degrade the matrix and natural fibers.

Screw Speed   The speed of the screw is another important parameter. Optimum speed is necessary to achieve a uniform distribution of fibers within the composites. Higher speed of the screw leads to higher heat generation and fiber attrition due to increased shear force. Sometimes, it can degrade the properties of the polymer also. In order to overcome these challenges, an optimum balance between the screw speed and back pressure is required.

Injection Speed   It plays a vital role during processing of thermoplastic-based composites. Injection speed has a direct influence on the viscosity of the melt compound. As the fiber loading increases, the viscosity of the melt compound also increases and to inject the melt compound and to fill the mold cavity totally, higher injection speed is required. Higher speed increases the shear force and excessive speed may degrade the properties of fibers and polymers. Higher injection speed is recommended for the processing of parts having thin cross section where high precision and accuracy are required.

Injection Pressure   It is one of the most important parameters during injection molding. It is the pressure by which the melt compound is forced to flow. Lower injection pressure leads to short shot type of defect in the molded part while higher pressure may lead to the formation of flash type of defect. Hence, optimum injection pressure is required to achieve the desired product without any defects. The operating window for injection molding process is depicted in Figure 2.8.

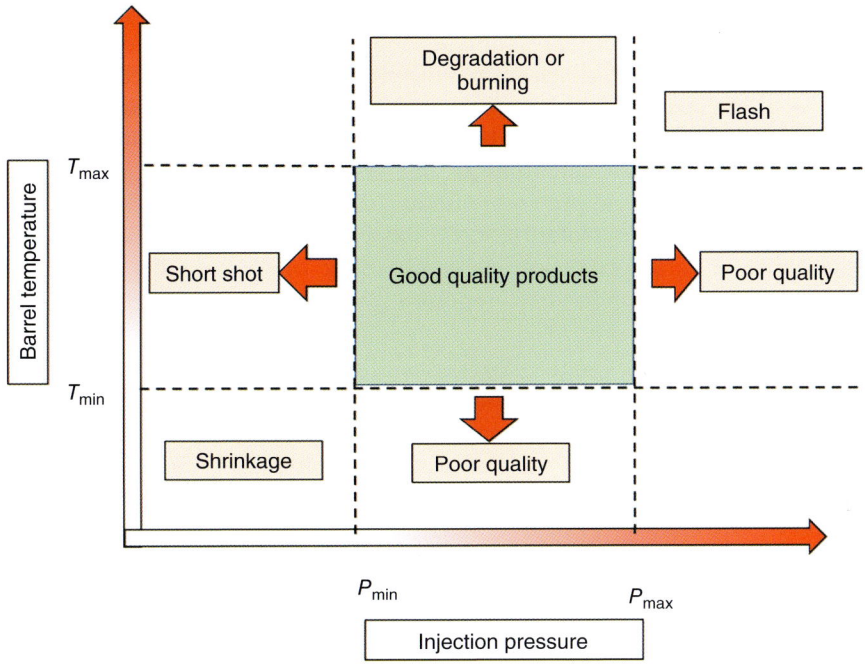

Figure 2.8 Processing window for injection molding process.

2.4.2.2 Challenges in Injection Molding of Natural Fiber-Based Composites

The challenges reported by several researchers during injection molding of polymeric composites have been discussed briefly in the following section.

Distribution and Alignment of Fibers The distribution and alignment of fibers have a direct influence on the behavior of the natural fiber-based composites. In case of short fibers, it becomes a challenging task to control the alignment of fibers. During processing, the melt compound is injected in the mold cavity through a nozzle. If the melt compound flows through the convergent section, the fibers tend to align in the flow direction, while in case of a divergent section, the fibers tend to align in the direction perpendicular to melt flow [23]. It has been reported that near the mold wall the fibers tend to align in a flow direction due to friction and shear force, while at the center (core region) the fibers tend to align in the direction perpendicular to the melt flow. It has been also reported that the pre-compounding (extrusion) of fibers and polymers prior to injection molding significantly improves the distribution and orientation of fibers resulting in the improvement of the mechanical properties of the composites [24].

Fiber Attrition The attrition of fibers is one of the important challenges associated with the injection molding process. The overall performance of the composites depends upon their load sharing capability with the fibers. If the fibers are damaged during processing stage itself, it may compromise the performance of the composites. Attrition of fibers takes place due to high shearing action, passage through narrow gates, and plasticizing during injection molding. The method (twin screw extrusion, pultrusion) of pre-compounding of fibers and polymers and subsequent pelletizing also leads to fiber attrition. Excessive injection speed, screw speed, and injection pressure are responsible for increased shearing action during processing, which results in the attrition of fibers. To overcome this challenge, careful selection of operating parameters and better tooling design are required. Better methods of pre-compounding can also be explored. In an investigation, the use of counterrotating screw has been used for compounding the fiber and matrix during extrusion. As compared to other methods, significant reduction in fiber attrition has been reported [25].

Residual Stresses Process induced stresses within the molded parts are called residual stresses; they can be either flow induced or thermally induced. During injection molding, these stresses are developed due to inadequate flow of melt compound and nonuniform cooling during solidification [26]. In case of composite materials, the stresses are developed due to difference in orientation of the polymeric chains, pressure gradient, and variation in the coefficient of thermal expansion of the constituents. These stresses can be minimized by designing a better cooling system that allows gradual cooling during solidification. These can also be reduced by heat treatment and optimizing the operating parameters such as screw and injection speed, barrel temperature, and injection pressure.

2.4.2.3 Fabrication of Polymeric Composites by Injection Molding Process

For processing of natural fiber-based polymeric composites, injection molding has been recommended as the most widely used process. In order to investigate the behavior of the natural fiber based composites, researchers have explored the injection molding route for the development of composites [27–29]. In an investigation, silk fiber-based PLA composites were developed using extrusion and injection molding processes. Good interfacial properties between the constituents were reported [30]. In a study, flax fiber reinforced PLA composites were developed by injection molding. The compounding of fibers and matrix was conducted using extrusion before injection molding. Good distribution of flax fibers within the biocomposites was reported without any fiber clusters [31]. In another investigation, the mechanical behavior of wood fiber-based polypropylene composites was developed by using three fabrication methods: high speed mixer, two roll mill, and twin screw extruder. The composite developed by twin screw extruder exhibited improved mechanical behavior [32].

2.4.2.4 Mechanical Performance of Injection Molded Composites

There are various factors (fiber loading, fiber aspect ratio, the distribution and orientation of fibers, interfacial characteristics between the constituents, processing routes) that govern the mechanical performance of natural fiber-based composites [33]. A brief review on the mechanical performance of injection molded composites is shown in Table 2.5.

Table 2.5 Mechanical properties of injection molded composites.

Materials	Findings	References
Abaca fibers (30 wt%) + PLA	Enhancement in tensile modulus and strength of the composite by 140% and 20%, respectively	[34]
Kenaf fiber (20 and 30 wt%) + PLA	As compared to neat PLA, the tensile modulus of the composites incorporating 20 and 30 wt% of kenaf fiber increased by 21% and 31%, respectively	[35]
Wood flour (20, 30, and 40 wt%) + PLA	The tensile modulus of the composite increased by 77%, 96%, and 133%, respectively	[36]
Cordenka (30 wt%) + PLA, flax fibers (30 wt%) + PLA	The impact strength of cordenka fiber reinforced composite was found to be higher	[37]
Kenaf fiber (20 wt%) + PLA, rice husk fiber (20 wt%) + PLA	The flexural and impact strength of the kenaf fiber-based composites exhibited superior properties	[38]
Banana fiber (10, 20, and 30 wt%) + PP	The tensile modulus, flexural strength, and modulus increased with respect to fiber loading	[21]

2.5 Conclusions

At present, the research community has been focusing on sustainable composites based on natural resources. Composite material can be called today's material as it possesses a plethora of opportunities. The application spectrum of composite materials has now spread and ranges from very common kitchen and sports goods to highly sophisticated automobile and aerospace components. As this spectrum is bound to increase further, there is a need for processing techniques for the fabrication of good quality composite products based on short natural fibers. The technique must have a fast processing cycle, should be easy to operate, and should be able to fabricate defect-free composite products. Considering these necessities, injection molding process has been recognized as an ideal fabrication technique. Much literature focusing on the performance of injection molded composites has been reported and is still continuing as the optimization and selection of these operating parameters during processing is a challenging task. In the current chapter, different processing techniques for the development of short fiber reinforced polymer composites have been studied. The various issues and challenges associated with the injection molding of short fiber-based composites have been discussed briefly. The effect of different processing parameters on the performance of injection molded composites has also been discussed.

It can be further concluded that the injection molding process possesses tremendous potential to be used for the fabrication of short fiber-based polymeric composites. Optimization and judicious selection of the injection molding parameters can lead to the development of defect-free sustainable composites, which will further increase the application spectrum of short natural fiber-based composites.

References

1 Najafi, A., Kord, B., Abdi, A., and Ranaee, S. (2012). The impact of the nature of nanoclay on physical and mechanical properties of polypropylene/reed flour nanocomposites. *Journal of Thermoplastic Composite Materials* 25: 717–727.

2 Joshi, S.V., Drzal, L.T., Mohanty, A.K., and Arora, S. (2004). Are natural fiber composites environmentally superior to glass fiber reinforced composites? *Composites. Part A, Applied Science and Manufacturing* 35: 371–376.

3 Ramamoorthy, S.K., Skrifvars, M., and Persson, A. (2015). A review of natural fibers used in biocomposites: plant, animal and regenerated cellulose fibers. *Polymer Reviews* 55: 107–162.

4 Faruk, O., Bledzki, A.K., Fink, H.P., and Sain, M. (2012). Biocomposites reinforced with natural fibers: 2000–2010. *Progress in Polymer Science* 37: 1552–1596.

5 Väisänen, T., Das, O., and Tomppo, L. (2017). A review on new bio-based constituents for natural fiber-polymer composites. *Journal of Cleaner Production* 149: 582–596.

6 Mohanty, A.K., Misra, M., and Hinrichsen, G. (2000). Biofibres, biodegradable polymers and biocomposites: an overview. *Macromolecular Materials Engineering* 276/277: 1–24.

7 Feyisetan Adekunle, K. (2015). Surface treatments of natural fibres—a review: part 1. *Open Journal of Polymer Chemistry* 5: 41–46.

8 Kabir, M.M., Wang, H., Lau, K.T., and Cardona, F. (2012). Chemical treatments on plant-based natural fibre reinforced polymer composites: an overview. *Composites. Part B, Engineering* 43: 2883–2892.

9 Huda, M.S., Drzal, L.T., Mohanty, A.K., and Misra, M. (2008). Effect of fiber surface-treatments on the properties of laminated biocomposites from poly(lactic acid) (PLA) and kenaf fibers. *Composites Science and Technology* 68: 424–432.

10 Vilay, V., Mariatti, M., Mat Taib, R., and Todo, M. (2008). Effect of fiber surface treatment and fiber loading on the properties of bagasse fiber-reinforced unsaturated polyester composites. *Composites Science and Technology* 68: 631–638.

11 Jandas, P.J., Mohanty, S., and Nayak, S.K. (2012). Renewable resource-based biocomposites of various surface treated banana fiber and poly lactic acid: characterization and biodegradability. *Journal of Polymers and the Environment* 20: 583–595.

12 Hu, R. and Lim, J.K. (2007). Fabrication and mechanical properties of completely biodegradable hemp fiber reinforced polylactic acid composites. *Journal of Composite Materials* 41: 1655–1669.

13 Komal, U.K., Verma, V., Ashwani, T. et al. (2018). Effect of chemical treatment on thermal, mechanical and degradation behavior of banana fiber reinforced polymer composites. *Journal of Natural Fibers* 00: 1–13.

14 Joseph, P., Joseph, K., and Thomas, S. (1999). Effect of processing variables on the mechanical properties of sisal-fiber-reinforced polypropylene composites. *Composites Science and Technology* 59: 1625–1640.

15 Petinakis, E., Yu, L., Edward, G. et al. (2009). Effect of matrix-particle interfacial adhesion on the mechanical properties of poly(lactic acid)/wood-flour micro-composites. *Journal of Polymers and the Environment* 17: 83–94.

16 Asaithambi, B., Ganesan, G.S., and Kumar, S.A. (2017). Banana/sisal fibers reinforced poly(lactic acid) hybrid biocomposites; influence of chemical modification of BSF toward thermal properties. *Polymer Composites* 1053–1062.

17 Asaithambi, B., Ganesan, G., and Ananda Kumar, S. (2014). Bio-composites: development and mechanical characterization of banana/sisal fibre reinforced poly lactic acid (PLA) hybrid composites. *Fibers and Polymers* 15: 847–854.

18 Chaitanya, S. and Singh, I. (2017). Sisal fiber-reinforced green composites: effect of ecofriendly fiber treatment. *Polymer Composites* 16: 101–113.

19 Mathew, A.P., Oksman, K., and Sain, M. (2006). The effect of morphology and chemical characteristics of cellulose reinforcements on the crystallinity of polylactic acid. *Journal of Applied Polymer Science* 101: 300–310.

20 Jagadeesh, D., Sudhakara, P., Prasad, C.V. et al. (2012). *A study on the mechanical properties of bio composites from renewable agro-waste/poly (lactic acid)*, vol. 6, 1–6.

21 Komal, U.K., Verma, V., Aswani, T. et al. (2018). Effect of chemical treatment on mechanical behavior of banana fiber reinforced polymer composites. *Materials Today: Proceedings* 5: 16983–16989.

22 Islam, M.S., Pickering, K.L., and Foreman, N.J. (2010). Influence of hygrothermal ageing on the physico-mechanical properties of alkali treated industrial hemp fibre reinforced polylactic acid composites. *Journal of Polymers and the Environment* 18: 696–704.

23 Ho, M.P., Wang, H., Lee, J.H. et al. (2012). Critical factors on manufacturing processes of natural fibre composites. *Composites. Part B, Engineering* 43: 3549–3562.

24 Chaitanya, S. and Singh, I. (2016). Processing of PLA/sisal fiber biocomposites using direct and extrusion–injection molding. *Materials and Manufacturing Processes*. https://doi.org/10.1080/10426914.2016.1198034.

25 Sykacek, E., Hrabalova, M., Frech, H., and Mundigler, N. (2009). Extrusion of five biopolymers reinforced with increasing wood flour concentration on a production machine, injection moulding and mechanical performance. *Composites. Part A, Applied Science and Manufacturing* 40: 1272–1282.

26 Wang, T.H. and Young, W.B. (2005). Study on residual stresses of thin-walled injection molding. *European Polymer Journal* 41: 2511–2517.

27 Adam, J., Korneliusz, B.A., and Agnieszka, M. (2013). Dynamic mechanical thermal analysis of biocomposites based on PLA and PHBV – a comparative study to PP counterparts. *Journal of Applied Polymer Science* 130: 3175–3183.

28 Bledzki, A.K., Mamun, A.A., and Faruk, O. (2007). Abaca fibre reinforced PP composites and comparison with jute and flax fibre PP composites. *Express Polymer Letters* 1: 755–762.

29 Mofokeng, J.P., Luyt, A.S., Tábi, T., and Kovács, J. (2012). Comparison of injection moulded, natural fibre-reinforced composites with PP and PLA as matrices. *Journal of Thermoplastic Composite Materials* 25: 927–948.

30 Cheung, H.Y., Lau, K.T., Tao, X.M., and Hui, D. (2008). A potential material for tissue engineering: silkworm silk/PLA biocomposite. *Composites. Part B, Engineering* 39: 1026–1033.

31 Rozite, L., Varna, J., Joffe, R., and Pupurs, A. (2013). Nonlinear behavior of PLA and lignin-based flax composites subjected to tensile loading. *Journal of Thermoplastic Composite Materials* 26: 476–496.

32 Bledzki, A.K., Letman, M., Viksne, A., and Rence, L. (2005). A comparison of compounding processes and wood type for wood fibre – PP composites. *Composites. Part A, Applied Science and Manufacturing* 36: 789–797.

33 Rao, M.V., Mahajan, P., and Mittal, R.K. (2008). Effect of interfacial debonding and matrix cracking on mechanical properties of multidirectional composites. *Composite Interfaces* 15: 379–409.

34 Bledzki, A.K. and Jaszkiewicz, A. (2010). Mechanical performance of biocomposites based on PLA and PHBV reinforced with natural fibres – a comparative study to PP. *Composites Science and Technology* 70: 1687–1696.

35 Pan, P., Zhu, B., Kai, W. et al. (2007). Crystallization behavior and mechanical properties of bio-based green composites based on poly(L-lactide) and kenaf fiber. *Journal of Applied Polymer Science* 105: 1511–1520.

36 Huda, M.S., Drzal, L.T., Misra, M., and Mohanty, A.K. (2006). Wood-fiber-reinforced poly(lactic acid) composites: evaluation of the physicomechanical and morphological properties. *Journal of Applied Polymer Science* 102: 4856–4869.

37 Bax, B. and Müssig, J. (2008). Impact and tensile properties of PLA/Cordenka and PLA/flax composites. *Composites Science and Technology* 68: 1601–1607.

38 Yussuf, A.A., Massoumi, I., and Hassan, A. (2010). Comparison of polylactic acid/kenaf and polylactic acid/rise husk composites: the influence of the natural fibers on the mechanical, thermal and biodegradability properties. *Journal of Polymers and the Environment* 18: 422–429.

3

Fabrication of Composite Laminates

Sandhyarani Biswas and Jasti Anurag

National Institute of Technology, Department of Mechanical Engineering, Rourkela 769008, India

3.1 Introduction

Over the last few decades, composite materials are used extensively in various emerging fields of engineering because of the advantages offered, particularly mechanical properties, with their attractive potential of saving weight. A substantial portion of the engineering materials ranging from everyday items to highly sophisticated applications have been taken over by the modern composite materials. Composite materials are much suited for weight sensitive applications, where the lighter members made of advanced fiber reinforced composites are used. Technologically, the most important composite materials are those in which the dispersed phase is in the form of a fiber. Fiber reinforced polymer (FRP) is a type of composite material made of a polymer matrix reinforced with fibers. FRP composite materials are very attractive for use in engineering applications due to their highly favorable material properties such as high strength to weight and stiffness to weight ratios, corrosion resistance, and light weight.

Laminate composites are one of the most commonly used structural composites that use laminated sheets of fabric materials embedded in the polymer matrix. Composite laminates are an assembly of layers of composite materials that are joined to meet the specific engineering requirements as represented (Figure 3.1). In a hybrid laminate, layers of different materials may be used. The individual layers in laminates are generally transversely isotropic (isotropic properties in the transverse plane) or orthotropic (normal properties in the orthogonal directions), resulting in the laminate exhibiting orthotropic, anisotropic, or quasi-isotropic properties. The laminate may exhibit coupling between out-of-plane and in-plane response depending on the stacking sequence of the laminates. The increased stiffness and strength for a specific weight, increased mechanical damping, toughness, and increased corrosion and chemical resistance in comparison to conventional metal are some of the main reasons that led to the improvement of laminated composites.

In the recent years, the significant growth of market share for FRP composites is due to their high performance and advances in fabrication techniques. An important aspect of composite part fabrication is the placement and direction

Reinforced Polymer Composites: Processing, Characterization and Post Life Cycle Assessment,
First Edition. Edited by Pramendra K. Bajpai and Inderdeep Singh.
© 2020 Wiley-VCH Verlag GmbH & Co. KGaA. Published 2020 by Wiley-VCH Verlag GmbH & Co. KGaA.

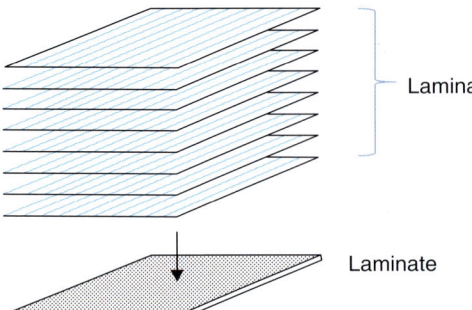

Figure 3.1 Composite laminate.

Lamina

Laminate

of fibers, which provides the required characteristics. Composite fabrication is a complex process involving various parameters such as matrix and reinforcement types, production volume, part geometry, and process and tooling requirements. Although, the properties of the matrix and reinforcing phase have significant effect on the final properties of the composites, the manufacturing technique selected to fabricate the composites also influences the performance of the composite materials. The availability of numerous choices makes it superior that the factors of design, economics, and manufacturing be integrated into the development process itself. Advances in the manufacturing process and technology of laminated composites have changed the use of the composites from secondary structural components to primary ones.

3.2 Fabrication Processes

Selection of composite fabrication process has emerged as one of the paramount challenges in the field of composite materials. The manufacturing technique for composite laminates depends on many factors such as the characteristics of matrix and reinforcement (fiber/matrix type, fiber content, fiber orientation, fiber length, etc.), geometry of the products (shape, size, etc.), and its end use. A brief outline of the different fabrication techniques for FRP laminate composites is described below.

3.2.1 Hand Lay-up Process

The simplest and most commonly used method for fabricating composite laminates is the hand lay-up process. It is the oldest method and generally uses continuous fibers in the form of unidirectional, woven, knitted, or stitched fabrics. Based on the application, the composite laminate can be fabricated by a combination of layers of laminae of different fiber orientation. In this method, alternative layers of matrix and reinforcement are placed over it. Generally, a releasing agent is used on the surface of the mold to prevent sticking and enable easy removal of the finished part. The matrix is pressed onto the reinforcements by a roller to remove the extra resin from each layer and to ensure uniform distribution of resin over the surface. The process is repeated for all layers of reinforcement until the

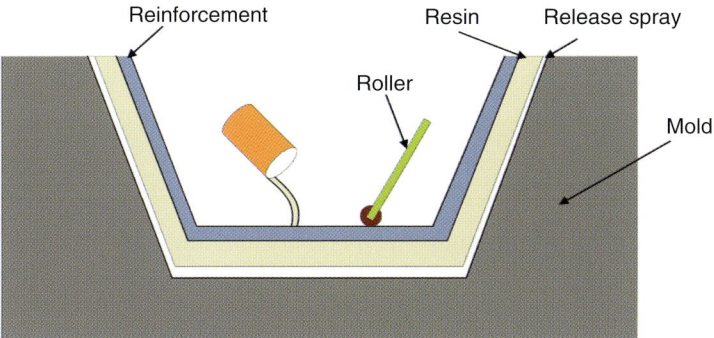

Figure 3.2 Hand lay-up process.

desired thickness is achieved. The whole process is done manually by hand. Curing of the composite laminates takes place at standard atmospheric temperature. Finally, the finished part is removed after complete cure. The performance and quality of composite products fabricated by the hand lay-up process are affected by many parameters such as fiber type, fiber content, fiber orientation, matrix type, pressure, and curing time. Figure 3.2 shows a typical hand lay-up process.

Advantages of this method include less capital investment, easy-to-change mold/design, and virtually no limit to the size of the part that can be produced. However, a few limitations of this process include the following: the method is suitable for low volume fraction/concentration of reinforcing phase; it is very time consuming; it has high content of voids and/or porosity; only one finished surface (which is in contact with the mold) is possible; it is labor intensive; control of thickness is not very accurate; it is difficult to achieve uniform fiber-to-resin ratio; and it is also not suitable for high volume production. This method is suitable for manufacturing wind turbine blades, tanks, vessels, boat hulls, etc.

A study on composite laminates fabricated using the hand lay-up technique has been made by many researchers [1, 2]. A study on yarn flax fibers for polymer-coated sutures and hand lay-up polymer composite laminates is made by Fong et al. [3]. The effect of moisture on the mechanical properties of yarn flax fibers as well as the possible dependence on knot geometry is investigated. An investigation on the fabrication and multifunctional properties of a hybrid laminate with aligned carbon nanotubes grown in situ is done by Garcia et al. [4]. The laminates are fabricated using the hand lay-up technique and mechanical and electrical tests are conducted. A study on the development of a telephone stand produced with woven banana pseudo-stem fiber/epoxy composites is performed by Sapuan and Maleque [5]. Similarly, a study on woven banana pseudo-stem fiber/epoxy composite-based multipurpose table fabricated using hand lay-up technique is made by Sapuan et al. [6]. Banakar et al. [7] investigated the influence of fiber orientation and thickness on tensile properties of glass FRP laminate composites. It is concluded from the study that the tensile strength mainly depends on the fiber orientation and thickness. Misri et al. [8] investigated the mechanical behavior of woven glass/sugar palm fiber reinforced unsaturated polyester hybrid composite based small boat made by hand lay-up technique. Investigations on the thermal and flexural properties of plain weave carbon/epoxy nanoclay composites fabricated by hand lay-up technique are performed by Chowdhury et al. [9].

3.2.2 Filament Winding Process

Filament winding is a process where composite parts are manufactured by winding of continuous fibers on a rotating mandrel in specific orientations. For high volume production of symmetrical composite parts, filament winding process is the most economical way. This process is primarily used for hollow, generally circular, or oval sectioned components, such as pipes and tanks. It has a wide range of applications ranging from a small gas cylinder to a large cryogenic tank. This method is well suited for automation where little or no human intervention is required. Before winding of fibers, they are passed through a resin bath, where the fibers are wetted by the resin. The number of layers and thickness of winding mainly depend on the desired properties of the composite parts. In order to compact the fibers on to the mandrel, the fibers are given the desired tension. The winding pattern can be changed with the movement of rotating mandrel and moving carriage. Curing is generally done at room temperature or at an elevated temperature. After curing is done, the mandrel is removed from the composite part and can be reused. Based on the desired applications, different winding patterns can be used. The most commonly used winding patterns are hoop winding, polar winding, and helical winding. Mandrel in the filament winding process plays a major role. The shape of the mandrel defines the shape of the composite part being produced. The material with which the mandrel is made mainly depends on the end use of the fabricated composite parts. Mandrels can be either removable or nonremovable. Removable mandrels can be classified as entirely removable, collapsible, breakable, or soluble. A typical filament winding process is shown in Figure 3.3.

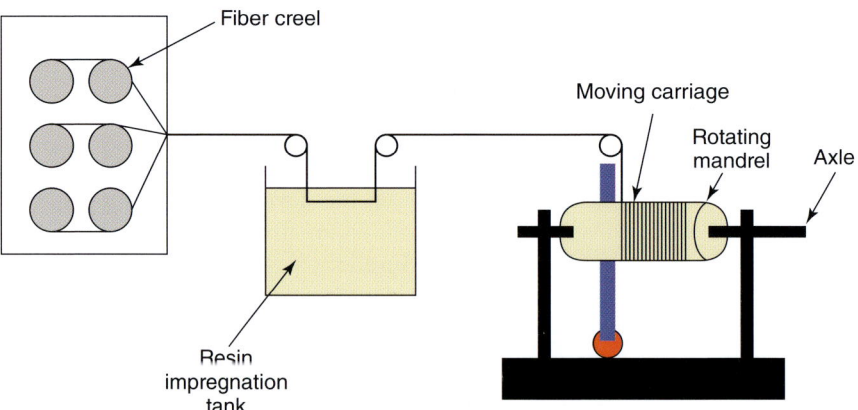

Figure 3.3 Filament winding process.

Advantages of filament winding include excellent mechanical properties due to the use of continuous fibers, fastness of the process, good thickness control, better control of fiber orientation and content, high volume fraction/concentration of reinforcing phase, and good internal finish. However, difficulty in winding complex shapes, which may require complex equipment; poor external finish; restriction to convex shaped components; high cost of mandrel; and need for low viscosity resins are the major limitations of this method. The application of filament winding process includes the manufacturing of open end structures such as gas cylinders and piping systems and closed end structures such as pressure vessels and chemical storage tanks.

Many researchers have been working on the composites fabricated with filament winding process. A filament wound graphite/thermoplastic composite ring-stiffened pressure hull model has been designed, analyzed, and manufactured, and hydrostatic testing done by Lamontia et al. [10]. Misri et al. [11] studied the mechanical behavior of kenaf fiber/unsaturated polyester composite hollow shafts fabricated using filament winding technique. Continuous kenaf fiber rovings were pulled through a drum-type resin bath and wound around an aluminum rotating mandrel. An experimental investigation on the influence of multi-angle filament winding on the strength properties of tubular composite structures has been carried out [12]. Cohen [13] investigated the effect of filament winding parameters on the strength and quality of the composite vessel. The impact of filament winding tension on the physical and mechanical behavior of glass fiber reinforced polymeric composite tubular has been studied by Mertiny and Ellyin [14]. Xia et al. [15] analyzed the stress and deformation of multilayered filament wound carbon fiber/epoxy composite pipes. In a study, the manufacturing of flat and cylindrical laminates and built up structure using automated thermoplastic tape laying, fiber placement, and filament winding is done by Lamontia et al. [16].

3.2.3 Compression Molding Process

One of the oldest methods for manufacturing composite laminates is compression molding. It is quite popular as it has the capability for mass production within a short time compared to other older methods. It is the process of molding a composite material into the desired shape by applying heat and pressure. The final shape of the product depends on the matched compression molds used. Compression molding is a two-step process, namely, preheating and pressurizing. Initially, the charge or preform is put in the cavity of matched mold while it is in the open position and the two halves are brought together, thereby closing the mold. Pressure is then applied to squeeze the resin, so that it fills the mold cavity. While under pressure, curing of the material takes place by the application of heat. After curing, the pressure is released and the mold is opened to remove the finished part. The parameters influencing the performance of composites made by compression molding are the pressure, temperature, quantity of molding compound, type of resin, cure time, etc. Figure 3.4 shows a typical compression molding process.

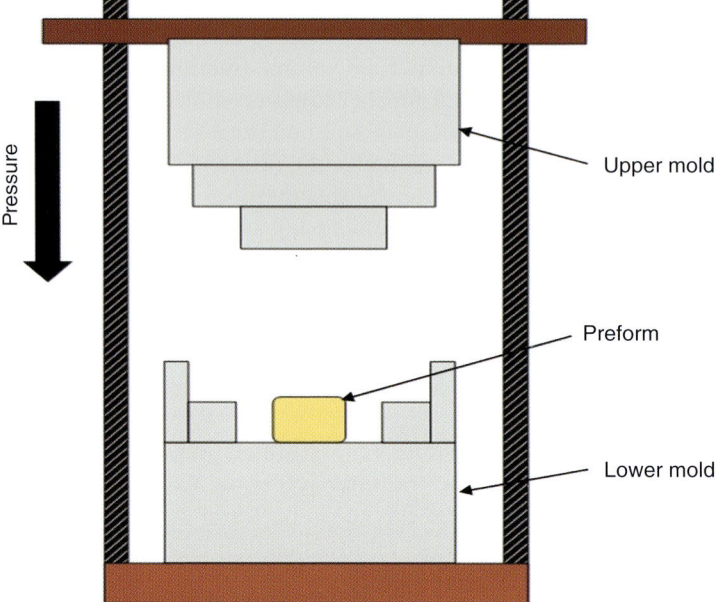

Figure 3.4 Compression molding process.

The primary advantage of compression molding is its ability to produce a large number of parts with little dimensional variations. Besides, other advantages include short cycle time, finished interior and exterior surfaces, better quality surface, uniformity in part shape, less maintenance cost, need for very little finishing operation, and better control of fiber content. However, the high initial capital investment, not being suitable for very large-sized parts, not being economical for low volume of production, and limitation on mold depth are the major disadvantages of this method. Typical products that can be produced using compression molding process include the bumper beams, road wheels, refrigerator doors, automobile panels, electrical fixtures, machine guards, door panels, hoods, kitchen bowls and trays, control boxes, etc.

Study on the compression molded composite products has been done by many researchers. An experimental investigation on the durability of compression molded sisal fiber reinforced mortar laminates is done [17]. The durability of laminates is investigated by determining the effects of accelerated gaining on the microstructure and flexural behavior of the composites. Prabu et al. [18] studied the mechanical properties of red mud filled sisal and banana fiber reinforced polyester composites fabricated by compression molding. Chen et al. [19] studied the structure and properties of composites made with polyurethane pre-polymer and various soy products fabricated using compression molding. The effect of fiber orientation on mechanical properties such as tensile and flexural strength of compression molded sisal fiber reinforced epoxy composites has been studied [20]. Correlo et al. [21] investigated the mechanical properties of scaffolds using chitosan–polyester blends and composites. Ismail et al. [22] made a review on the compression molding of natural fiber reinforced thermoset composites in terms of their thermal and mechanical properties. Holbery and Houston [23] did a comparative study on the compression molded unsaturated polyester composites reinforced with glass fiber and flax fibers, respectively. A study on the compression molding of the long chopped fiber thermoplastic composites is done by Howell et al. [24].

3.2.4 Vacuum Bagging Process

Vacuum bagging is a composite manufacturing process in which vacuum pressure is used during the resin cure cycle. In this process, atmospheric pressure is generally used to hold the resin and fibers in the desired place, which consolidates the layers within the laminate. The laminate is mainly sealed in an airtight bag, and a vacuum pump then evacuates all the air out of the bag, resulting in an even atmospheric pressure over the entire composite laminate. In vacuum bagging process, different layers include mold, release agent, composite laminate, peel ply, bleeder, release film, breather, and vacuum bag. Peel ply is used to create a clean surface for bonding purpose. Release agent is essential for preventing the resin from sticking to the mold surface when laminating a part. To provide a vacuum tight seal between the mold surface and the bag, sealant tapes are generally used on both sides of the bag. A bleeder layer is generally used to absorb the excess resin from the laminate. The release film is a perforated film that allows the entrapped air and volatiles to escape. The breather is used to create uniform pressure around the part. Pressure and heat are applied for a specific time. Pressure helps in distribution of matrix evenly and allows bonding between matrix and fibers. The fabricated composite laminates can be cured at room temperature or at an elevated temperature. The quality of composite laminates fabricated by vacuum bagging process is influenced by many parameters such as type of reinforcement, quality of sheets used, viscosity of the resin, and defined pressure. A typical vacuum bagging process is shown in Figure 3.5.

This process has many advantages: good adhesion between layers leading to better quality parts; high fiber volume fraction achievable; uniform distribution of matrix; very little emissions because the entire laminate is sealed in a bag; increased part consolidation; and reduced mold costs. However, the disadvantages include that the process is not suitable for high production volume, breather cloth needs to be replaced frequently, and that the process needs expensive

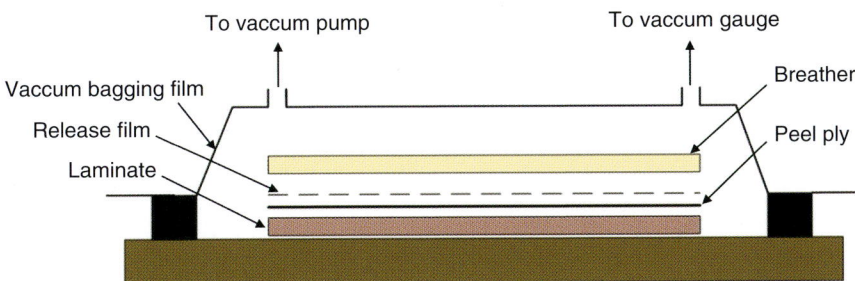

Figure 3.5 Vacuum bagging process.

curing ovens. Vacuum bagging process is mainly used for manufacturing parts such as large boat hulls, aircraft structures, racing car components, and bathtubs.

A few researchers have studied the performance of composite laminates made by vacuum bagging process. Aparna et al. [25] studied the continuous glass FRP composites fabricated by vacuum bag molding. Mariatti and Abdul Khalil [26] studied the mechanical behavior of bagasse fiber reinforced polyester composites.

3.2.5 Autoclave Molding

Autoclave molding is similar to the vacuum bagging process with some modifications. The need for autoclave in composite fabrication is due to the high quality requirements of composite laminates for very demanding industries such as aerospace. This advanced process generally produces compact and void-free parts because this method uses an autoclave to provide high heat and pressure to the composite products during curing. In autoclave molding, the material most often employed is composite prepreg. Autoclave curing gives uniform consolidation of prepreg laminates with lesser voids. The materials used in this process include mold, peel ply, release agent, bleeder, breather, and vacuum bag. In this process, layers of prepreg are piled with different fiber orientation to form the desired thickness above the molding plate. By the application of pressure, the prepreg is pressed down to the molding plate. Subsequent heating hardens the matrix and the laminate gets the desired shape. Generally, the peel plies are attached on the top and bottom layers to provide better surface finish. Layers of porous bleeder and breather cloth are laid to absorb excess resin during curing. The whole assembly is vacuum bagged to remove any air entrapped in between the layers. Finally, the mold and composite part with vacuum applied is moved into an autoclave for curing. This process is very versatile and it provides uniform quality products as the heat and pressure can be regulated precisely. The process is mainly used for making aircraft parts, spacecraft, military applications, etc. The low production rate and the limitation on part size, which mainly depend on the autoclave size, are the major drawbacks of this process. Also, the process is very expensive, time consuming and labor intensive.

Investigations on the composite laminates fabricated by autoclave process have been done by a few researchers. Experimental and statistical study on drilling of carbon fiber reinforced plastics manufactured by autoclave is done by Davim and Reis [27]. Tarsha-Kurdi and Olivier [28] analyzed the effects of autoclave curing pressure and cooling rate on the room temperature curvature of carbon fiber reinforced epoxy laminated strips.

3.2.6 Resin Transfer Molding (RTM) Process

Resin transfer molding (RTM) is a widely used technique for manufacturing high performance thermosetting composite laminates. The pre-shaped reinforcement in the shape of the final product is generally kept in the lower half of the mold. The upper mold is closed on to the lower mold. Catalyzed, low-viscosity resin is injected into the mold with high pressure and temperature. High pressure helps

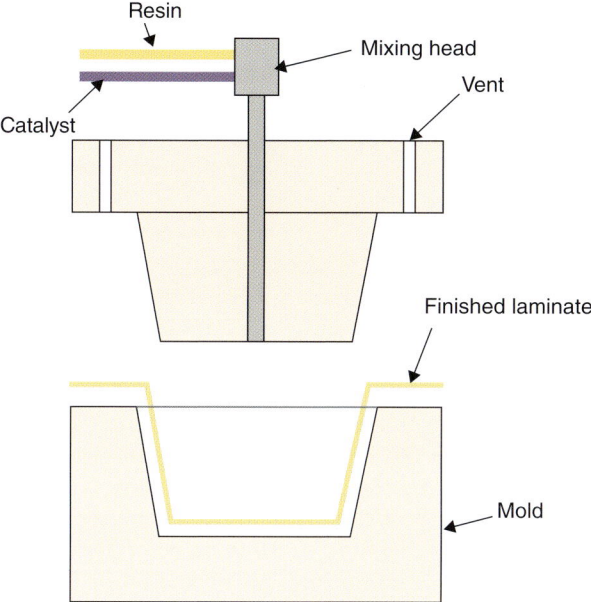

Figure 3.6 Resin transfer molding process.

the resin to impregnate into the reinforcement and the gases escape out of the mold through the vents provided in the mold. After curing at room or elevated temperature, the mold is opened and the laminate is removed. Curing depends on the thickness of the laminate, the type of resin used, and the temperature and pressure in the mold. RTM process has the capability of rapid manufacture of large, complex, high performance composite structures with good surface finish on both sides. The parameters influencing the quality of the final composite product are fiber geometry, fiber content, resin viscosity, applied pressure, mold temperature, etc. A typical RTM process is depicted in Figure 3.6.

This process yields increased laminate compression, a high fiber-to-resin ratio, and outstanding strength to weight characteristics. Other advantages include the good surface finish on both sides, better control over the product thickness, short cycle time, low volatile emissions, and better utilization of resin and fiber. However, a few limitations of the process are that the tooling cost is high, mold cavity generally limits the size of the part, and that the process is limited to low viscosity resins. The process is commonly used in aerospace, automotive, and sporting industries.

A great deal of work has been done on the composites fabricated using RTM process by many researchers. Tari et al. [29] studied the rapid prototyping of composite parts using RTM and laminated object manufacturing. Salim et al. [30] analyzed the influence of stitching density of nonwoven fiber mat on the mechanical behavior of kenaf fiber reinforced epoxy composites fabricated by RTM. The effect of fiber surface modification on the mechanical and water absorption characteristics of sisal fiber reinforced polyester composites prepared by RTM is studied by Sreekumar et al. [31]. Pothan et al. [32] investigated the tensile and flexural behavior of sisal fabric reinforced polyester textile composites made by RTM. Rouison et al. [33] studied the resin transfer molded natural fiber reinforced composites. The thermo-physical behavior of hemp/kenaf fiber reinforced unsaturated polyester composites is analyzed in their study. Cairns et al. [34] had done a modeling of composite materials with oriented unidirectional plies. In another study, porosity reduction using optimized flow velocity in RTM is investigated by Leclerc and Ruiz [35].

One attractive manufacturing technique to fabricate low cost, high performance composite components is vacuum-assisted resin transfer molding (VARTM). The main difference between VARTM and RTM is that in VARTM, resin is drawn into a preform through the use of a vacuum only, rather than pumped in under pressure. VARTM does not require high heat or pressure like RTM. For that reason, VARTM operates with low cost tooling, making it possible to inexpensively produce large, complex parts in one shot. In addition, complex shapes with unique fiber architectures allow the fabrication of large parts that have a high structural performance. A great deal of work has been done on the fabrication of composites with VARTM techniques. An experimental investigation on the use of carbon nanofibers to improve the interlaminar fracture properties of polyester/glass fiber composites fabricated by VARTM is done by Sadeghian et al. [36]. In another study, the characterization of layered silicate/glass fiber/epoxy hybrid nanocomposites prepared using VARTM is done by Lin et al. [37].

3.2.7 Pultrusion Process

Pultrusion process is a highly automated continuous fiber laminating process producing high fiber volume profiles with a constant cross section. This process creates a continuous composite profile by pulling raw composites through a heated die. This process is similar to extrusion; however, the main difference is that the composite part is pulled from the die, whereas in the case of extrusion process it is pushed through the die. Fiber is wetted or impregnated with resin and is organized and then the excess resin removed. Finally, the curing of composite part takes place inside the heated dies. The shape of the final composite product generally follows the shape of the dies. At the end of the pultrusion machine there is a cut-off saw to cut out parts of the desired size. The most common shapes that can be obtained using this process are square, circular, rectangular, and I-shaped sections. The end products are generally in the form of bars and rods. The parameters influencing the quality of composite products fabricated by pultrusion are resin viscosity, resin polymerization, fiber content, pulling speed, die temperature, etc. A typical pultrusion process is shown in Figure 3.7.

Pultrusion is a very fast and highly automated process. Other advantages of this process are as follows: possibility of high volume fraction, not being labor intensive, better strength properties, consistent quality, easy control of resin content, good structural properties, better surface finish, etc. However, the main limitation of this process is that it is suitable to produce the components only with uniform cross sections. Pultrusion is used to manufacture constant cross-section shapes, viz. I-beam, box, channels, and tubings.

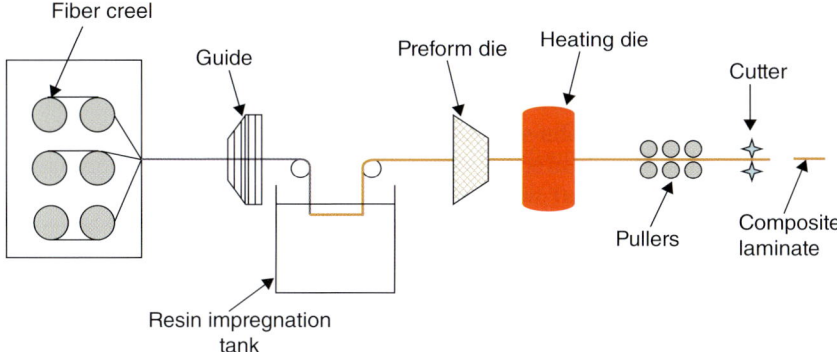

Figure 3.7 Pultrusion process.

Research on the FRP composites fabricated by pultrusion has been done by many researchers. Akil et al. [38] analyzed the flexural and indentation behavior of pultruded jute/glass and kenaf/glass fiber reinforced hybrid polyester composites monitored using acoustic emission. Van de Velde and Kiekens [39] investigated the possibility of the use of flax fiber as reinforcement in thermoplastic-based pultruded composites. Round flax fiber reinforced polypropylene and rectangular glass fiber reinforced polypropylenes profiles are fabricated using pultrusion process and various properties are tested. Fairuz et al. [40] studied the mechanical behavior of kenaf fiber/vinyl ester composites fabricated by pultrusion process using Taguchi's design of experiment. The effect of three different parameters such as speed, temperature, and filler loading has been analyzed in their study. Moschiar et al. [41] have done a theoretical modeling of pulling speed for pultruded epoxy composites. It is concluded from the study that the high pulling velocity reduces the thermal stress and pressure inside the heated die. Davalos et al. [42] have performed the analysis and design of pultruded FRP composite beams under bending. It is concluded from the study that the material architecture of pultruded composite shapes can be efficiently modeled as a layered system. Zhu et al. [43] analyzed the mechanical behavior of glass FRP composites fabricated by pultrusion using chemically modified soy-based epoxy resins. It is concluded from the study that pultruded composites with soy-based co-resin systems show comparable or improved structural performance characteristics. A study on the high speed pultrusion of thermoplastic matrix composites is done by Miller et al. [44].

3.3 Conclusions

Recently, advances in the manufacturing process and technology of laminated composites have changed the use of the composites from secondary structural components to primary ones. According to the end-item design requirements, a variety of fabrication techniques can be used for composite laminates. Each technique requires different kinds of material systems, different tools, and also different processing conditions. Depending on the property, cost, quality, and quantity of the product, a suitable technique can be selected. One attractive manufacturing technique to fabricate low cost, high performance composite components is VARTM. Some of the recent improvements include more advanced and sophisticated equipment that has been used with precise control of pressure, temperature, and other parameters to achieve better quality products. Nowadays, filament winding equipment is available with numerous levels of sophistication. The machine generally ranges from the simple mechanically controlled equipment to sophisticated computer controlled equipment. Advanced composites for aircraft and aerospace applications as well as other highly demanding end-use application have encouraged development of a number of specialized composite processing methods such as vacuum bagging and autoclave molding. Similarly, many variations in compression molding have been developed that are suitable for many engineering applications. Significant technological developments have greatly enhanced the quality of molded parts and the efficiency of the composite laminate fabrication processes.

References

1 Biswas, S., Deo, B., Patnaik, A., and Satapathy, A. (2011). Effect of fiber loading and orientation on mechanical and erosion wear behaviors of glass-epoxy composites. *Polymer Composites* 32 (4): 665–674.
2 Mishra, V. and Biswas, S. (2016). Three-body abrasive wear behavior of short jute fiber reinforced epoxy composites. *Polymer Composites* 37 (1): 270–278.
3 Fong, T.C., Saba, N., Liew, C.K. et al. (2015). Yarn flax fibres for polymer-coated sutures and hand layup polymer composite laminates. In: *Manufacturing of Natural Fibre Reinforced Polymer Composites*, 155–175. Cham: Springer.
4 Garcia, E.J., Wardle, B.L., John Hart, A., and Yamamoto, N. (2008). Fabrication and multifunctional properties of a hybrid laminate with aligned carbon nanotubes grown in situ. *Composites Science and Technology* 68 (9): 2034–2041.
5 Sapuan, S.M. and Maleque, M.A. (2005). Design and fabrication of natural woven fabric reinforced epoxy composite for household telephone stand. *Materials and Design* 26 (1): 65–71.
6 Sapuan, S.M., Harun, N., and Abbas, K.A. (2007). Design and fabrication of a multipurpose table using a composite of epoxy and banana pseudostem fibres. *Journal of Tropical Agriculture* 45 (1–2): 66–68.
7 Banakar, P., Shivananda, H.K., and Niranjan, H.B. (2012). Influence of fiber orientation and thickness on tensile properties of laminated polymer composites. *International Journal of Pure and Applied Sciences and Technology* 9 (1): 61.
8 Misri, S., Leman, Z., Sapuan, S.M., and Ishak, M.R. (2010). Mechanical properties and fabrication of small boat using woven glass/sugar palm fibres reinforced unsaturated polyester hybrid composite. *IOP Conference Series: Materials Science and Engineering* 11 (1): 1–13.

9 Chowdhury, F.H., Hosur, M.V., and Jeelani, S. (2007). Investigations on the thermal and flexural properties of plain weave carbon/epoxy-nanoclay composites by hand-layup technique. *Journal of Materials Science* 42 (8): 2690–2700.

10 Lamontia, M.A., Gruber, M.B., Smoot, M.A. et al. (1995). Performance of a filament wound graphite/thermoplastic composite ring-stiffened pressure hull model. *Journal of Thermoplastic Composite Materials* 8 (1): 15–36.

11 Misri, S., Sapuan, S.M., Leman, Z., and Ishak, M.R. (2015). Torsional behaviour of filament wound kenaf yarn fibre reinforced unsaturated polyester composite hollow shafts. *Materials and Design* 65: 953–960.

12 Mertiny, P., Ellyin, F., and Hothan, A. (2004). An experimental investigation on the effect of multi-angle filament winding on the strength of tubular composite structures. *Composites Science and Technology* 64 (1): 1–9.

13 Cohen, D. (1997). Influence of filament winding parameters on composite vessel quality and strength. *Composites Part A: Applied Science and Manufacturing* 28 (12): 1035–1047.

14 Mertiny, P. and Ellyin, F. (2002). Influence of the filament winding tension on physical and mechanical properties of reinforced composites. *Composites Part A: Applied Science and Manufacturing* 33 (12): 1615–1622.

15 Xia, M., Takayanagi, H., and Kemmochi, K. (2001). Analysis of multi-layered filament-wound composite pipes under internal pressure. *Composite Structures* 53 (4): 483–491.

16 Lamontia, M.A., Funck, S.B., Gruber, M.B. et al. (2003). Manufacturing flat and cylindrical laminates and built up structure using automated thermoplastic tape laying, fiber placement, and filament winding. *Sampe Journal* 39 (2): 30–43.

17 Toledo Filho, R.D., Silva, F.d.A., Fairbairn, E.M.R., and de Almeida MeloFilho, J. (2009). Durability of compression molded sisal fiber reinforced mortar laminates. *Construction and Building Materials* 23 (6): 2409–2420.

18 Prabu, V.A., Manikandan, V., Uthayakumar, M., and Kalirasu, S. (2012). Investigations on the mechanical properties of red mud filled sisal and banana fiber reinforced polyester composites. *Materials Physics and Mechanics* 15 (2): 173–179.

19 Chen, Y., Zhang, L., and Du, L. (2003). Structure and properties of composites compression molded from polyurethane prepolymer and various soy products. *Industrial and Engineering Chemistry Research* 42 (26): 6786–6794.

20 Kumaresan, M., Sathish, S., and Karthi, N. (2015). Effect of fiber orientation on mechanical properties of sisal fiber reinforced epoxy composites. *Journal of Applied Science and Engineering* 18 (3): 289–294.

21 Correlo, V.M., Boesel, L.F., Pinho, E. et al. (2009). Melt-based compression-molded scaffolds from chitosan–polyester blends and composites: morphology and mechanical properties. *Journal of Biomedical Materials Research Part A* 91A (2): 489–504.

22 Ismail, N.F., Sulong, A.B., Muhamad, N. et al. (2015). Review of the compression moulding of natural fiber-reinforced thermoset composites: material processing and characterisations. *Journal of Tropical Agricultural Science* 38 (4): 533–547.

23 Holbery, J. and Houston, D. (2006). Natural-fiber-reinforced polymer composites in automotive applications. *JOM* 58 (11): 80–86.

24 Howell, D.D, and Fukumoto, S. (2014). Compression molding of long chopped fiber thermoplastic composites In *Composites and Advanced Materials Expo (CAMX) Conference produced by The American Composites Manufacturers Association (ACMA) and the Society for the Advancement of Material Process Engineering (SAMPE) at Orange County Convention Center in Orlando, FL, USA*, October 13–16, 2014.

25 Aparna, M.L., Chaitanya, G., Srinivas, K., and Rao, J.A. (2016). Fabrication of continuous GFRP composites using vacuum bag moulding process. *International Journal of Advanced Science and Technology* 87: 37–46.

26 Mariatti, J. and Abdul Khalil, H.P.S. (2009). Properties of bagasse fibre-reinforced unsaturated polyester (USP) composites. In: *Research on Natural Fibre Reinforced Polymer Composites* (ed. S.M. Sapuan), 63–83. Serdang: UPM Press.

27 Davim, J.P. and Reis, P. (2003). Drilling carbon fiber reinforced plastics manufactured by autoclave—experimental and statistical study. *Materials and Design* 24 (5): 315–324.

28 Tarsha-Kurdi, K.E. and Olivier, P. (2002). Thermoviscoelastic analysis of residual curing stresses and the influence of autoclave pressure on these stresses in carbon/epoxy laminates. *Composites Science and Technology* 62 (4): 559–565.

29 Tari, M.J., Bals, A., Park, J. et al. (1998). Rapid prototyping of composite parts using resin transfer molding and laminated object manufacturing. *Composites Part A: Applied Science and Manufacturing* 29 (5–6): 651–661.

30 Salim, M.S., Ishak, Z.A.M., and Abdul Hamid, S. (2011). Effect of stitching density of nonwoven fiber mat towards mechanical properties of kenaf reinforced epoxy composites produced by resin transfer moulding (RTM). *Key Engineering Materials* 471–472: 987–992.

31 Sreekumar, P.A., Thomas, S.P., Saiter, J.M. et al. (2009). Effect of fiber surface modification on the mechanical and water absorption characteristics of sisal/polyester composites fabricated by resin transfer molding. *Composites Part A: Applied Science and Manufacturing* 40 (11): 1777–1784.

32 Pothan, L.A., Mai, Y.W., Thomas, S., and Li, R.K.Y. (2008). Tensile and flexural behavior of sisal fabric/polyester textile composites prepared by resin transfer molding technique. *Journal of Reinforced Plastics and Composites* 27 (16–17): 1847–1866.

33 Rouison, D., Sain, M., and Couturier, M. (2004). Resin transfer molding of natural fiber reinforced composites: cure simulation. *Composites Science and Technology* 64 (5): 629–644.

34 Cairns, D.S., Humbert, D.R., and Mandell, J.F. (1999). Modeling of resin transfer molding of composite materials with oriented unidirectional plies. *Composites Part A: Applied Science and Manufacturing* 30 (3): 375–383.

35 Leclerc, J.S. and Ruiz, E. (2008). Porosity reduction using optimized flow velocity in resin transfer molding. *Composites Part A: Applied Science and Manufacturing* 39 (12): 1859–1868.

36 Sadeghian, R., SudhirGangireddy, B.M., and Hsiao, K.-T. (2006). Manufacturing carbon nanofibers toughened polyester/glass fiber composites using vacuum assisted resin transfer molding for enhancing the mode-I delamination resistance. *Composites Part A: Applied Science and Manufacturing* 37 (10): 1787–1795.

37 Lin, L.-Y., Lee, J.-H., Hong, C.-E. et al. (2006). Preparation and characterization of layered silicate/glass fiber/epoxy hybrid nanocomposites via vacuum-assisted resin transfer molding (VARTM). *Composites Science and Technology* 66 (13): 2116–2125.

38 Akil, H.M., De Rosa, I.M., Santulli, C., and Sarasini, F. (2010). Flexural behaviour of pultruded jute/glass and kenaf/glass hybrid composites monitored using acoustic emission. *Materials Science and Engineering: A* 527 (12): 2942–2950.

39 Van de Velde, K. and Kiekens, P. (2001). Thermoplastic pultrusion of natural fibre reinforced composites. *Composite Structures* 54 (2): 355–360.

40 Fairuz, A.M., Sapuan, S.M., Zainudin, E.S. et al. 2014a. Study of pultrusion process parameters. *Proceedings of the Postgraduate Symposium on Composites Science and Technology 2014 & 4th Postgraduate Seminar on Natural Fibre Composites*, Putrajaya, Malaysia, (28 January). pp. 116–120.

41 Moschiar, S.M., Reboredo, M.M., Larrondo, H., and Vazquez, A. (1996). Pultrusion of epoxy matrix composites: pulling force model and thermal stress analysis. *Polymer Composites* 17: 850–858.

42 Davalos, J.F., Salim, H.A., Qiao, P. et al. (1996). Analysis and design of pultruded FRP shapes under bending. *Composites Part B: Engineering* 27 (3): 295–305.

43 Zhu, J., Chandrashekhara, K., Flanigan, V., and Kapila, S. (2004). Manufacturing and mechanical properties of soy-based composites using pultrusion. *Composites Part A: Applied Science and Manufacturing* 35 (1): 95–101.

44 Miller, A.H., Dodds, N., Hale, J.M., and Gibson, A.G. (1998). High speed pultrusion of thermoplastic matrix composites. *Composites Part A: Applied Science and Manufacturing* 29 (7): 773–782.

4

Processing of Polymer-Based Nanocomposites

Ramesh K. Nayak[1], Kishore K. Mahato[2], and Bankim C. Ray[2]

[1] *Maulana Azad National Institute of Technology, Department of Materials and Metallurgical Engineering, Bhopal, Madhya Pradesh, 462003, India*
[2] *National Institute of Technology, Department of Metallurgical and Materials Engineering, Rourkela, Odisha, 769008, India*

4.1 Introduction

Nanotechnology is engineering at the molecular or atomic level. It is the common or combined word for an extensive variety of measurements, processing techniques, and technologies. This involves handling of matter at the minimum scale, i.e. in nanometer level. The word nano has been invented from the Latin word nannusand and Greek word $ν\tilde{α}νoς$, both having the same meaning as dwarf. The officially adopted SI unit is 10^{-9}. Usually, nanotechnology deals with structures and components having size between 1 and 100 nm at least in any one direction. It is a technology that develops particles and materials at the nanoscale level. Eventually, these materials and particles are said to be nanomaterials and nanoparticles, respectively. E. Drexler said that "Nanotechnology is the source of handling of the structure of matter at the molecular level. It entails the ability to build molecular systems with atom-by-atom precision, yielding a variety of nanomachines" [1]. Nanotechnology is the employment of nanostructures into suitable nanoscale products [2]. The various laws related to science were not sufficient to manage all engineered nanostructures or nanomaterials. Generally, the nanoparticles possess a very large specific surface area, a high surface to mass ratio, and a high aspect ratio. The other factors that affect nanoparticles can be the chemical, physical, biological, electrical, and mechanical properties.

The nanomaterials used commonly for engineering applications as consumer products are carbon nanotubes, metal oxides at nano sizes (titanium dioxide, ferrous oxides, and zinc oxide), nanosilver, gold, and silica. Other engineered nanomaterials used in industrial, consumer, and medical products are aluminum oxide, nanoclays, nanocarbon, iron oxide, cerium oxide, nickel and copper oxide, and quantum dots. Polymer matrix-based nanocomposites comprising mainly inorganic fillers (filler/polymer composites) have been gaining substantial consideration recently for their outstanding and advantageous characteristics, such as good mechanical properties, chemical resistance, and thermal resistance [3].

Owing to recent developments in the field of nanotechnology, there has been growing interest in polymer matrix composites in which nano-sized fillers are distributed homogeneously (known as filler/polymer nanocomposites), due to their unique mechanical, electric, optical, and magnetic properties, as well as their thermal and dimensional stability [4–10]. Since the surface movement of the nanoparticles is exceptionally high, the nanoparticles subsequently have an inclination to aggregate closely, generating micron-sized filler clusters. Therefore, agglomeration of nanoparticles is one of the main problems in the fabrication of polymer/filler nanocomposites. In regard to this immense problem, there have been countless efforts to disperse nanoparticles homogeneously in polymer matrices by using procedures with organic modification at the surface or interlayer of nano-fillers and a variation of sol–gel and/or polymerization reactions. This chapter reviews the conventional and developed methods for the fabrication of nano-fillers/polymer nanocomposites.

One of the probable methods to improve the interface strength of matrix and fiber is by adding nano-fillers into glass fiber reinforced polymer (GFRP) composites [11–13]. Figure 4.1 shows the illustration of damage developed in the multiscale glass fabric/epoxy/CNT–Al_2O_3 nanocomposites. Microcrack propagation can be hindered through nano-fillers. This is due to the high toughness of nano-fillers in the polymer matrix. The crack propagation hinders, bypasses the nanoparticles, or stops at the nanoparticles due to the high toughness of the nano-fillers, resulting in improvement of overall strength of the composites.

Figure 4.2 shows a schematic of the observed crack propagation mechanism in thermally reduced graphene oxide (TRGO)/graphite nano-platelets (GNPs) epoxy nanocomposites [15]. While considering the failure mode (A) – crack pinning, the schematic shows that as the crack propagates into the composite it encounters one of the agglomerates of nano-fillers (GNP/TRGO) oriented perpendicular to the crack direction. At this point, the crack face has to bifurcate and one part goes around the sheet and another beneath the particle (A1). This continues to grow with difference in height between the two separated

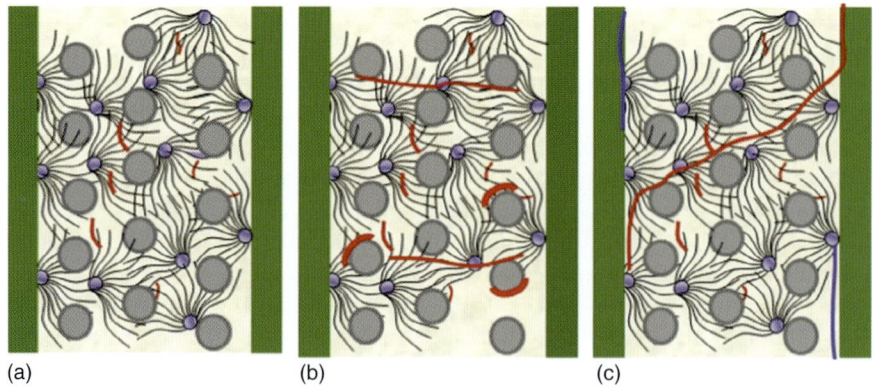

(a)   (b)   (c)

Figure 4.1 Illustration of damage development in the multiscale glass fabric/epoxy/ CNT–Al_2O_3 composites: (a) matrix-dominated cracking, (b) transverse cracks and fiber/matrix debonding, and (c) delamination. Source: Li et al. 2014 [14]. Reproduced with permission of Elsevier.

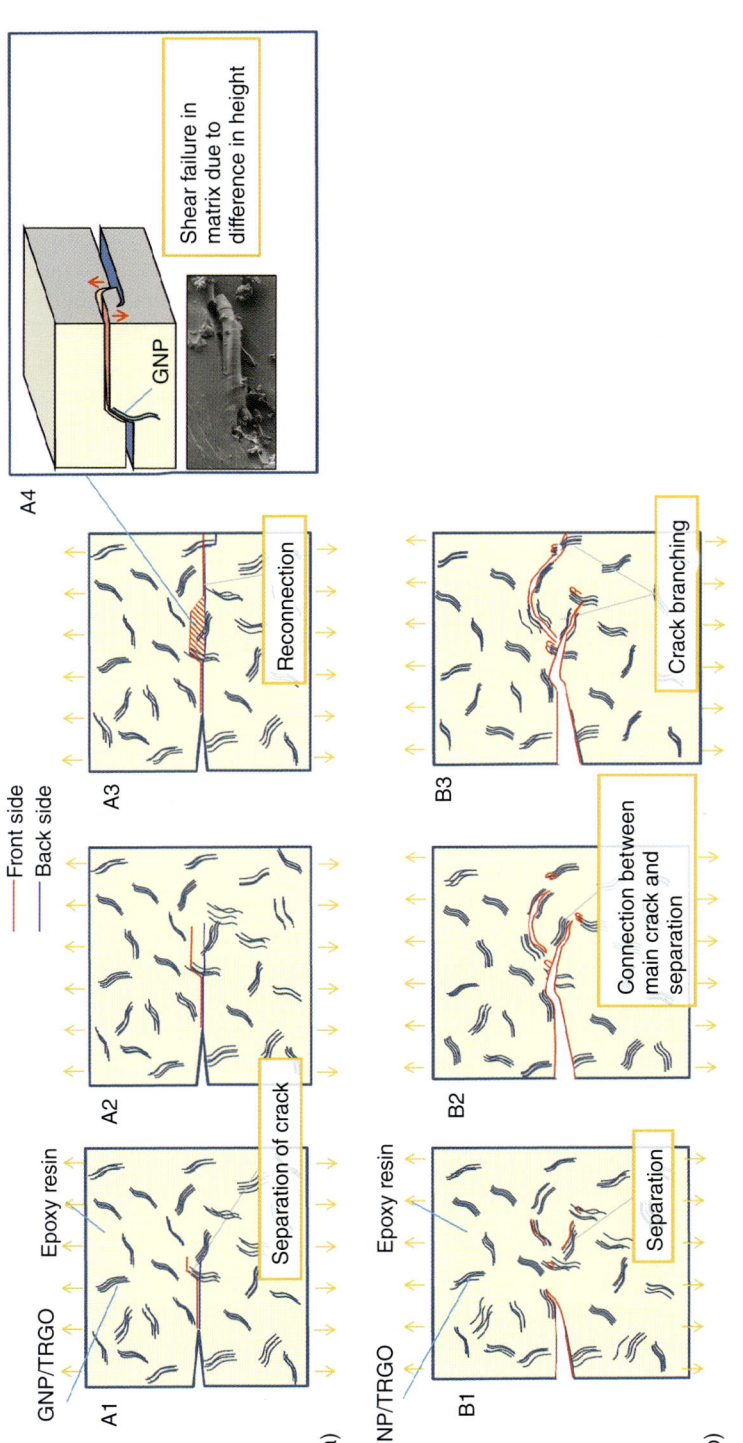

Figure 4.2 Schematic diagram of crack propagation mechanism of GNP/TRGO and epoxy matrix. (a) Failure process A: crack pinning/bifurcation and (b) failure process B: crack deflection + separation between layers [15].

crack faces (A2) and then joins after some distance (A3), leaving behind a narrow band formation. In the scanning electron microscopic (SEM) image (A4) at the right corner, we see one such narrow band formation. Here, the flow pattern on the narrow band deformation area is aligned in the perpendicular direction. This implies that the cracks grew from the outside of the narrow band deformation area to the inside of the narrow band deformation area. In other words, the crack runs along the surface (B1), that is, along the TRGO or GNP/epoxy interface. A similar phenomenon has also been reported in the literature through modeling studies [16]. In certain places, depending on the orientation of the nanoparticles relative to the crack face, separation between the graphitic layers initiates (B2). Since the force between the sheets is a secondary force (van der Waals) [17], the separation of sheets is more facilitated with more easy crack propagation. This separation occurs simultaneously at several particles and, in a few instances the main crack propagates through this particle (B3).

From the strengthening mechanism point of view, nano-fillers have great potential to enhance the mechanical properties of the composites. Therefore, fabrication of nanocomposites is a challenge for researchers and scientists. Different fabrication techniques of nanocomposites are required to understand the full potential of the new class of materials.

4.2 Classification of Nanomaterials

Nanomaterials can be classified based upon their dimensions, which normally lie in the nano range (≤ 100 nm). Zero-dimensional nanomaterials have measurements in the nanoscale, that is, the dimension does not exceed more than 100 nm. The nanoparticles are one of the general examples. These nano-shaped particles could be amorphous or crystalline, ceramic, polymeric, or metallic. One-dimensional nanomaterial has a minimum of one dimension in the nano level. Some examples of one-dimensional nanomaterials are nanoclays, nanosheets, nanoplatelets, and nanorods. Two-dimensional nanomaterials have at least two dimensions in nano level and these include nanotubes, nanofibers, nanorods, and whiskers. Three-dimensional nanomaterials have all measurements in nanoscale in all the three directions and these include nanoclays, nanogranules, and equiaxed nanoparticles. The nanomaterials can be amorphous or crystalline or polycrystalline. They can be composed of single or multiphase chemical elements. They could be in numerous forms and shapes, ceramic, metallic, or polymeric.

4.2.1 Nanocomposites

Nanocomposites are said to be composite materials having any one of the phases in nanometric scale. These composites came into existence due to their superlative properties as compared to traditional and microcomposites. Furthermore, the preparation methods and fabrication of nanocomposites reveal various challenges because of the stoichiometry in the nanophased and elemental composition. In order to improve the various properties of the nanocomposites, nanophased filler materials are embedded into them [18].

Nanocomposites consist of two or more separate elements or phases having dissimilar chemical and physical properties and are parted by an interface. The exclusive properties are not described by any one of the constituents. The constituent whose percentage is usually more in the nanocomposites is called the matrix. The reinforcements or nanomaterials are incorporated in the matrix phase of the nanocomposites to improve the physical, mechanical, thermal, and electrical properties. These reinforcements are usually in the form of nano-shaped fillers. Most commonly, the nanocomposites exhibit anisotropic behavior due to the different properties of the constituents materials and inhomogeneous scattering of the reinforcement. The different advantages of nanocomposites over conventional composites are as follows [19]:

- Enhancement in properties of the matrix phase in nanocomposites can be accomplished by the accumulation of lower quantity of nano-filler materials compared to conventional composites that need higher concentration of microparticles imperative to enhance various properties.
- The density to weight ratio of nano-fillers embedded nanocomposites is quite low as compared to conventional composites.
- The shape and size dependence of nanomaterials certainly increases the chemical, mechanical, thermal, magnetic, optical, and electrical properties to a higher range as compared to conventional composites.

Therefore, nanocomposites possess exceptionally better properties than conventional composite materials, thus finding a broad range of applications across various areas. Research and development in the field of nanocomposites increased with the introduction of carbon nanotubes in 1991 by Lijima. Nanocomposites are also present in nature in the form of abalone shell and bone. The difference in nanocomposites is the high surface area to volume ratio of reinforcement nano-fillers and higher aspect ratio than conventional composites. The reinforcement material consists of nanoparticles (CNT, metal oxide nano-fillers such as Al_2O_3, TiO_2, SiO_2), fibers (carbon nanofibers), or sheets (graphene, exfoliated clay stacks).

The classification of nanocomposites in accordance to the different kinds of reinforcement and matrix materials is required in the applications. Nanocomposites with various matrix materials are commonly divided into the following three categories [20]: However, polymer matrix nanocomposites have been discussed elaborately.

- Polymer matrix nanocomposites
- Ceramic matrix nanocomposites
- Metal matrix nanocomposites

4.2.2 Polymer Matrix Nanocomposites

The polymer-based nanocomposite materials consist of polymer as matrix material and nano-fillers as reinforcement. The nano-fillers can be one dimensional (fibers and nanotubes), two dimensional (layered materials like clay), or three dimensional (spherical particles). These polymer-based nanocomposites have been achieving significant consideration both in industries and academia, due to their exceptional mechanical properties such as high strength and elastic stiffness with addition of small amounts of nano-fillers. The additional outstanding properties of these nanocomposites include better electrical and optical resistance; flame retardance; wear resistance; and magnetic properties. Polymeric nanocomposites usually contain a polymer called the matrix phase and nano-fillers called the reinforcement.

Polymers possess exceptional properties such as easy processing, light weight, corrosion resistance, and high durability, ductility, and low cost. As related to metals and ceramics, polymers have relatively reduced thermal, electrical, and mechanical properties. Polymers also have lower heat resistance, gas barrier, and fire performance properties. Polymers are less dense than metals and ceramics, and have a lower coordination number. This eventually leads to lower weight of atoms of hydrogen and carbon, which finds applications in various structural materials and construction materials in lightweight applications such as aerospace, automobile, defense, and electronics industries [21]. Nowadays, research in polymer nanocomposites has been increasing rapidly due to the involvement of carbon nanotubes, exfoliated graphite, exfoliated clay, and nanocrystalline metals. The "effect of nanoparticles" is certainly a crucial factor affecting the different properties of the nanocomposites [22].

Different experimental research has indicated that nano-filler addition leads to the development of new phenomena, which eventually changes the properties of the material. The addition of nanoparticles with high surface energy, high surface area, and anisotropic properties in the polymer matrix composites certainly decreases the interparticle distance and increases the strength of the composite [23]. The micromechanics concepts rely on various properties of the composites mainly influenced by different constituents such as volume percentage of the matrix and reinforcement, and shape and size. These concepts help in calculating the properties of composites but are independent of the shape and size of the addition particles. The statement related to micromechanics is true for micro size reinforcement but not for nanoscale reinforcement. Therefore, it is a great challenge to fabricate good dispersion of nano-fillers in polymer matrix nanocomposites. Investigators around the globe have been using several methods to produce polymer matrix nanocomposites such as mixing filler materials during electro spinning, in situ polymerization, melt mixing, and many other methodologies. In the literature, several processes have been designated for the fabrication of polymer matrix nanocomposites that include layered structured materials with different nano-fillers. The common techniques are direct mixing of polymer and fillers, sol–gel process, in situ polymerization, melt intercalation, template synthesis, and intercalation of polymer from solution. The melt mixing process is one of the newer processes for nanocomposites fabrication.

4.3 Fabrication Techniques of Polymer Matrix Nanocomposites

Usually, polymer matrix nanocomposites were fabricated by mechanical or chemical processes. The main aim of various processing techniques is to get homogeneous and uniform distribution of nanoparticles in the polymer matrix. But proper dispersion of nanoparticles in the polymer matrix is one of the challenging problems in these nanocomposites. The nanoparticles have an affinity to agglomerate among themselves and form clusters of micron size. This results in improper distribution of nanoparticles in the polymer matrix, resulting in decreasing properties of the nanocomposites. Various researchers have tried different techniques to disperse nanoparticles homogeneously and uniformly in the polymer matrix by various methods such as polymerization reactions or surface modification, and chemical reaction of filler materials [24]. The following methods are mainly used for fabricating polymer matrix nanocomposites.

4.3.1 Ultrasonic and Dual Mixing

This method is used to disperse the different nano-fillers in polymer matrix by an ultrasonic apparatus. In this technique high impact energy is transferred by ultrasonic equipment that produces lower shear energy in the ultrasonic chamber. The advantage of using a sonicator is that agglomerated nanoparticles start to de-agglomerate. Qian et al. [25] have successfully achieved the dispersion of multiwalled carbon nanotubes (MWCNTs) in a polystyrene polymer by applying high energy sonication method. Primarily, the acetone is taken as solvent and CNTs are dispersed in it and this solution is further sonicated for some period of time. Sonication helps in de-agglomeration of the nanoparticles. The affinity toward agglomeration arises due to higher specific surface area. Therefore, it is difficult to get good dispersion in the case of single-walled carbon nanotubes (SWCNT). Also, during sonication the generation of high energy damages the shape and size of the nanoparticles. Finally, for a specific period of sonication, the polymer and the suspension are mixed until complete evaporation of the solvent. Song et al. [26] fabricated successfully and achieved a good dispersion of MWCNT/epoxy composites by this technique. The CNTs were sonicated in ethanol solution for a period of two hours. Further, sonication of CNT/ethanol mixture was done for one hour at 80 °C and the suspension was kept in vacuum for a duration of five days to ensure complete removal of air bubbles. Researchers have reported that mechanical stirring followed by sonication is a good method to disperse nano-fillers in polymer matrix [27–29]. A. Chatterjee and M.S. Islam have fabricated and characterized TiO_2–epoxy nanocomposites [30]. They have used a Vibracell ultrasonic processor for nano-TiO_2 filler dispersion in polymer matrix. Figure 4.3 shows the Vibracell ultrasonic processor.

Nayak et al. [10, 31] fabricated the epoxy with nano-TiO_2 particles stirred by a magnetic stirrer for one hour, followed by sonication with a high frequency sonicator at 60 °C, as shown in Figure 4.4. The shape and size of nano-TiO_2 particles has been examined by field emission scanning electron microscope (FESEM).

Figure 4.5a,b shows the nano-TiO_2 particles shape and X-ray diffraction (XRD) plot respectively. It is observed that the shape of the nano-TiO_2 particles is nearly spherical and the purity level of the nanoparticles is good. Nanoparticles are mixed with epoxy through magnetic stirring and sonication. Distribution of nanoparticles in epoxy matrix is observed through FESEM and reasonably uniform distribution of nanoparticles was found in the epoxy matrix of the

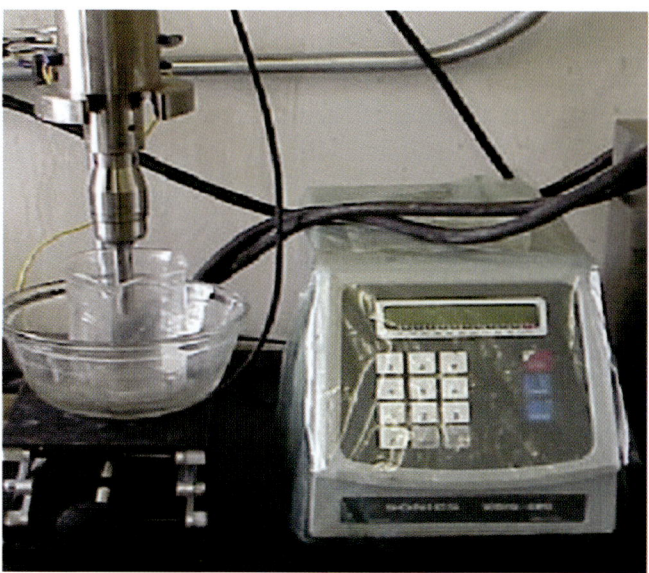

Figure 4.3 Vibracell ultrasonic processor for fabrication of nano-TiO$_2$ enhanced polymer nanocomposites.

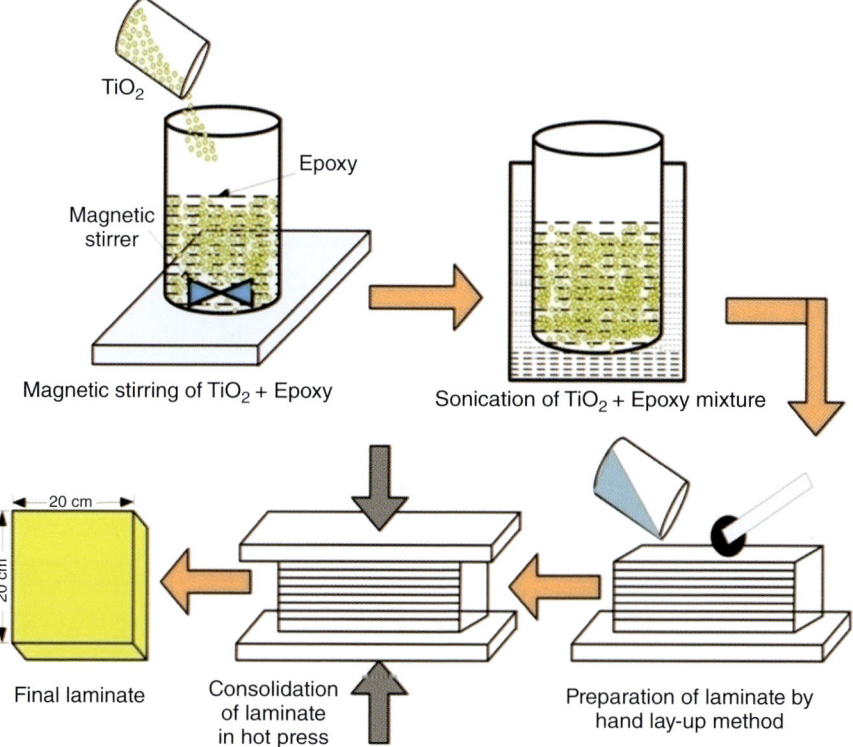

Figure 4.4 Schematic diagram of fabrication method of nanocomposite laminates.

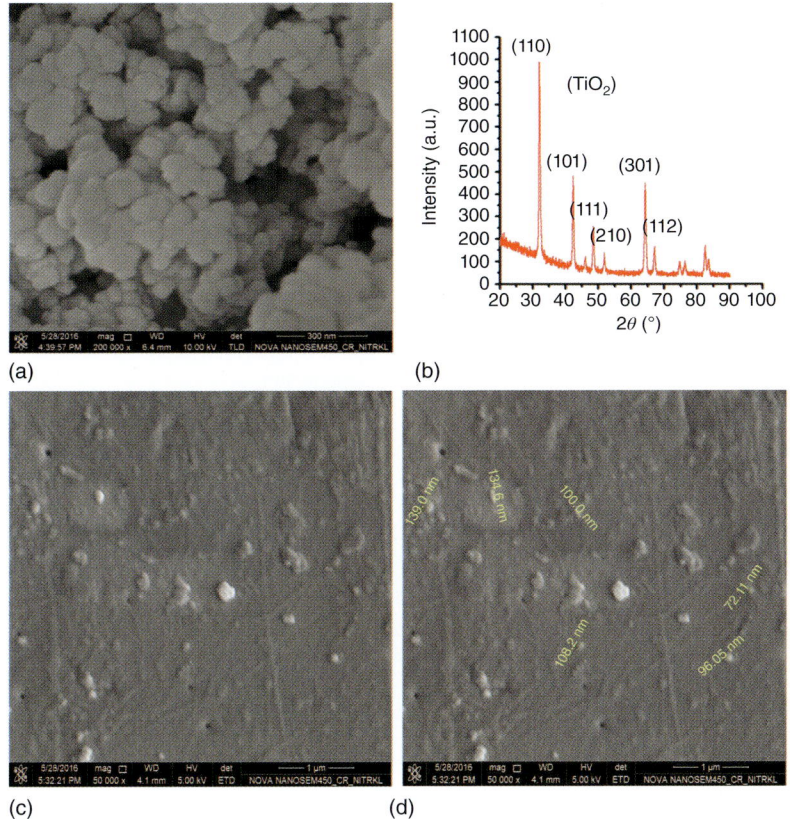

Figure 4.5 FESEM image of nano-TiO$_2$ particles (a) shape, (b) intensity versus 2θ and (c,d) the distribution and size of nanoparticles in epoxy matrix at 0.7 wt% TiO$_2$.

composites having 0.7 wt% of nano-TiO$_2$ as shown in Figure 4.5c,d. Figure 4.6 shows the three filling configurations of ceramics particles in ceramic/polymer composites: (a) dispersed, (b) mechanically contacted, and (c) chemically bonded. Kumar et al. [33] used a dual mixing method to increase the nano-TiO$_2$ content in the epoxy matrix and successfully fabricated the nanocomposites at higher concentration of nano-TiO$_2$ content. Ash et al. [34] have conducted sonication process to disperse the nano-Al$_2$O$_3$ particles in polymethyl methacrylate polymer to improve the mechanical properties of the nanocomposites.

4.3.2 Three-Roll Mixing of Nano-fillers in PMC

Organic or inorganic nano-fillers improve the mechanical properties of micro/macro scale fiber reinforced polymer composites. However, researchers have started investigating the effect of multiple nano-fillers into the polymer matrix to understand the synergistic effect of both types of nano-fillers on the mechanical properties of the hybrid composites. Li et al. [35] studied the effect of hybrid fillers comprised of CNTs directly grown on alumina microspheres by chemical vapor deposition incorporated into the epoxy matrix, which was then reinforced

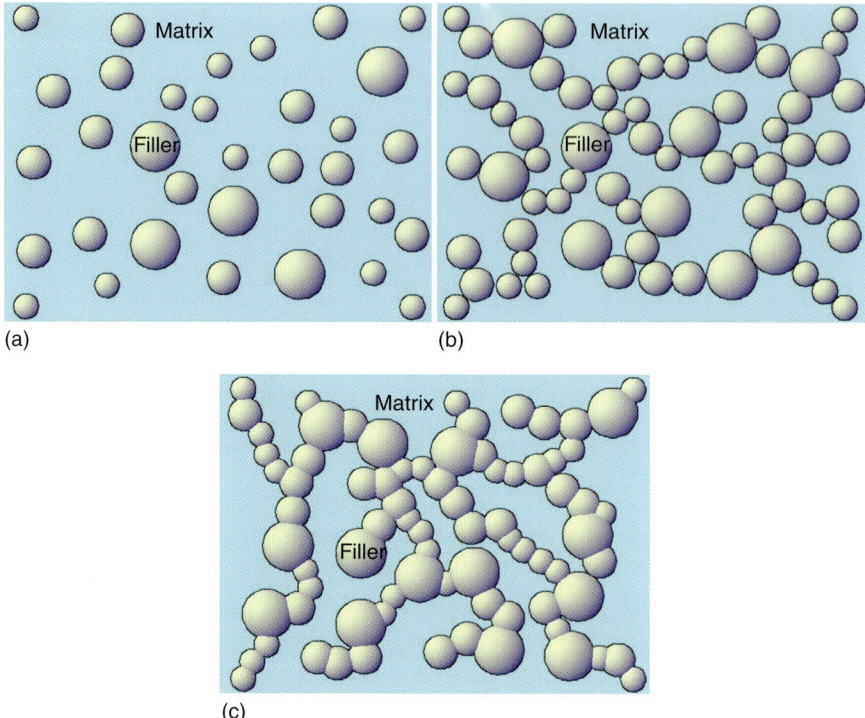

(a)

(b)

(c)

Figure 4.6 Three filling configurations of ceramics particles in ceramic/polymer composites: dispersed (a), mechanically contacted (b), and chemically bonded (c). Source: Hu et al. 2016 [32]. Reproduced with permission of Elsevier.

with woven glass fibers. The hierarchical composite with 0.5 wt% hybrid loading was observed to exhibit an improvement of 19% and 11% in flexural modulus and interlaminar shear strength, respectively. Figure 4.7 shows the fabrication method for multiple fillers into the polymer matrix. In this fabrication method, three rolls are used to disperse the nano-fillers by shear action.

The gap size between the adjacent rollers was set to $50\,\mu m$ and rotation speed was set to 80 rpm. The dwell time of the obtained suspension on the rolls was about 10 minutes. The filler content was set as 0.5 wt% of epoxy resin-hardener mixture, which was close to the percolation threshold (0.49 wt%) of the CNT–Al_2O_3/epoxy system. The weight fraction of CNTs and Al_2O_3 in the CNT + Al_2O_3 mixture was the same as that of CNT–Al_2O_3 hybrids. After collecting the suspensions, the curing agent was added at a mass ratio of $3:1$ (epoxy:hardener) and then mixed for 10 minutes under mechanical stirring. The glass fabric/epoxy composites were prepared using a combination of hand lay-up and hot compression.

4.3.3 Intercalation Method

This method normally deals with the dispersion of different types of nanoplatelets into the polymer matrix nanocomposites. The exfoliation of the silicates layered

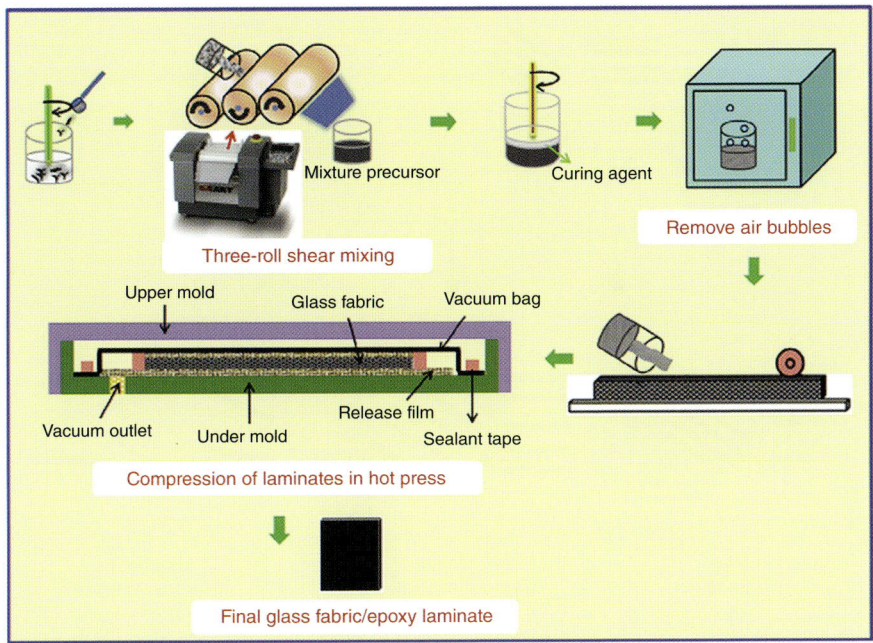

Figure 4.7 Schematic of filler dispersion in the epoxy matrix and fabrication process of glass fabric/epoxy composite.

used as the inorganic filler produce by intercalating an organic compound into the interlayer space of the silicate, resulting in the even dispersion of plate-like nanoparticles [35–38]. The layered silicate should be organically altered by organic surfactants comprising quaternary cation functionality, such as alkyl ammonium, amino acids, phosphonium salts, and imidazolium, to attain sufficient hydrophobicity to be miscible with the organic compounds since the silicate is hydrophilic in nature but the organic compound is hydrophobic [39–42]. The intercalation of polymeric materials into the organically altered layered silicates and the successive exfoliation of the silicates are normally done by employing a mechanical or chemical method. Figure 4.8 shows clay modification and intercalation of polymer to form polymer nanocomposites.

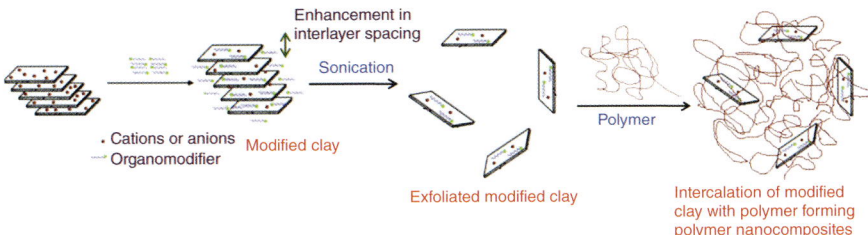

Figure 4.8 Schematic diagram showing clay modification and intercalation of polymer to form polymer nanocomposites.

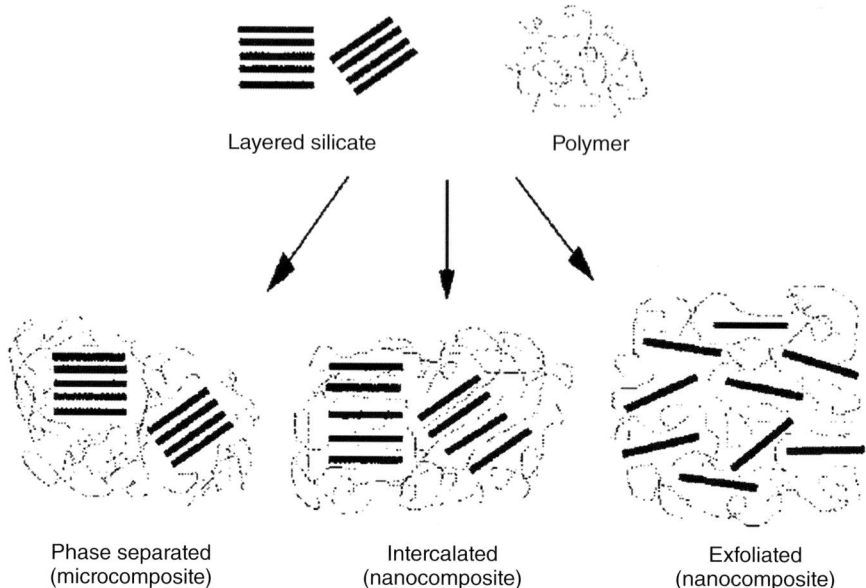

Layered silicate   Polymer

Phase separated
(microcomposite)

Intercalated
(nanocomposite)

Exfoliated
(nanocomposite)

Figure 4.9 Schematic diagram of three main types of layered silicates in polymer matrix.

Figure 4.9 shows the schematic representation of three main types of layered silicates in polymer matrix. For example, fusion of nanoclays into the polymer matrix up to an optimum level will improve the bulk behavior of the composites. Intercalation is a method that involves surface modification of nanoplatelets for better dispersion of plate-like nanoparticles in the polymer matrix. Intercalated morphology is created when polymer chains are converted into layered structure. There are broadly two interaction methods in which the nanoplatelets are homogeneously dispersed [43–45]. The first intercalation method is about chemical polymerization. This includes in situ polymerization technique where nano-fillers are distributed all over the monomer. In this technique, nanoplatelets were distributed into the polymer matrix followed by polymerization reaction. Generally, the nanoplatelets are enlarged from their usual size in the monomer and the polymeric reaction takes place between the intercalated sheets by the chemical method. The chemical method is in situ polymerization of the monomers within the silicate layers (in situ intercalative polymerization method) [4, 36, 42, 46, 47]. Shioyama et al. have investigated that in situ polymerization of the monomers could also happen in the interlayer spacing of graphite with a layered structure similar to that of the abovementioned silicate [37, 38]. The above results show the prospect of the fabrication of graphite/polymer nanocomposites by the in situ intercalative polymerization technique.

The second method is direct intercalation. This method deals with direct intercalation of nanoplatelets with polymer matrix by the solution mixing process [36, 47, 48]. In this process, the polymer matrix is dissolved in a solvent while the nanoplatelet sheets are swollen in the solvent. The above two solutions are mixed together and the polymeric chains intercalated in between the layers of nanoplatelets. One more intercalation method is broadly used in industry, called melt intercalation. This technique comprises mixing of the polymer matrix into the nano-fillers at a molten temperature. The nanofibers and polymer are mixed thoroughly either under shear or statically [36, 47, 49, 50]. This technique is well established in various industrial processes as in injection and extrusion molding. Further, it permits the usage of polymers that are not appropriate for in situ solution or intercalation polymerization.

4.3.4 Sol–Gel Method

This synthesis method is a very practical approach to produce nanocomposites as compared to other methods. In this process, the word sol–gel is related to two steps – one is sol and the other is gel. Generally, the colloidal suspension of solid nanoparticles in monomer is related to sol and the three-dimensional connection forms between the phases are called gel [4, 5]. Usually, the solid nanoparticles are spread all over the monomer, establishing a colloidal suspension of nanoparticles, i.e. sol. Similarly, by the polymerization reaction, interconnecting regions between phases are developed, called the gel. The three-dimensional polymer nanoparticles network spreads throughout the liquid phase. The polymer matrix acts as the nucleation media and develops the growth of layered crystals. The development of crystals leads to the formation of nanocomposites. There are a number of trials that have been carried out to synthesize several kinds of polymer/filler nanocomposites, where the inorganic particle phases ranging in dimensions from an angstrom to nanometers are uniformly and homogeneously distributed, i.e. inorganic/organic molecular hybrid materials, and have been done by this process using reactions of metal alkoxides [4, 5, 51–53] as given below.

$$Si(OR)_4 + 4H_2O \rightarrow [Si(OH)_x(OR)_{4-x} + xROH] \rightarrow SiO_2 + 4ROH + 2H_2O$$

$$(4.1)$$

Here, Eq. (4.1) states the polycondensation and hydrolysis reactions of tetra alkoxy silane as one of the usual sol–gel reactions used for synthesizing silica/polymer molecular hybrid materials. With the current progress of this sol–gel technology, molecular hybrid materials and nanocomposites have been widely considered by many researchers and organizations. By the use of this technology, it is quite possible to dissolve inorganic fillers with a dimension smaller than the molecular chain length of the polymer matrix. However, the different polymers used in the organic area of the hybrid materials in this sol–gel method is constrained to polymers having hydrogen bond acceptor groups that can form hydrogen bonds with the hydroxyl groups on the inorganic filler surface, such as alcohol and water-soluble polymers. Furthermore, the conditions of the sol–gel reactions employed have an intense effect on the structure of the inorganic network formed. At the molecular level, it is quite difficult to control the arrangement and size of the inorganic field in the hybrid materials. As this technology is noble, a synthesis process of hybrid materials of segmented polymer and silica (site selective molecular hybrid method) has been established by using silane-modified polymers where oligomers of alkoxysilane were introduced selectively into suitable places in each polymer [54].

This technique permits the usage of polymer matrix having no interactions with metal alkoxides, which was formerly not proper for the conventional sol–gel process. The ultimate benefit of this process is that it is promising to regulate the microstructure in hybrid material, permitting its objective application. Goda et al. used this technology for the development of the properties of polyurethane [54]. In general, elastic properties of the rubber in segmented urethane block copolymers are affected by their two-phase microstructure in which hard parts detach from the soft parts to form areas. By the molecular hybrid technique, a hybrid material having dispersed hybrid areas of the hard part interpenetrating with silica in the soft zone constant phase could be synthesized. It has been stated that usually, hybrid material reveal good thermal and mechanical behaviors because the hybrid regions are much tougher and more heat resistant than the original hard part, while sustaining the flexibility of the soft part [54].

From the above discussions, it is seen that several complex reactions usually occur during the synthesis process. Therefore, intricate amenities and process control for the polymerization reactions and the clearance of chemical wastes discharged from the process are vital for the useful operation of these procedures on an industrial scale.

4.3.5 Direct Mixing of Polymer and Nano-fillers

Direct mixing of nano-fillers and polymer matrix is a suitable method for fabrication of nanocomposites and it is the breaking of agglomerated nano-fillers during the mixing process [24]. This method includes two techniques in the mixing of nano-fillers and polymer. The first method is the mixing of a polymer without the presence of any solvent with nano-fillers above the glass transition temperature of the polymer, mostly called melt compounding technique [24, 55–58]. The second way of mixing includes mixing of nano-fillers and polymer in the presence of a solvent, usually called solution mixing or solvent method [55, 58–62].

4.3.5.1 Melt Compounding

Melt compounding includes addition of nanofibers to the polymer matrix above the glass transition temperature. In this case, the shear stress (hydrodynamics force) is applied in the polymer matrix melt by dragging viscous force. The shear force helps in breaking the nano-filler agglomerates. This breaking of agglomerates helps in the development of uniform and homogeneous dispersion of nano-filler in the nanocomposites [63, 64].

The mechanism of organoclay dispersion and exfoliation during melt processing is shown in Figure 4.10. A rupture model revealed that rupture arises along a cross section of the nano-filler agglomerates, where the amount of contact of each primary element with its neighbors is very small. The nanoparticle agglomerates split into almost two equal portions. According to the "onion peeling" model, the pressures produced at the surface of agglomerate are quite enormous enough at any point on the agglomerate surface to eliminate a primary particle or an

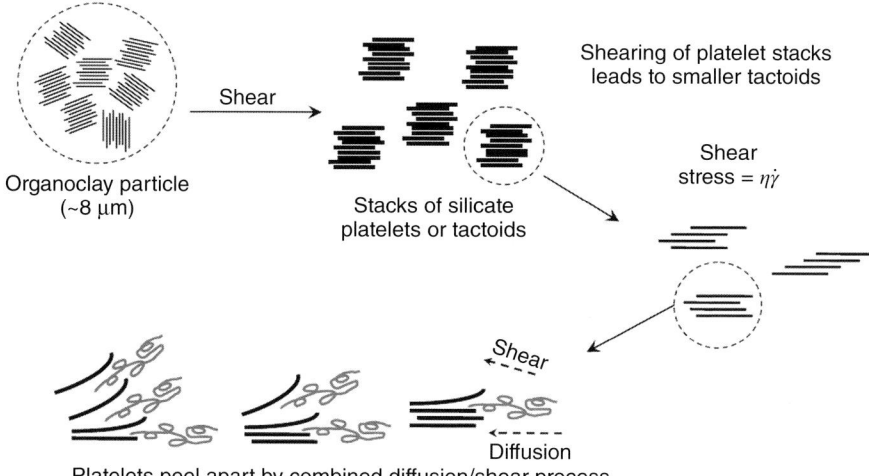

Figure 4.10 Organoclay dispersion and exfoliation during melt processing.

aggregate of primary particles from the surface of the larger agglomerate. The removal of the nanoparticles aggregates forms a cloud near the initial agglomerate, partially protecting it from further reduction in size. When the aggregates are swept from the cloud of agglomerates, reduction in size occurs and fresh aggregates from the agglomerate replace them. In both the dispersion models, when the shear stress is higher than some critical threshold value for breaking down agglomerates of fillers, a dispersive action will occur inside the kneaded polymer melt [56, 65].

4.3.5.2 Solvent Method

Solvent method is a process where the nanoparticles are dispersed in the solvent and a co-solvent is used to dissolve the polymer. The subsequent nanocomposites are recovered from the solvent through solvent coagulation or by solvent evaporation method. This technique comprises lower shear stress into the polymer matrix than that of melt compounding. Here, the nanoparticles are previously dispersed in the solvent using sonication to de-agglomerate the nanoparticles. The polymer nanocomposites are fabricated by the above process by conventional manufacturing procedures such as extrusion molding, injection molding, casting, calendaring, rotational molding, compression molding, blow molding, and thermoforming. Wang et al. [66] have developed homogenous organic dispersion of graphene oxide (GO) sheets that were prepared by a solvent exchange method. The schematic diagram of the process is shown in Figure 4.11. This method enabled the simultaneous achievement of full exfoliation and high concentration of GO in several organic solvents such as dimethyl sulfoxide, which would facilitate the fabrication of individual graphene reinforced polymer composites through a solution-based process. To this end, poly[2,2′-(*p*-oxydiphenylene)-5,5′-bibenzimidazole] (OPBI)/GO composites were fabricated. They found that the composites showed a 17% increase in

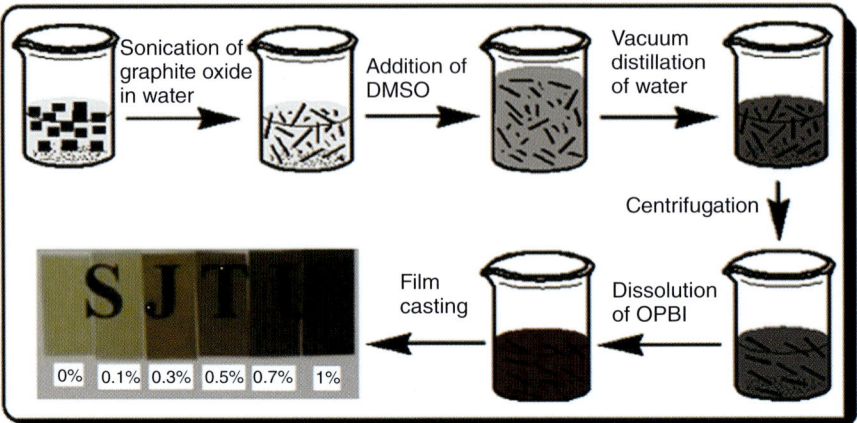

Figure 4.11 Fabrication technique for GO/OPBI composites.

Young's modulus, 33% increase in tensile strength, and 88% improvement in toughness by the addition of only 0.3 wt% of GO. ABS/montmorillonite nanocomposites have been prepared by a solvent/non-solvent method and found a good interaction between them [67].

4.4 Future Perspective and Challenges

FRP composites have been steadily replacing their metallic counterparts either fully or partially for the last few decades and the replacement and repair of metallic materials with FRP composites is in full swing. However, unfortunately there are very rare cases of unprecedented failure of FRP composite materials before reaching the predicted service life period of the components/parts. It is reasonably believed that the environment is the biggest issue of concern to trigger the nucleolus zones of early failure of FRP component.

At present, different nano-fillers have been added to epoxy polymer matrix to enhance the damage tolerance of the composites. The effects of nano-fillers on all possible failure modes are to be evaluated through micro characterization such as FESEM, TEM, FTIR, and Raman spectroscopy. If the investigation is carried out in different loading mode and water uptake kinetics theory, the mechanics and effect of nano-fillers on diffusivity at different temperatures at the time of conditioning and time of testing are important parameters to optimize the nano-fillers GFRP composites to attain the highest possible potential of the material. The micro characterization will certainly make the composites reliable, sustainable, and predictable in different service conditions. Extensive work in different directions and dimensions of these new generation composites may meet the emerging demands of present day technological evolution.

Different fabrication methods have been developed to fabricate nanocomposites to meet the design requirement of different applications. However, there is a limit to the quantity of the nano-fillers being mixed in the polymer matrix of every fabrication process. To take this further into the next level, commercialization of the nanocomposite's fabrication is necessary. The weight percentage of nano-fillers into the polymer matrix needs to be increased to achieve the full potential of the nanocomposites. However, the environmental effect of nano-powders/fillers needs to be taken care of considering better human health.

Acknowledgment

We would like to thank Maulana Azad National Institute of Technology, Bhopal and NIT, Rourkela, for supporting infrastructure facilities to write the book chapter.

References

1 Drexler, K.E. (1996). *Engines of Creation*. Fourth Estate.
2 Ratner, M. and Ratner, D. (2002). *Nanotechnology: A Gentle Introduction to the Next Big Idea*. Prentice Hall.
3 Nielsen, L.E. and Landel, R.F. (1994). *Mechanical Properties of Polymers and Composites*, 2e. New York, NY: Marcel Dekker.
4 Schadler, L.S. (2004). Polymer-based and polymer-filled nanocomposites. *Nanocomposite Science and Technology* 77–153. https://doi.org/10.1002/3527602127.ch2.
5 Luther-Davies, B., Samoc, M., and Woodruff, M. (1996). Comparison of the linear and nonlinear optical properties of poly(p-phenylenevinylene)/sol–gel composites derived from tetramethoxysilane and methyltrimethoxysilane. *Chemistry of Materials* 8: 2586–2594. https://doi.org/10.1021/cm9504448.
6 Nayak, R.K., Mahato, K.K., Routara, B.C., and Ray, B.C. (2016). Evaluation of mechanical properties of Al_2O_3 and TiO_2 nano filled enhanced glass fiber reinforced polymer composites. *Journal of Applied Polymer Science* 133: https://doi.org/10.1002/app.44274.
7 Nayak, R.K., Mahato, K.K., and Ray, B.C. (2016). Water absorption behavior, mechanical and thermal properties of nano TiO_2 enhanced glass fiber reinforced polymer composites. *Composites Part A: Applied Science and Manufacturing* 90: 736–747. https://doi.org/10.1016/j.compositesa.2016.09.003.
8 Mahato, K.K., Dutta, K., and Ray, B.C. (2017). Static and dynamic behavior of fibrous polymeric composite materials at different environmental conditions. *Journal of Polymers and the Environment* 1–27. https://doi.org/10.1007/s10924-017-1001-x.
9 Mahato, K.K., Rathore, D.K., Prusty, R.K. et al. (2017). Tensile behavior of MWCNT enhanced glass fiber reinforced polymeric composites at various crosshead speeds. *IOP Conference Series: Materials Science and Engineering* 178: 012006. https://doi.org/10.1088/1757-899X/178/1/012006.
10 Nayak, R.K. and Ray, B.C. (2017). Water absorption, residual mechanical and thermal properties of hydrothermally conditioned nano-Al_2O_3 enhanced glass

fiber reinforced polymer composites. *Polymer Bulletin* 74: 4175–4194. https://doi.org/10.1007/s00289-017-1954-x.

11 Fan, X.J., Lee, S.W.R., and Han, Q. (2009). Experimental investigations and model study of moisture behaviors in polymeric materials. *Microelectronics Reliability* 49: 861–871. https://doi.org/10.1016/j.microrel.2009.03.006.

12 Dikobe, D.G. and Luyt, A.S. (2010). Comparative study of the morphology and properties of PP/LLDPE/wood powder and MAPP/LLDPE/wood powder polymer blend composites. *Express Polymer Letters* 4: 729–741.

13 Jongsomjit, B., Chaichana, E., and Praserthdam, P. (2005). LLDPE/nano-silica composites synthesized via in situ polymerization of ethylene/1-hexene with MAO/metallocene catalyst. *Journal of Materials Science* 40: 2043–2045. https://doi.org/10.1007/s10853-005-1229-z.

14 Li, W., He, D., Dang, Z., and Bai, J. (2014). In situ damage sensing in the glass fabric reinforced epoxy composites containing $CNT-Al_2O_3$ hybrids. *Composites Science and Technology* 99: 8–14.

15 Chandrasekaran, S., Sato, N., Tölle, F. et al. (2014). Fracture toughness and failure mechanism of graphene based epoxy composites. *Composites Science and Technology* 97: 90–99.

16 Parashar, A. and Mertiny, P. (2013). Multiscale model to study of fracture toughening in graphene/polymer nanocomposites. *International Journal of Fracture* 179: 221–228.

17 Gong, L., Kinloch, I.A., Young, R.J. et al. (2010). Interfacial stress transfer in a graphene monolayer nanocomposites. *Advanced Materials* 22: 2694–2697.

18 Asmatulu, R., Khan, W.S., Reddy, R.J., and Ceylan, M. (2015). Synthesis and analysis of injection-molded nanocomposites of recycled high-density polyethylene incorporated with graphene nanoflakes. *Polymer Composites* 36: 1565–1573. https://doi.org/10.1002/pc.23063.

19 Nanocomposites: properties and applications. Posted by Nano on January 4 2008 at 1:13pm in Polymer Nanocomposite Group (PNC), Discussions B to PNG (PNC). The International Nanoscience Community, Nanopaprika. http://www.nanopaprika.eu/forum/topics/1612324:Topic:6001?groupUrl=polymernanocompositegrouppnc&xg_source=activity (accessed 27 September 2018).

20 Chung, D.D.L. (2003). *Composite Materials: Functional Materials for Modern Technologies*. London: Springer-Verlag.

21 Thostenson, E.T., Li, C., and Chou, T.-W. (2005). Nanocomposites in context. *Composites Science and Technology* 65: 491–516. https://doi.org/10.1016/j.compscitech.2004.11.003.

22 Paul, D.R. and Robeson, L.M. (2008). Polymer nanotechnology: nanocomposites. *Polymer* 49: 3187–3204. https://doi.org/10.1016/j.polymer.2008.04.017.

23 Winey, K.I. and Vaia, R.A. (2007). Polymer nanocomposites. *MRS Bulletin* 32: 314–322. https://doi.org/10.1557/mrs2007.229.

24 Tanahashi, M. and Tanahashi, M. (2010). Development of fabrication methods of filler/polymer nanocomposites: with focus on simple melt-compounding-based approach without surface modification of nanofillers. *Materials* 3: 1593–1619. https://doi.org/10.3390/ma3031593.

25 Qian, D., Dickey, E.C., Andrews, R., and Rantell, T. (2000). Load transfer and deformation mechanisms in carbon nanotube–polystyrene composites. *Applied Physics Letters* 76: 2868–2870. https://doi.org/10.1063/1.126500.

26 Song, Y.S. and Youn, J.R. (2005). Influence of dispersion states of carbon nanotubes on physical properties of epoxy nanocomposites. *Carbon* 43: 1378–1385. https://doi.org/10.1016/j.carbon.2005.01.007.

27 Chang, L.N. and Chow, W.S. (2010). Accelerated weathering on glass fiber/epoxy/organo-montmorillonite nanocomposites. *Journal of Composite Materials* 44: 1421–1434. https://doi.org/10.1177/0021998309360944.

28 Soundararajah, Q.Y., Karunaratne, B.S.B., and Rajapakse, R.M.G. (2009). Mechanical properties of poly(vinyl alcohol) montmorillonite nanocomposites. *Journal of Composite Materials* https://doi.org/10.1177/0021998309347040.

29 Barbezat, M., Brunner, A.J., Necola, A. et al. (2009). Fracture behavior of GFRP laminates with nanocomposite epoxy resin matrix. *Journal of Composite Materials* 43: 959–976. https://doi.org/10.1177/0021998308100799.

30 Chatterjee, A. and Islam, M.S. (2008). Fabrication and characterization of TiO_2–epoxy nanocomposites. *Materials Science and Engineering: A* 487: 574–585. https://doi.org/10.1016/j.msea.2007.11.052.

31 Nayak, R.K. and Ray, B.C. (2018). Influence of seawater absorption on retention of mechanical properties of nano-TiO_2 embedded glass fiber reinforced epoxy polymer matrix composites. *Archives of Civil and Mechanical Engineering* 18: 1597–1607. https://doi.org/10.1016/j.acme.2018.07.002.

32 Hu, Y., Du, G., and Chen, N. (2016). A novel approach for Al_2O_3/epoxy composites with high strength and thermal conductivity. *Composites Science and Technology* 124: 36–43. https://doi.org/10.1016/j.compscitech.2016.01.010.

33 Kumar, K., Ghosh, P.K., and Kumar, A. (2016). Improving mechanical and thermal properties of TiO_2-epoxy nanocomposites. *Composites Part B: Engineering* 97: 353–360. https://doi.org/10.1016/j.compositesb.2016.04.080.

34 Ash, B.J., Rogers, D.F., Wiegand, C.J. et al. (2002). Mechanical properties of Al_2O_3/polymethylmethacrylate nanocomposites. *Polymer Composites* 23: 1014–1025. https://doi.org/10.1002/pc.10497.

35 Li, W., Dichiara, A., Zha, J. et al. (2014). On improvement of mechanical and thermo-mechanical properties of glass fabric/epoxy composites by incorporating CNT–Al_2O_3 hybrids. *Composites Science and Technology* 103: 36–43.

36 Sinha Ray, S. and Okamoto, M. (2003). Polymer/layered silicate nanocomposites: a review from preparation to processing. *Progress in Polymer Science* 28: 1539–1641. https://doi.org/10.1016/j.progpolymsci.2003.08.002.

37 Shioyama, H. (2000). The interactions of two chemical species in the interlayer spacing of graphite. *Synthetic Metals* 114: 1–15. https://doi.org/10.1016/S0379-6779(00)00222-8.

38 Shioyama, H., Tatsumi, K., Iwashita, N. et al. (1998). On the interaction between the potassium—GIC and unsaturated hydrocarbons. *Synthetic Metals* 96: 229–233. https://doi.org/10.1016/S0379-6779(98)00098-8.

39 Usuki, A., Kawasumi, M., Kojima, Y. et al. (1993). Swelling behavior of montmorillonite cation exchanged for ω-amino acids by $\in$-caprolactam. *Journal of Materials Research* 8: 1174–1178. https://doi.org/10.1557/JMR.1993.1174.

40 Fudala, Á.Á., Pálinkó, I., and Kiricsi, I. (1999). Preparation and characterization of hybrid organic-inorganic composite materials using the amphoteric property of amino acids: amino acid intercalated layered double hydroxide and montmorillonite. *Inorganic Chemistry* 38: 4653–4658.

41 Lin, J.-J., Cheng, I.-J., Wang, R., and Lee, R.-J. (2001). Tailoring basal spacings of montmorillonite by poly(oxyalkylene)diamine intercalation. *Macromolecules* 34: 8832–8834. https://doi.org/10.1021/ma011169f.

42 Bottino, F.A., Fabbri, E., Fragalà, I.L. et al. (2003). Polystyrene-clay nanocomposites prepared with polymerizable imidazolium surfactants. *Macromolecular Rapid Communications* 24: 1079–1084. https://doi.org/10.1002/marc.200300054.

43 Yang, F., Ou, Y., and Yu, Z. (1998). Polyamide 6/silica nanocomposites prepared by in situ polymerization. *Journal of Applied Polymer Science* 69: 355–361. https://onlinelibrary.wiley.com/doi/abs/10.1002/%28SICI%291097-4628%2819980711%2969%3A2%3C355%3A%3AAID-APP17%3E3.0.CO%3B2-V.

44 Alexandre, M. and Dubois, P. (n.d.). Polymer-layered silicate nanocomposites: preparation, properties and uses of a new class of materials. *ScienceDirect* 28: 1–63. https://www.sciencedirect.com/science/article/pii/S0927796X00000127.

45 Jannapu Reddy, R. (2010). Preparation, characterization and properties of injection molded graphene nanocomposites. MS thesis. Wichita State University. https://soar.wichita.edu/handle/10057/3726.

46 Usuki, A., Kojima, Y., Kawasumi, M. et al. (1993). Synthesis of nylon 6-clay hybrid. *Journal of Materials Research* 8: 1179–1184. https://doi.org/10.1557/JMR.1993.1179.

47 Pinnavaia, T.J. and Beall, G.W. (2000). *Polymer-Clay Nanocomposites*. Chichester: John Wiley.

48 Ogata, N., Jimenez, G., Kawai, H., and Ogihara, T. (1997). Structure and thermal/mechanical properties of poly(L-lactide)-clay blend. *Journal of Polymer Science Part B: Polymer Physics* 35: 389–396. https://onlinelibrary.wiley.com/doi/full/10.1002/(SICI)1099-0488(19970130)35:2%3C389::AID-POLB14%3E3.0.CO;2-E.

49 Vaia, R.A., Ishii, H., and Giannelis, E.P. (1993). Synthesis and properties of two-dimensional nanostructures by direct intercalation of polymer melts in layered silicates. *Chemistry of Materials* 5: 1694–1696. https://doi.org/10.1021/cm00036a004.

50 Vaia, R.A. and Giannelis, E.P. (1997). Lattice model of polymer melt intercalation in organically-modified layered silicates. *Macromolecules* 30: 7990–7999. https://doi.org/10.1021/ma9514333.

51 Morikawa, A., Iyoku, Y., Kakimoto, M., and Imai, Y. (1992). Preparation of a new class of polyimide-silica hybrid films by sol–gel process. *Polymer Journal* 24: 107.

52 Claude, C., Garetz, B., Okamoto, Y., and Tripathy, S. (1992). The preparation and characterization of organically modified silicates that exhibit nonlinear optical properties. *Materials Letters* 14: 336–342. https://doi.org/10.1016/0167-577X(92)90049-P.

53 Popall, M., Meyer, H., Schmidt, H., and Schulz, J. (1990, 1990). Inorganic-organic composites (Ormocers) as structured layers for microelectronics. *MRS Proceedings* 180: 995. https://doi.org/10.1557/PROC-180-995.

54 Goda, H. and Frank, C.W. (2001). Fluorescence studies of the hybrid composite of segmented-polyurethane and silica. *Chemistry of Materials* 13: 2783–2787. https://pubs.acs.org/doi/abs/10.1021/cm000711w.

55 Hashimoto, M., Takadama, H., Mizuno, M., and Kokubo, T. (2006). Enhancement of mechanical strength of TiO_2/high-density polyethylene composites for bone repair with silane-coupling treatment. *Materials Research Bulletin* 41: 515–524. https://doi.org/10.1016/j.materresbull.2005.09.014.

56 Ess, J.W. and Hornsby, P.R. (1987). Twin-screw extrusion compounding of mineral filled thermoplastics: dispersive mixing effect. *Plastics and Rubber Processing and Applications* 8 (3): 147–156.

57 Yang, F. and Nelson, G.L. (2006). Polymer/silica nanocomposites prepared via extrusion. *Polymers for Advanced Technologies* 17: 320–326. https://doi.org/10.1002/pat.695.

58 Kalaitzidou, K., Fukushima, H., and Drzal, L.T. (2007). A new compounding method for exfoliated graphite–polypropylene nanocomposites with enhanced flexural properties and lower percolation threshold. *Composites Science and Technology* 67: 2045–2051. https://doi.org/10.1016/j.compscitech.2006.11.014.

59 Chae, D.W. and Kim, B.C. (2005). Characterization on polystyrene/zinc oxide nanocomposites prepared from solution mixing. *Polymers for Advanced Technologies* 16: 846–850. https://doi.org/10.1002/pat.673.

60 Carotenuto, G., Her, Y.-S., and Matijević, E. (1996). Preparation and characterization of nanocomposite thin films for optical devices. *Industrial and Engineering Chemistry Research* 35: 2929–2932. https://doi.org/10.1021/ie950721k.

61 Andrews, R., Jacques, D., Qian, D., and Rantell, T. (2002). Multiwall carbon nanotubes: synthesis and application. *Accounts of Chemical Research* 35: 1008–1017. https://doi.org/10.1021/ar010151m.

62 Preghenella, M., Pegoretti, A., and Migliaresi, C. (2005). Thermo-mechanical characterization of fumed silica-epoxy nanocomposites. *Polymer* 46: 12065–12072. https://doi.org/10.1016/j.polymer.2005.10.098.

63 Manas-Zloczower, I., Nir, A., and Tadmor, Z. (1982). Dispersive mixing in internal mixers–a theoretical model based on agglomerate rupture. *Rubber Chemistry and Technology* 55 (5): 1250–1285.

64 Shiga, S. and Furuta, M. (1985). Processability of EPR in an internal mixer (II)–morphological changes of carbon black agglomerates during mixing. *Rubber Chemistry and Technology* 58 (1): 1–22.

65 Palmgren, H. (1975). Processing conditions in the batch-operated internal mixer. *Rubber Chemistry and Technology* 48: 462–494. https://doi.org/10.5254/1.3547462.

66 Wang, Y., Shi, Z., Fang, J. et al. (2011). Graphene oxide/polybenzimidazole composites fabricated by a solvent-exchange method. *Carbon* 49: 1199–1207. https://doi.org/10.1016/j.carbon.2010.11.036.

67 Pourabas, B. and Raeesi, V. (2005). Preparation of ABS/montmorillonite nanocomposites using a solvent/non-solvent method. *Polymer* 46: 5533–5540. https://doi.org/10.1016/j.polymer.2005.04.055.

5

Advances in Curing Methods of Reinforced Polymer Composites

Ankit Manral[1], Furkan Ahmad[1], and Bhasha Sharma[2]

[1]*Netaji Subhas University of Technology (Formerly NSIT-University of Delhi), Division of Manufacturing Processes and Automation Engineering, Azad Hind Fauj Marg, Sector-3, Dwarka, New Delhi, 110078, India*
[2]*Netaji Subhas University of Technology (Formerly NSIT-University of Delhi), Chemistry Department, Azad Hind Fauj Marg, Sector-3, Dwarka, New Delhi, 110078, India*

5.1 Introduction

High specific strength to weight ratio, noncorrosive nature, and high toughness are some of the advantages of fiber reinforced polymer composites (FRPCs) over conventional materials. Demand for FRPCs is increasing day by day in the field of aerospace, sports, and infrastructure industries. In aircrafts (Boeing-787), around half of the total number of components are made up of FRPCs [1]. Even in some of aircrafts such as Airbus-A350, the percentage of FRPCs was predicted to be up to 53 [2]. It has become very easy to make highly complex products using FRPCs. Aircraft wings, structural components of aircraft, wind turbine blades, and car bodies are some examples of products that are being used under different environmental conditions. Polymer matrix of FRPCs must be completely cured to increase the interfacial bonding between fiber and matrix so that the product can serve its defined life satisfactorily. Curing of thermoplastic is quite different from that of thermoset polymers used as matrix in FRPCs. Contrary to thermoset polymers, thermoplastic polymer matrix does not require any chemical reaction for polymerization. Heating up to a defined temperature followed by cooling is required for curing of thermoplastic-based matrix materials. The curing time of thermoplastic-based matrix is very low in comparison to that of thermoset polymers. Long curing time results in limited production volume of FRPCs products. Hence, curing time is a governing factor for mechanical properties of FRPCs by controlling the residual stresses generated during curing of FRPCs [3]. Therefore, temperature, pressure, and curing time must be at an optimized level for proper curing. Curing energy must be sufficient for the proper polymerization of resin. If curing temperature is lower than the optimum level, longer time will be required for proper polymerization. On the other hand, curing at higher temperature may result in generation of residual stresses in the composite material. The generation

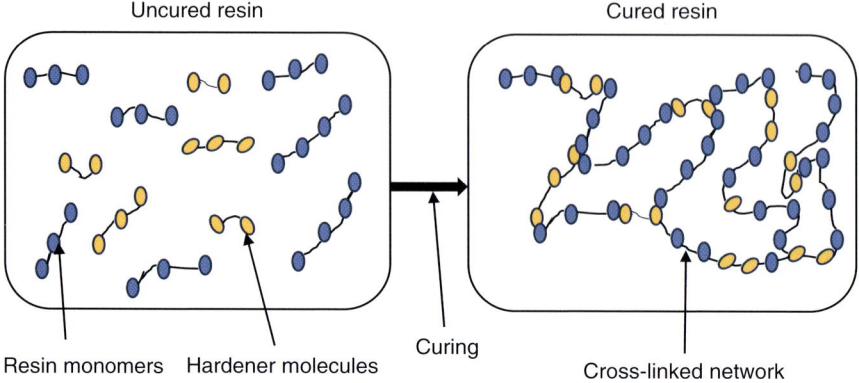

Uncured resin

Cured resin

Resin monomers   Hardener molecules

Curing

Cross-linked network

Figure 5.1 Curing of resin.

of residual stress reduces the load bearing capacity of the material, which may cause sudden failure of the product before completion of life.

Autoclave, microwave, and direct electric and radiant energy are some of the prominent curing techniques that are being used for the development of FRPCs. Conventional curing techniques are based on thermal heating processes such as microwave and autoclave. In conventional curing process, if heat is not enough to penetrate the layers of FRPCs, improper polymerization of matrix will result in low quality composite products. The lower heat penetration may be because of low thermal conductivity of the reinforcement material, which lowers the flow of heat from one layer to another. Therefore, radiating and convective heat curing processes are suggested to be used for proper penetration of heat throughout the material for better polymerization (interlocking of monomers units of resin) [4]. Figure 5.1 shows the resin before and after curing, and the monomer unit of resin forming a cross-linked structure after curing.

Convective heat transfer during curing of FRPC is also time consuming and requires high energy input; still it is better than conductive heat transfer. The contribution of convection curing reduces the generation of residual stresses in FRPC, which was the main drawback of conduction curing processes. On the other hand, high energy input in convection curing may degrade the fiber with lower thermal stability such as natural fiber. The curing time energy consumption also depends on the size of the developed composites. The thickness of product influences the heat transfer rate and distribution of thermal energy, which in turn results in generation of residual stresses. With oven and autoclave curing, limited size composite components can be cured. The world's largest autoclave chamber of diameter 9 m with length of 25 m was designed as per the demand of "Autoclave Systems for Aerospace Composites (ASC) processes system" (USA) for the production of Boeing-787 components [5]. Size limitation is still a big problem for curing of polymer composite products as in aerospace and wind industries [4]. Hence, industries and researchers are still exploring new curing processes for highly efficient curing methods with no size limitations. At present, autoclave technique is the most popular technique among all available options.

The lower thermal conductivity of the composite material constituent retards the flow of heat, which becomes a big problem for both convection and conduction type of curing. To overcome this problem, microwave curing comes into the picture. Microwave of a certain wavelength travels irrespective of the thermal conductivity of the composite material and penetrates the component easily throughout the layers of composite materials. The polymerization done by microwave heating is more efficient and effective than conduction and convection curing techniques. Curing time is also lower in microwave curing method than other methods due to instant propagation of heat, which in turn accelerates the rate of polymerization. Microwave heating also minimizes the generation of void percentage by reducing the viscosity of resin during curing [6, 7]. However, microwaves are hazardous to human health. Hence, the size of microwave chamber is limited. Size limitation of the microwave chamber limits its application in the field of composite production for aerospace and power generation sectors. Limitation with conventional curing techniques (autoclave, microwave, etc.) leads to the development of advanced curing techniques such as radiation, electric, and thermal additives-based heating. In radiation curing, radiation rays ionize resin molecules with the help of high electromagnetic radiation (ultraviolet, X-ray, and γ rays) [4]. In radiation curing, pure resin can be polymerized as it does not require any hardener for curing. High beam radiation ionizes the molecules of resin, forming cations and anions, which accelerates the bonding of resin molecules instantly and hence reduces the curing cycle time. This special curing mechanism is superior than other conventional curing processes. All curing mechanisms can be categorized on the basis of mechanisms used for heat generation, which is required for polymerization. In the present chapter, various advanced curing techniques used for polymerization of different polymers have been discussed. Figure 5.2 shows the classification of curing techniques on the basis of radiation and thermal energy.

A number of curing techniques are being used for curing of polymer-based composites used in different applications. Advanced curing processes have lower curing cycle due to which the application spectrum is very wide starting from household products to aerospace and automobile sector. Figure 5.3 shows some of the important applications of conventional and advanced curing processes. In the present chapter, various conventional and advanced curing techniques for thermosets and thermoplastics have been discussed.

5.2 Curing Method

Curing of polymer composites refers to the polymerization or solidification of the matrix material used for holding up the reinforced composites. Thermal curing techniques are more popular in the field of automobile and aerospace industries but these techniques are time consuming. Large volumetric components of aircraft industries are not cured with normal curing processes. Therefore, advance curing techniques are mostly preferred to cured large volumetric components. Microwave curing is mostly used by aerospace industries

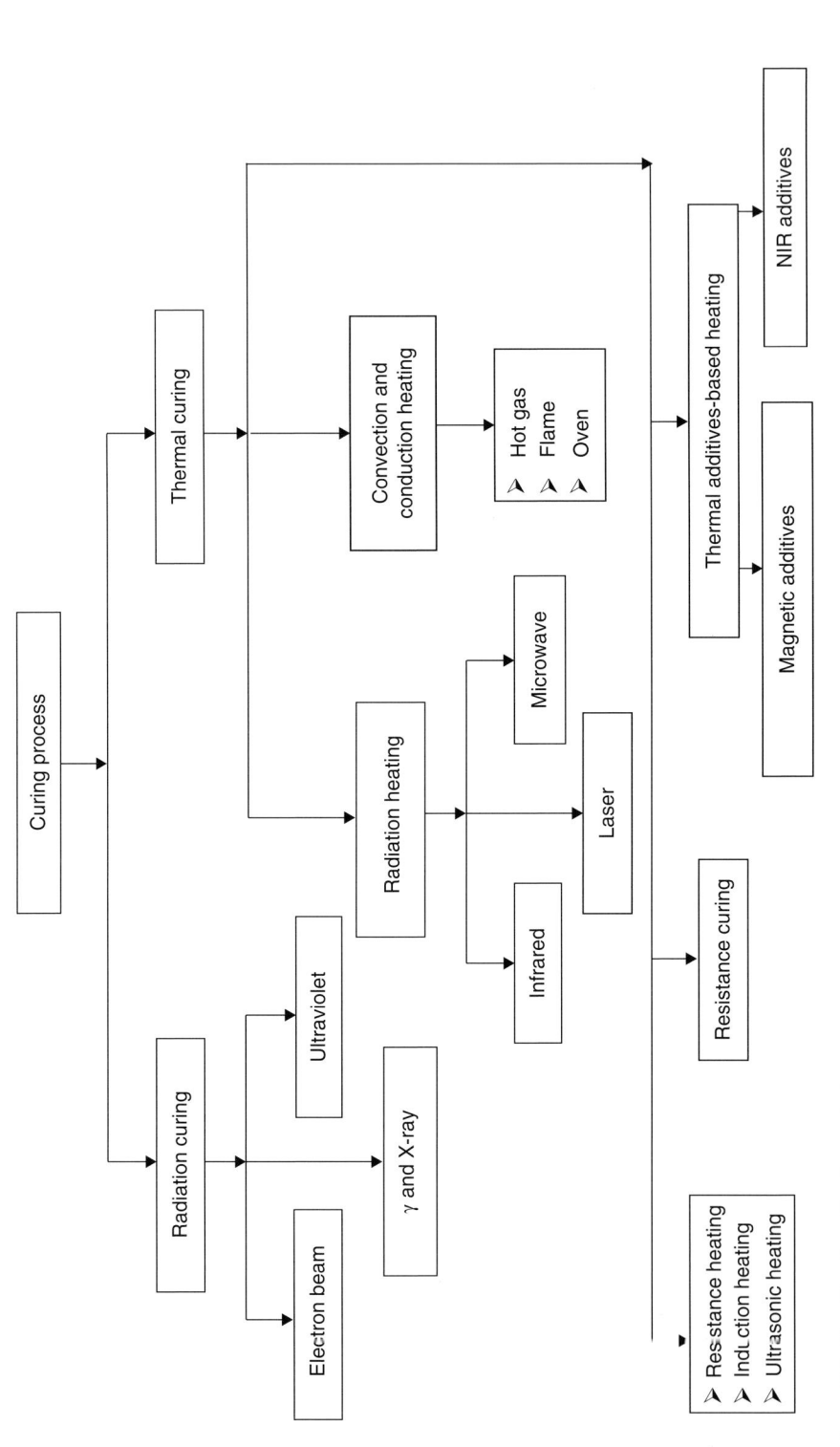

Figure 5.2 Classification of curing techniques.

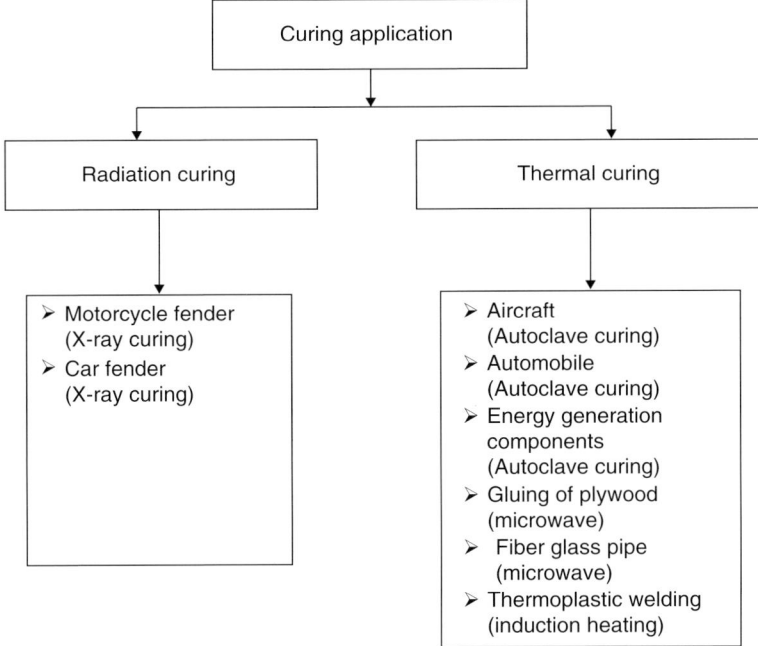

Figure 5.3 Conventional and advanced curing application.

due to the lower time required for curing and low energy requirement without compromising on the quality of FRPCs product. All curing processes fall under the category of conventional and advanced curing techniques. Each process has some disadvantages such as long curing time, high energy input, and limitation of size while some of the curing methods have some advantages such as lower curing cycle and lower energy input for proper polymerization of resin.

5.3 Thermal Curing of FRPC

In thermal curing, polymerization of resin is done by the application of heat. Although polymerization of resin can be easily completed at room temperature, it takes around 24–48 hours for curing in the case of epoxy and polyester. So, to accelerate the rate of polymerization, thermal energy is used. Classification of different thermal processes is done on the basis of heat input mechanisms during the curing of FRPCs.

5.3.1 Autoclave Curing

Autoclave curing is a convective heat transfer process used for the curing of FRPCs. In autoclave a closed vessel maintained at certain temperature and pressure for a definite time depending upon the type of resin under curing. The composite cured by autoclave technique can be prepared by the hand lay-up process or vacuum bagging process using prepregs. After preparation of the laminate

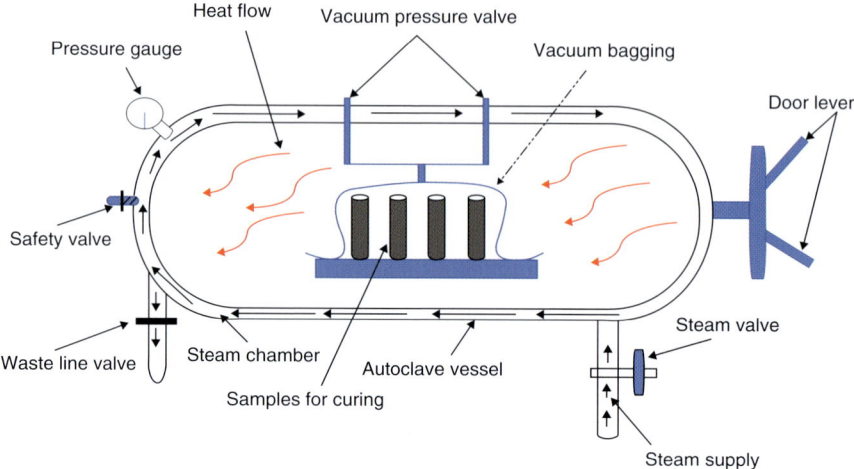

Figure 5.4 Autoclave curing.

stack, the component is placed in the autoclave chamber at a particular temperature and pressure for a particular time. Using certain optimized parameters, the
components get cured/solidified. Mostly all types of composite laminates can be
cured using cylindrical autoclaves. Large volumetric components of aircraft and
wind energy generation wings can be easily cured using autoclave chambers.

The autoclave curing chamber contains a pressure chamber in which components get cured under the required pressure and heat. Figure 5.4 shows an
autoclave pressure vessel. The cylindrical shape of the vessel provides for both
a flat and a cylindrical body for curing. For proper curing of components, the
pressure should be maintained at a sustainable limit. The pressure vessel is made
leakproof and the door is properly sealed after closing. The required pressure in
the chamber is achieved using an air compressor mounted on the outer body
of the chamber. For vacuum bagging, vacuum is maintained by two hose pipes
connected to a vacuum compressor. Temperature sensors and thermocouples
are placed inside the chamber for detecting temperature. Vessel pressure can be
maintained by a safety valve that is mounted on the chamber to release excess
pressure above the required level. For heating up the chamber, gas firing and electric heaters are used to complete polymerization/curing of polymer composites.
Direct gas firing is mostly preferred for large volumetric components (aircraft
and wind turbine blade), although direct heating systems are preferred for small
components (automotive components). Large components require high thermal
energy to spread over the surface of the components that can be only achieved
by the gas firing method, whereas for small components the polymerization of
components can be easily achieved by direct heating. The design of the gas firing tube should be done according to prescribed measurements to avoid leakage
during operating hours.

For prepregs curing, initially vacuum is generated in vacuum bags to remove
the air bubbles present in the laminate before placing it in the autoclave

chamber. The vacuum developed during polymerization removes the void contents, which were developed because of mobility of molecules in regions with viscosity difference [8]. The reduction in void percentage directly influences the mechanical properties of the developed composites. The vacuum level created in the vacuum bag must be at an optimized limit so that it only releases the air and volatiles. The excess vacuum pressure releases the resin from the prepegs, forming a low resin region. A very high vacuum may even displace the reinforcement.

The curing process occurs in three different phases, namely heating phase, holding phase, and cooling phase. Every phase of curing must have its own definite limit of parameters. Proper time should be provided for proper and efficient curing of the polymer. The cure cycle varies according to the type of resin to be cured in the autoclave chamber. The curing cycle starts with the first phase, the heating phase, where the temperature linearly increases with respect to time. In this phase, polymerization of resin just starts by binding the molecules of the resin and hardener. In the second phase, the holding phase, the temperature is constant with respect to time while polymerization is under progress. With increasing temperature, it becomes very hard to maintain the stability of the fiber. If the curing temperature is higher, it may degrade the reinforced fiber. The third phase of curing is cooling, where the composite material component is cooled at a certain rate of heat transfer till it reaches room temperature. A variety of reinforcement fibers (carbon fiber, glass fiber, aramid, etc.) are used to incorporated in the polymer matrix (epoxy, phenolic resin, polyester, polyvinyl ester, unsaturated polyester, and thermoplastic resin) using autoclave curing. The curing cycles for different thermoset polymers are shown in Figure 5.5. The intensity of heat input varies according to the type of resin

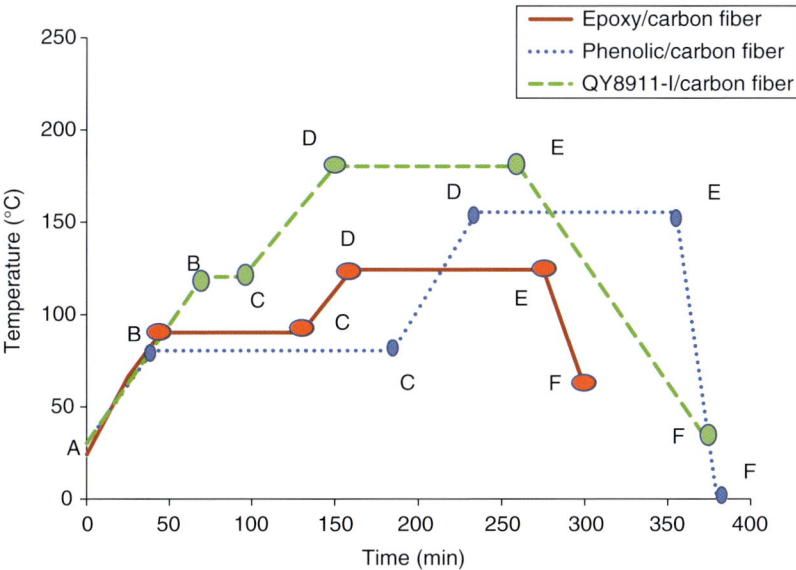

Figure 5.5 Curing curves for different resin and carbon fibers [9–11].

used as matrix in FRPCs. It is evident from Figure 5.5 that all other polymers except epoxy start polymerizing at higher temperature. Even the holding time varies according to the type of resin. Phenolic resin requires high holding time for curing. The variation of phases of curing curve is highlighted by different points as follows:

A–B → Linear increase in temperature with time
B–C → Maintained at low temperature (initial polymerization)
C–D → Temperature rise with respect to time
D–E → Maintained at higher temperature (constant time)
E–F → Reduction in heat input (cooling)

The cooling rate at the end phase of curing is also different according to the type of resin cured. From Figure 5.5 it is clear that cooling of phenolic/carbon composites is sudden as compared to other combinations of resin and carbon composites.

The intensity of curing may vary along the thickness of components due to the generation of a temperature gradient across the thickness of components. The temperature gradient varies from layer to layer, which generates a problem of inefficient curing of components [12]. Variation of temperature in all layers starting from the upper surface to the lower surface depends on the fabric architecture and thermal conductivity of the fabric material. Low thermal conductivity of the constituents of composite material also reduces the propagation of heat from surface to core and then from core to surface. Variations of temperature that exist between layers are shown in Figure 5.6. The average temperature of the autoclave chamber is generally 100 °C, thus leading to heat flow from surface to core due to temperature difference. The temperature changes from surface to core of the material depending on the thermal conductivity of the constituents. Difference in thermal conductivity of constituents leads to a situation where curing

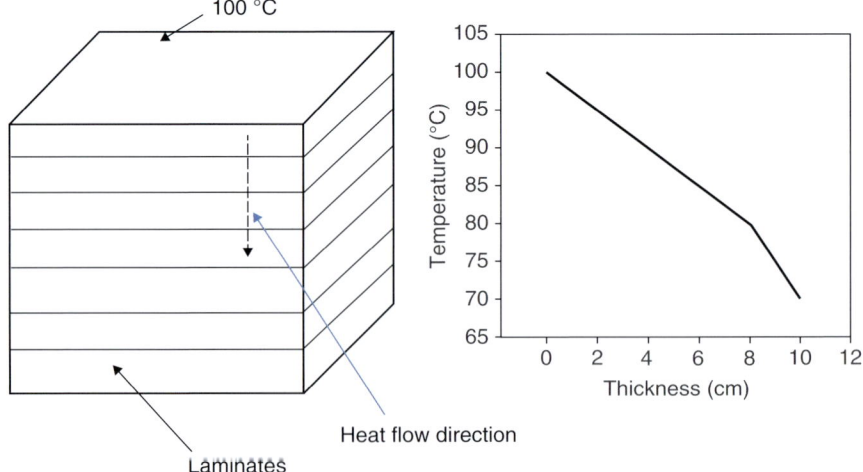

Figure 5.6 Variation of temperature across thickness.

Table 5.1 Advantages and disadvantages of autoclave curing.

S. No.	Advantages	Disadvantages
1.	Higher fiber reinforcement fraction can be used in composites	Size limitation of components
2.	Applicable for both thermoset and thermoplastic resin	High initial and running cost
3.	Better surface quality and high interfacial adhesion can be achieved	Not good for high production rate due to long curing cycle
4.	Curing by autoclaves reduces the void formation	Costly technique
5.	Improved wettability of reinforced fibers	Require highly skilled labor

is not efficient or complete and temperature distribution is not uniform. The above problem of temperature distribution can be solved by proper selection of reinforcement and matrix material. Increasing the time of curing cycle may also solve the problem to some extent. At the same time, increasing the time of curing results in low volume production of composite products.

Irrespective of long curing cycles, presently autoclave curing is being used more than any other curing technique in automobile and aircraft industries. Autoclave curing has some advantages for FRPCs products along with some disadvantages. The following are some of the advantages and disadvantages of autoclave curing as discussed in Table 5.1.

5.3.1.1 Properties of FRPCs Influenced by Autoclave Parameters

Autoclave curing cycle parameters (time, temperature, and pressure) influence the mechanical and thermal properties of the developed FRPCs. Optimized temperature is required to obtain better flowability of resin and for efficient polymerization of resin. Flowability of polymer improves the dispersion of molten resin over the dry surface of fiber/reinforcement. Optimized pressure also influences the wettability of fiber and optimized vacuum pressure reduces the void percentage during curing of composites due to change in viscosity of resin. For proper polymerization, the resin is required to sustain at constant temperature for certain time. A number of studies have been performed to study the effect of autoclave parameters on the mechanical (tensile, flexural, and impact strength) and thermal properties. Comparative studies of different parameters have been performed to achieve optimized results in terms of better mechanical and physical properties of the developed composites.

Chamber Temperature The first and foremost parameter of autoclave curing is the temperature of the chamber. If the temperature of the autoclave chamber is below the optimized level it may affect the storage modulus, glass transition temperature (T_g), and some other viscoelastic properties of the developed composites [13]. Lower temperature reduces the polymerization rate due to less penetration of heat inside the samples. Improper penetration results in the generation of internal thermal stresses and reduction in internal laminar strength. Generation of internal stresses and lower interfacial adhesion minimize the properties of the developed composites. Most of the mechanical properties (tensile and flexural strength) of the developed composite depend upon the interfacial adhesion. On other hand, higher temperature of the autoclave chamber than the optimized temperature results in fiber/reinforcement burn. Hence, it is advisable to use chamber temperature within the optimized limit to obtain better mechanical properties of the developed FRPCs.

Chamber Pressure Autoclave chamber pressure is responsible for generation of voids in the developed FRPCs. Voids present in the laminate result in reduced strength of the component and the product may fail earlier than its life at normal loading conditions. If the chamber pressure decreases from its optimum value, the chances of generation of voids increase due to insufficient force to resist the mixing of air with resin. Formation of voids in composites may be because of any of the reasons such as chemical reaction excreting volatile content due to resin system or entrapped air during stacking or lay-up of prepregs [14]. Hence, curing pressure and vacuum should be properly selected so that it helps in squeezing out all the entrapped air or gases from the laminate. There are some mathematical models that were developed to correlate the effect of void percentage on composites properties [15]. These models help predict the effect of curing parameter on the properties of FRPCs without any actual curing of samples. Huang and Talrej [15] developed a quantitative relation between the properties of FRPCs and the void percentage. Therefore, it is always good to find the optimized curing parameters before performing the actual autoclave curing.

5.3.2 Induction Curing

Induction curing is based on the principle that when a nonmagnetic, highly electric conductive material is placed in a magnetic field eddy current losses result in the rise of temperature of the nonmagnetic material. In the past years, it has been seen that induction heating is being widely used for welding and heating of polymer composites [4]. The working principle of induction curing is explained in Figure 5.7. The generation of heat by magnetic hysteresis is lower than heat generated by eddy currents [16]. Arrangement for induction heating varies according to the type of fiber incorporated in the polymer. For glass FRPC, an additional electric conductive material should be placed between the fiber and the induction coil for the generation of heat. The thermal conductivity of glass fiber is very low, which resists generation of heat from the induced magnetic field. On the other hand, carbon fiber is an electrically conductive material and hence it does not require any conductive material for induction of heat from the developed magnetic field [17]. Induction setup depends upon the coil geometry, coil current, and input electric power. Coupling distance and frequency of coil input current are also some additional factors that influence induction quality. Current input frequency in the coil decides the limit of

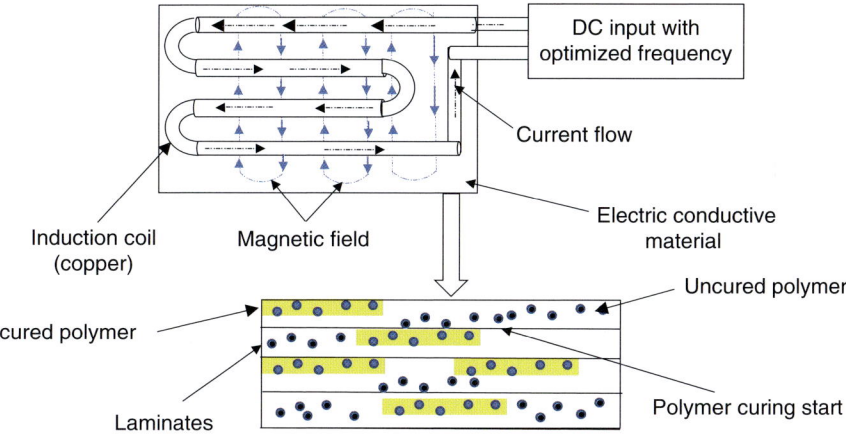

Figure 5.7 Principle of induction curing.

penetration of electromagnetic field. The penetration of magnetic field can be found by Maxwell relation as given in Eq. (5.1) [16]:

$$\delta = \sqrt{\frac{\rho}{\mu \pi f}} \qquad (5.1)$$

δ	=	penetration depth
μ	=	magnetic permeability
ρ	=	electric resistivity
f	=	frequency

Furthermore, coupling distance also influences the induction heating, and lower coupling distance improves the local temperature. Heating dispersion is inhomogeneous in nature, which influences the variation in temperature of the surface of the composite material. For homogeneous heating of the component, the optimized coupling distance should be used. The following are some of the parameters that should be considered during the design of coil geometry. The shape of coil used in induction heating depends upon the surface or component that is to be heated. There are some common induction coil geometries that are usually preferred in heating, joining, and curing of FRPCs such as pancake coil, conical coil, single-turn coil, Helmholtz coil, and helical coil [16].

There are a number of contributions from various authors on the use of induction method in the field of polymer composites as shown in Figure 5.8.

Induction heating is widely used for joining and welding of nonhomogeneous materials. The spectrum of research in polymer curing by induction heating has spread to the field of composites applications [18–21]. In induction heating, the rate of heat generation is very high as compared to other conventional heating processes. Instant heat generation reduces curing time, which accelerates the rate of polymerization. Tay et al. [22] compared induction curing with other

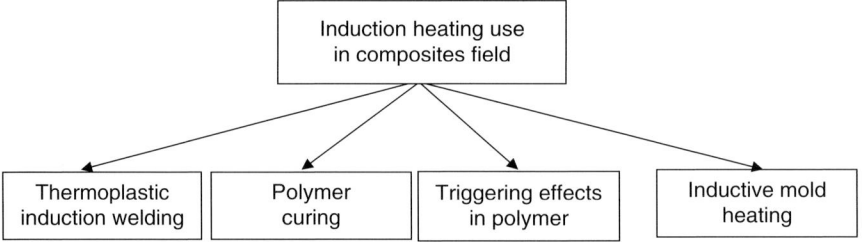

Figure 5.8 Use of induction heating in the field of polymer composites.

curing techniques and concluded that induction curing reduces the curing time from 15 minutes to 1 hour in comparison to oven heating. There are two different mechanisms of heat generation required for curing of FRPCs by induction. For an electrically conductive fiber, the eddy current induces in the fiber an electric current flow over and between the fibers layers, which causes the generation of heat by Joule's law of heating [23]. Looping is required for better heating of composite components; hence, for a unidirectional fiber mat, induction heat is not sufficiently developed. On the other hand, some authors claimed that induction heating can be generated in composites curing by dielectric losses between fibers in the polymer [24]. Induction curing was found to be efficient for small scale composites industries. The size of copper coil used in induction curing depends upon the size of laminates to be cured. Induction curing has some limitations related to the size and therefore large volumetric components are not cured with induction heating efficiently. This limitation of induction heating makes it applicable for aerospace industries. An induction coil of copper material is used through which current flow generates a magnetic field around the coil. When nonmagnetic and high electric conductive material comes across the magnetic field, it accelerates the electron present in the conductive material. The eddy current induced in the conductive material in the form of eddy loss results in the generation of heat on the surface of the conductive plate instantly. The generated thermal energy ignites the curing of polymer present in the laminates kept under the conductive plate. The use of conductive plate above the laminates is necessary for natural fiber reinforced or low thermal conductive reinforced composites. Curing of carbon fiber reinforced laminates by induction techniques does not require any conductive plate because of the electrically conductive nature of the carbon fibers that allows the electron to flow. Induction curing reduces the curing heating time, but for cooling a special arrangement is required with the apparatus so that cooling can be done in the desired conditions. In the whole curing process, cooling is time consuming. Overall, the time required for induction curing is lower than that in other curing processes. Induction curing process has been proved to be efficient for curing of FRPCs with some advantages and disadvantages as discussed in Table 5.2.

There are some parametric analyses of induction curing for proper curing of FRPCs. The magnetic field induced during induction must overlap all the circumferential surfaces of the electrically conductive material for homogeneous heating of all the laminae or prepregs of FRPCs. The main influencing factors for quality of

Table 5.2 Advantages and disadvantages of induction curing.

S. No.	Advantages	Disadvantages
1.	Lower time required for curing	Require special arrangement for non-electrically conductive fiber/reinforcement
2.	Lower initial setup cost	Limitation of size for curing
3.	No special arrangement required for curing of high electric conductivity fiber	Frequency variation and coupling distance affect penetration of heating and heat distribution respectively
4.	Lower energy input required	Curing of limited fiber reinforced polymer composites
5.	Used for curing, joining and welding of FRPC	

FRPCs are location of coil over the cured surface and area ratio of coil to specimen [25]. Location of coil during induction is most important for proper curing of composite layers. The location of coil influences the spectrum of the generated magnetic field required for curing of the layers efficiently. Proper optimized position of the coil over the conductive material should be ensured for proper curing of the composite. Also, the ratio of coil area and specimen should be nearly equal to unity, i.e. the coil area must cover the whole area of the specimen. Both factors should be maintained during heating for proper heat input inside the composite layer so that the generation of induced stresses can be eliminated [26].

5.3.3 Resistance Curing

Resistance curing is based on the electrical resistance of conductive materials and the generation of heat using the resistance. When electric current passes through an electric or semi-electric conductive material the resistance offered by the material causes the generation of heat under pressure. Resistance curing is not applicable for natural fiber reinforced composites because of the nonconductive nature of natural fiber. In the case of autoclave curing, heating takes significant amount of energy and time to achieve the temperature at which curing can take place. Furthermore, some extra time is needed for cooling, which makes autoclave curing a more time consuming and costlier method for composites, which limits its use in mass production.

The electric conductive nature of a number of fibers provides the potential to use conductive fibers as a heating element by applying electric current across fibers [27, 28]. Generation of heat around conductive fibers initiates the polymerization of resin surrounding the surface of the conductive fiber. Curing time depends upon the intensity of current and resistance offered by conductive fibers/reinforcement. A schematic diagram of resistance curing is shown in Figure 5.9. The ends of prepregs are separated by copper blocks. Copper blocks are held in place with the help of locking pins, which hold through the height

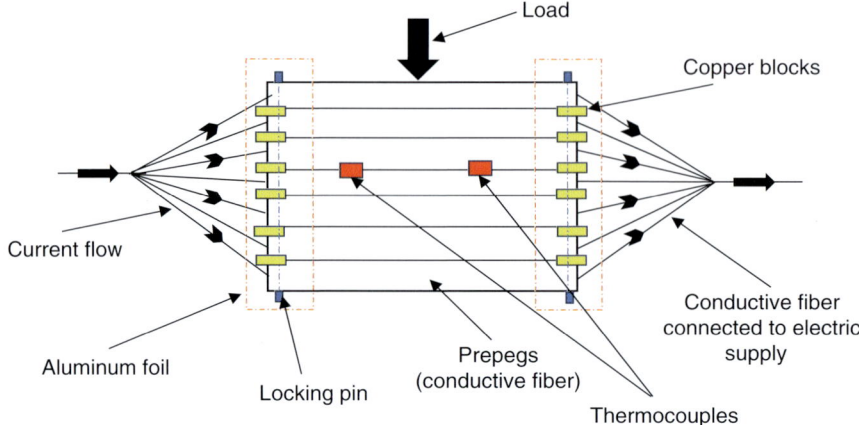

Figure 5.9 Layout diagram of the working principle of resistance curing.

of the stacks. Aluminum foil is wrapped around the ends of the copper block to promote good electric connection and to reduce the stickiness of epoxy with copper blocks. Current flows across the stack of prepregs and they are placed under some applied load. The polymerization of resin starts under some load and applied voltage across the conductive fiber. Thermocouples are placed at the center of each layer of prepregs to confirm whether the curing temperature is achieved. One of the advantages with resistance curing is the generation of heat across the whole surface of the conductive fiber, which helps in polymerizing the whole product at the same rate. The generation of thermal stresses also reduces due to uniform curing of the product, which directly influences the mechanical properties of the developed composites.

Proper resistance curing can only be achieved when prepregs are properly arranged with all other accessories. Improper arrangement leads to improper contact resistance that affects the rate of heat generation during curing. Many researchers have worked in the field of resistance curing of FRPCs. Hayes et al. [29] compared the different curing techniques of carbon/epoxy and concluded that resistance cured samples achieved higher flexural modulus as compared to oven and autoclave cured specimens. Sancaktar et al. [30] concluded that resistance curing improved the interfacial strength and reduced the ability of initiation of crack in the polymer matrix. Resistance curing has some advantages and disadvantages as shown in Table 5.3.

5.3.4 Microwave Curing

In the present scenario, thermoset polymer resins are the most used matrix material for the fabrication of FRPCs. Furthermore, among all the thermoset polymer resins, epoxy is widely used as a matrix material for the fabrication of composites. Epoxy resin exhibits good physical and mechanical properties after complete curing. The incorporation of matrix with reinforcement results in a different property than the individual constituent. Excellent properties of thermoset polymers

Table 5.3 Advantages and disadvantages of induction curing.

S. No.	Advantages	Disadvantages
1.	Resistance curing takes lower time and lower running cost for curing of composites	Arrangement of prepregs or laminates is time consuming and requires skilled labor
2.	Heating rate is constant, which reduces the induced thermal stresses. Thick prepegs can easily be cured.	Applicable only for electrically conductive fibers such as carbon, aramid, etc.
3.	Can be used for mass production due to short curing cycle	Limitation of size of components and also requires high energy input
4.	Can also be used for joining of composites	
5.	Less controlling parameters	

open a new door of opportunities in the field of aerospace, marine, automobile, and other industries. The ever-growing demand for thermoset polymers presents a challenge to the engineers to find a better curing technique than other available conventional processes. Conventional processes are time consuming and require high energy input, which affects the production output in a negative manner. As a result, nonconventional techniques are being introduced in the market such as electron beam (EB) curing, ultraviolet curing, and microwave curing.

Microwave curing is done by propagation of microwaves through the thickness of the laminate, which transfers momentum to the molecules of the material. Transfer of energy takes place when the molecules collide with the microwave, resulting in the generation of heat. This heat helps in starting and completing the polymerization of the matrix. In microwave curing, heat penetrates deep inside the composite material resulting in complete polymerization of the matrix material. Dielectric properties of the material decide the microwave curing efficiency [31, 32]. The curing time can be minimized by changing the input parameters of microwave (frequency and amplitude). Lower curing time of composites can be beneficial for mass production of polymer composite components. The common frequency used for microwave curing is around 2.5 GHz. This technique is inexpensive and easily available [33]. Waves propagate during curing in two different modes, namely transverse magnetic and transverse electric. Transverse magnetic has longitudinal and transverse components while transverse electric has electric intensity only in the transverse direction. A schematic diagram of microwave curing is shown in Figure 5.10. During curing, microwaves strike the molecules of material and transfer the momentum to the molecule of the matrix. This momentum gain by the molecule results in the displacement of the molecules and hence, molecules of the matrix start colliding with each other. The collision of molecules results in the generation of heat. This generated heat energy is enough to raise the temperature of the material up to a level where polymerization starts in the resin and propagates. A dead weight is placed on the top of laminates stacking layer to maintain the thickness, to squeeze out excess amount of resin, and to reduce void

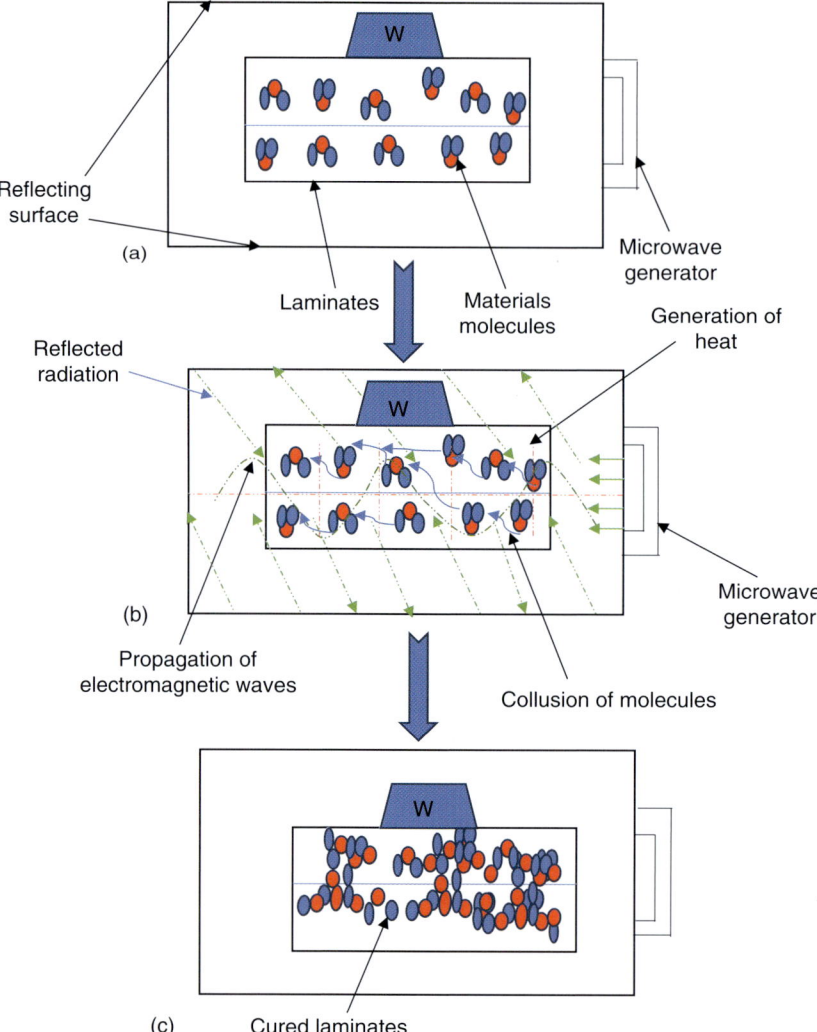

Figure 5.10 Schematic diagram of microwave curing. (a) Before microwave curing, (b) during microwave curing, and (c) after microwave curing.

formation. Similarly, distribution of temperature across the thickness of material helps in curing various complicated shape products.

5.3.4.1 Properties Influenced by Microwave Curing of FRPC

The properties of developed FRPCs, polymerized by microwave curing technique, are directly influenced by the input parameters. In microwave curing, the parameters are time, temperature, and heat input. Curing time is the time at which laminates sustain the heating and cooling phase. Properties of FRPCs depend upon the interfacial adhesion of reinforcement and matrix material, which can be achieved by curing parameters. In microwave curing, increase in power input increases the generated temperature, which degrades the interfacial adhesion. With increase in temperature, it is observed that the relaxation time of dipole also decreases, resulting in reduced interfacial adhesion, and hence the mechanical properties decrease [34]. Hang et al. [35] studied the effect of microwave curing temperature and curing time on the properties of carbon fiber reinforced composites and concluded that increasing the heating rate degrades the tensile properties of microwave cured composites although the interlaminar and flexural strengths increase with increment in heat input. The effect of input parameters of microwave curing and a comparative study of microwave curing with other processes are summarized in Table 5.4.

5.3.5 Ultrasonic Curing

Ultrasonic curing is another unconventional technique for curing of FRPCs. Ultrasonic curing is based on the generation of heat using high frequency mechanical vibration. The ultrasonic mechanical vibrations are generally produced using two methods, namely magnetostriction and piezoelectric effect. Using magnetostriction method, ultrasonic waves of the frequency range of 20–100 kHz can be produced while piezoelectric method can generate ultrasonic waves of frequency more than 100 kHz. Ultrasonic vibration gives a small displacement to the chain segments and atoms around their position [40] whereas in polymers, the force acts between the molecular and segment chain for creating a displacement in the neighboring zones [41]. The displacement of particles in the resin initiates the polymerization. A wide range of ultrasonic waves can be used depending upon the type of application for which ultrasonic waves are being used. The types of waves are lamb, shear, and longitudinal waves. Shear waves are generally used for curing of liquid and soft gel resins [42]. Ultrasonic waves can be characterized on the basis of propagation velocity, displacement amplitude, and wavelength of the developed waves. Ultrasonic waves are produced with the help of transducers, which convert electric energy into ultrasonic waves. A schematic diagram of ultrasonic curing is shown in Figure 5.11.

A monitoring unit gives electrical input to the transducer, which converts it into ultrasonic waves. These ultrasonic waves strike the laminates, which starts the displacement in molecules and chain segment starting the polymerization. This displacement of molecules also removes the air bubbles developed during pouring of resin by minimizing the viscosity of resin. For monitoring the curing quality of laminates, sensors are mounted inside the unit. Feedback signals control the input signal for proper curing of laminates. Ultrasonic curing is mostly preferred for surface curing of polymers in packaging industries. Ultrasonic curing has some advantages and disadvantages over conventional curing process as shown in Table 5.5.

5.4 Radiation Curing of FRPCs

In radiation curing, high electromagnetic radiation is used for the generation of heat required for curing of FRPCs. Various types of radiation such as ultraviolet,

Table 5.4 Effect of curing techniques parameters on properties of polymer composite.

S. No.	Matrix/ reinforcement	Conventional curing technique/ (fabrication technique)	Optimized parameters			Tensile strength (MPa)	Flexural strength (MPa)	Impact strength (kJ/m²)	Conclusion	References
			Power input (W)	Cure time (m)	Frequency (GHz)					
1.	PP/Sisal PP/Grewia optiva EVA/Sisal EVA/Grewia	Microwave/ (Hand-lay-up)	900	12.5 8.5	2.45	40-45 19-20 23-25 5-10	100 58-60 70-80 40-50	40 25-28 20-25 5-6	• At given optimized time and energy input in microwave curing enhanced the properties of developed composites	[36]
2.	DGEBA (epoxygrade)/ carbon LY/HY5052 epoxygrade)/ carbon	Microwave, Conventional thermal/ (RTM)	MW-250 CT-2000	MW-40 CT-60 MW-40 CT-180	—	—	—	—	• In MW curing the cure cycle for (DGEBA-carbon) reduce from 60 to 40min. • And for LY/HY5052-carbon cure with MW reduces cure time from 3hrs to 40min • The reduction in cure cycle highlighted the improve efficiency of MW curing	[37]
3.	Glass-epoxy	Microwave/ Thermally cured/(Vacuum bag mold)	MW-.258 KWh TC-6kWh	20 4hr	2.45 —	— —	— —	— —	• The power consumption during microwave curing is less compared to thermal curing	[38]

No.	Material	Process	Power	Time	Freq.			Findings	Ref.
4.	Epoxy/carbon	Microwave, conventional/ (RTM)	MW-.250 TC-4KW	90 3hr	2.45 —	— —	— —	• Microwave cured sample have higher tensile strength as compared to conventional cured samples • The Tg of microwave cured sample is higher as compared to thermal cured samples • Microwave curing improved flexural strength as compared to conventional cured samples. • Microwave reduce 50% cured cycle time. • The interlaminar strength of microwave cured sample increase 9% as to conventional cured samples.	[6]
5.	Carbon fiber pre-pegs	Microwave, Autoclave/ (Vacuum bag mold)	MW-3KW AC	30 180	2.45 —	62.4 110.5Ksi	— —	• Autoclave composite show higher tensile strength. The curing time for autoclave and microwave were significant effect on tensile strength of composite.	[39]

Note- PP- polypropylene, EVA-ethylene vinyl acetate, MW-microwave, CT- conventional thermal, RTM-resin transfer mold, TC- thermally cured, AC -auto Clave

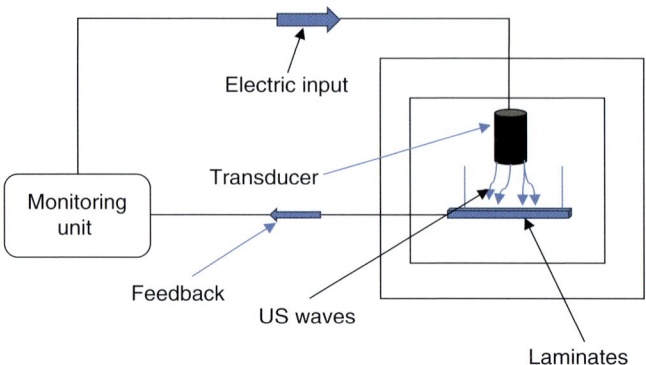

Figure 5.11 Ultrasonic curing of laminates.

Table 5.5 Advantage and disadvantage with US curing.

S. No.	Advantages	Disadvantages
1.	Time saving as compared to conventional curing techniques	Initial setup cost is high
2.	Waves are not harmful to human being and environment	High input required for high thick cured laminates
3.	Curing cost is low	Highly skilled labor is required
4.	Displacement in chain segment reduces the air bubble formed during pouring	Initially setup time for input parameters is high

X-rays, and γ rays are being used in radiation curing. The radiation incident on laminates can ionize the resin molecules and splits the molecules into cations and anions. These negatively and positively charged ions attract each other and start polymerization of the resin. Curing agent is not required for curing of polymer using radiation, thus reducing the cost of the composite product. Polymerization of resin takes lower time in comparison to conventional curing processes. The advanced curing mechanism of radiation curing makes it superior to other conventional and thermal curing processes. Classification of radiation curing can be done on the basis of the type of radiation used for curing of FRPCs.

5.4.1 Electron Beam Radiation Curing

Electron beam radiation (EBR) curing uses high intensity electron beams for the ionization of polymer resin. Curing of polymer composites using EBR method was first investigated by France for missile fuel tank aerospace applications in early 1970 [43]. EBR curing is very helpful for the curing of various aircraft components such as wing skin and bulkhead [44, 45]. EBR method has various advantages over the conventional curing techniques. EBR is not only used for complete curing of composite but it can also be used for surface curing of polymeric materials. The application of EBR in various fields depends upon the

Table 5.6 Applications according to the EB intensity.

S. No.	Intensity of radiation	Electron penetration	Application field
1.	10 MeV	24 mm or less	Composites (carbon fiber)
2.	4–10 MeV	38 mm	Sterilization
3.	0.4–3 MeV	11 mm	Cables and wire
4.	300–800 keV	2 mm	Shrink film
5.	80–300 keV	0.4 mm	Surface curing

Source: Berejka et al. 2010 [46].

intensity of electron beams. The classification of EBR application on the basis of intensity of electron beams is shown in Table 5.6.

EBR curing technique is an alternative to autoclave and other conventional curing techniques that take more time for curing of polymer composites. In conventional curing, the curing/polymerization of monomers starts at a certain temperature. More time is consumed in achieving and maintaining the particular temperature necessary for curing, limiting the production volume. There are basically two mechanisms for the polymerization known as stepwise curing and chain mechanism [47]. In stepwise curing, cross-linking of the monomer units occurs step by step. The linked monomer units link with others to form dimers, trimers, and so on. On the other hand, chain mechanism first activates a single species, which will then initiate chain polymerization.

Proper curing of thick laminated composites requires high energy input, which in turn requires large capital cost. In order to resolve this problem, layer by layer curing of laminates is used, which requires less energy input. Much research has been done in the field of curing of composites using the layer by layer curing method. In this technique, first, the laminates may be cured on single sided or double sided irradiation. Once a single laminated sheet is cured, it is placed on module and then tape placement roller is used for proper stacking of cured prepeg sheets over the module. The stack of cured laminates after irradiation is placed in the oven at a certain temperature for proper curing until composites sheets of the desired thickness are achieved. Oven curing is an additional process with layer by layer radiation curing of composites laminates. This technique is generally adopted for minimizing the radiation input, which directly affects the processing cost of curing. Lower range of energy input (in keV) can be used for curing of higher thickness composite components used in aerospace and automotive industries using layer by layer method. Layer by layer curing by irradiation is time consuming due to the additional process of oven curing and separate curing of each lamina of composites. Oven curing used after layer by layer curing of laminates is sometimes optional and it depends on the type of fabrication process used for the development of composites. Some fabrication processes such as filament winding do not required oven curing as at every winding layer it is already cured with low energy EBR [48]. The schematic diagram for low energy layer by layer curing of laminates is shown in Figure 5.12. Guasti and Rosi [48]

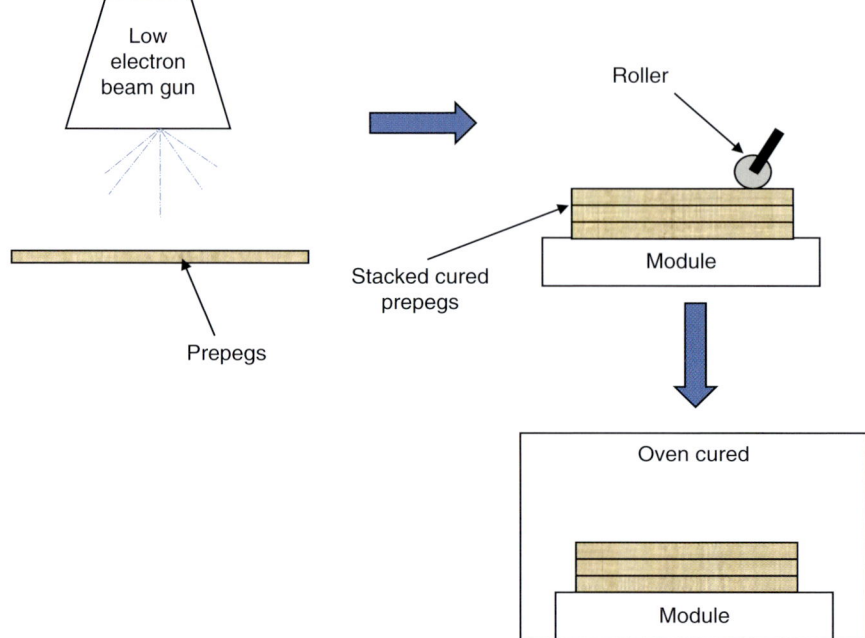

Figure 5.12 Radiation curing of higher thickness composites laminates.

examined the curing of filament wound composites by low energy electron beam layer by layer curing technique and achieved successful curing by this process. Although for proper penetration of radiation through the higher thickness composite laminates, irradiation was required on both sides of the laminate surface [49], both side irradiation curing of composites laminates reduces the radiation energy inputs. Zhao et al. [49] evaluated the low energy input (125 keV) for the curing of 125 mm thick carbon fiber/epoxy laminates. Both side irradiation of the order of 125 keV energy input is enough for better curing of 125 mm thick composites laminates. The electron beam curing technique has some advantages and disadvantages as tabulated in Table 5.7.

5.4.2 Ultraviolet Curing (UV)

UV curing also falls under the category of radiation curing. It is a cost-efficient, environment-friendly and energy-saving curing technique mostly used for thin film coated material curing abliz-[4, 50, 51]. UV curing takes less time for curing of material than other thermal and conventional curing processes. The curing time can be considered in minutes compared to hours of unconventional methods of curing. UV radiation can be generated by a number of sources but most of the time UV is generated from mercury arc lamps. The UV cured polymer properties can be easily predicted by the cross-linking density [52]. The penetration of UV radiation in the material is limited; hence, it can cure only limited thickness of composites laminates. There are two mechanisms of UV curing, namely free radicals and cations, which start polymerization. The radicals develop new radicals by the light irradiation; this leads to cross-linking

Table 5.7 EB curing advantages and disadvantages.

S. No.	Advantages	Disadvantages
1.	Low curing time	High energy input required for curing of thick laminated composites
2.	Generates no thermal stresses in comparison to conventional curing method	Achieves poor interfacial bonding between fiber and matrix
3.	Low curing cost for thin composite laminate	Curing makes the matrix material brittle
4.	Can be used for surface curing in packaging industries	High investment cost
		Highly skilled labor is required to operate

of monomers due to polymerization. UV curing is very fast or instant in thin film laminates converting liquid or gel form of resin to solid material by polymerization [53]. The reduction in curing time and the overall energy input are lower than with other conventional techniques for the same thickness of laminate [52]. Low energy input reduces the chances of thermal stress during curing. UV curing is mostly used for curing of thin film applications such as paints, adhesives, and varnishes. The wavelength of UV radiation is lower than that of visible light radiation. Proper curing of material can be achieved by proper penetration of radiation inside the material. High radiation energy with lower wavelength can be used for curing of thick laminates.

UV curing of FRPCs can be done on single side or both side irradiation on the surface of laminates. A number of authors have investigated the field of radical polymerization of FRPCs by UV curing with single and both side techniques. Cured samples with one side curing and both side curing were found to have distinct properties. Pogany and Vancso-Szmercsanyi [54] compared mechanical properties of chopped glass/unsaturated polyester resin composite cured on the basis of single and both side UV curing. The flexural strength of one side cured samples was higher than that of both side cured samples. UV curing is the best for low thickness laminates. To overcome the problem of inefficient curing of high thickness laminates, layer by layer UV curing was introduced similar to electric beam curing. Some authors combined UV curing with filament winding and then investigated the layer by layer curing of FRPCs [55]. Schematic of FRPCs fabricated with filament winding technique combined with UV curing is shown in Figure 5.13. Advantages and disadvantages of UV curing are mentioned in Table 5.8.

5.5 Conclusion

Advanced curing processes are more efficient than conventional curing process for mass production of FRPCs. Requirement of less curing time and efficient curing of complicated shapes used in aerospace and aircraft industries demand advanced curing methods. Curing cycle time is generally lower for

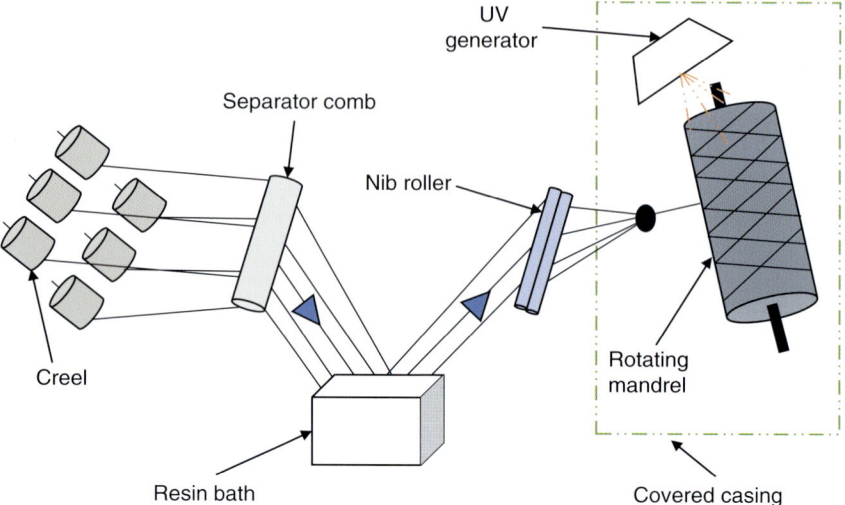

Figure 5.13 Layer by layer UV curing in filament winding.

Table 5.8 UV curing advantages and disadvantages.

S. No.	Advantages	Disadvantages
1.	Laminates get instantly cured, which helps in increasing the production efficiency	Curing can be done up to a limited thickness of laminate
2.	Cured samples have good strength, wear resistance, and solvent resistance ability	Penetration of UV is weak as compared to other radiation and conventional curing techniques
3.	Lower energy input required in comparison to conventional curing process for same laminate	Initial setup cost is very high
4.	UV radiation is not harmful	

advanced curing techniques in the case of FRPCs mass production. Complicated shapes/thick laminated FRPCs products can easily and effectively be cured using unconventional curing techniques. Uniform curing reduces the chances of development of thermal stresses. The selection of curing technique should be based on the type of reinforcement used in FRPC along with the nature of the polymer to be cured.

References

1 Compositesworld (2005). Boeing sets pace for composite usage in large civil aircraft. http://www.compositesworld.com/articles/boeing-sets-pace-for-compositeusage-in-large-civil-aircraft (accessed 2 October 2018).

2 Risikoblog (2011). Schwer ist beim Flugzeug-Leichtbauleicht was. http://www .risikoblog.de/hintergrund/schwer-ist-beimflugzeug-leichtbau-leicht-was/ (accessed 12 November 2018).

3 Kim, Y. and Daniel, I. (2002). Cure cycle effect on composite structures manufactured by resin transfer molding. *Journal of Composite Materials* 36 (10): 1725–1741.

4 Abliz, D., Duan, Y., Steuernagel, L. et al. (2013). Curing methods for advanced polymer composites – a review. *Polymers and Polymer Composites* 21 (10): 341–348.

5 ASC (2006). Completes "World's Largest Autoclave". http://www.aschome .com/WorldsLargestAutoclave2.htm.

6 Papargyris, D., Day, R., Nesbitt, A., and Bakavos, D. (2008). Comparison of the mechanical and physical properties of a carbon fiber epoxy composite manufactured by resin transfer moldings using conventional and microwave heating. *Composites Science and Technology* 68: 1854–1861.

7 Davies, L., Day, R., Bond, D. et al. (2007). Effect of cure cycle heat transfer rates on the physical and mechanical properties of an epoxy matrix composite. *Composites Science and Technology* 67 (9): 1892–1899.

8 Boey, F. (1989). Development of an autoclave curing system for fiber reinforced polymer composites. *Polymer Testing* 8: 375–384.

9 Guliherme, F., Cassu, S., Diniz, M. et al. (2017). Evaluation of Out-of-Autoclave (OOA) epoxy system. *Polymers* 27 (4): 353–361.

10 Adeodu, A., Anyaeche, C., Oluwole, O., and Afolabi, S. (2015). Modeling of conventional autoclave curing of unsaturated polyester based composite materials as production process guide. *International Journal of Materials Science and Applications* 4 (3): 203–208.

11 Sun, J., Gu, Y., Li, Y. et al. (2012). Role of tool-part interaction in consolidation of l-shaped laminates during autoclave process. *Applied Composite Material* 19: 583–597.

12 Zhang, J., Xu, Y., and Huang, P. (2009). Effect of cure cycle on curing process and hardness for epoxy resin. *Express Polymer Letters* 3 (9): 534–541.

13 Kashani, P. and Minaie, B. (2011). An ex-situ state-based approach using rheological properties to measure and model cure in polymer composites. *Journal of Reinforced Plastics and Composites* 30: 123–133.

14 Olivier, P., Cottu, J.P., and Ferret, B. (1985). Effects of cure cycle pressure and voids on some mechanical properties of carbon/epoxy laminates. *Composites* 26: 509–515.

15 Huang, H. and Talrej, R. (2005). Effects of void geometry on elastic properties of unidirectional fiber reinforced composites. *Composites Science and Technology* 65 (13): 1964–1981.

16 Rudnev, V., Loveless, D., Cook, R., and Black, M. (2003). *Handbook of Induction Heating*. Basel: Marcel Dekker AG.

17 Bayerl, T., Duhovic, M., Mitschang, P., and Bhattacharyya, D. (2017). The heating of polymer composites by electromagnetic induction – a review. *Composites: Part A* 57: 27–40.

18 Fink, B.K., McKnight, S.H., Yarlagadda, S., Gillespie, Jr. J.W. (1999). Non-polluting composites repair and remanufacturing for military applications: induction-based repair of integral Armor. *ARL-TR-2121*. Aberdeen Proving Grounds: Army Research Laboratory.

19 A process for the accelerated adhesive curing. Patent DE 10 037 884 A1 (2000).

20 Ferromagnetic resonance excitation and their use for heating substrates teilchengefullter. Patent DE 10 037 883 A1 (2000).

21 Wetzel, E.D., Sand, J.M., and Yungwirth, C.J. (2003). Induction curing of a phase-toughened adhesive. *ARL- TR-2999*, Aberdeen Proving Grounds: Army Research Laboratory.

22 Tay, T.E., Fink, B.K., McKnight, S.H. et al. (1999). Accelerated curing of adhesives in bonded joints by induction heating. *Journal of Composite Material* 33 (17): 1643–1664.

23 Miller, A.K., Chang, C., Payne, A. et al. (1990). The nature of induction-heating in graphite-fiber polymer-matrix composite materials. *Sampe Journal* 26: 37.

24 Fink, B.K., Mccullough, R.L., and Gillespie, J.W. (1992). A local theory of heating in cross-ply carbon-fiber thermoplastic composites by magnetic induction. *Polymer Engineering Science* 32 (5): 357–369.

25 Fink, B.K., Mccullough, R.L., Gillespie, J.W. (1999). Induction heating of carbon-fiber composites: electrical potential distribution model. *ARL-TR-2130*, Aberdeen Proving Grounds, MD 21005-5069: Army Research Laboratory.

26 Mahdi, S., Kim, H.J., Gama, B.A. et al. (2003). A comparison of oven-cured and induction-cured adhesively bonded composite joints. *Journal of Composite Materials* 37 (6): 519–541.

27 Colak, Z.S., Sonmez, F.O., and Kalenderoglu, V. (2002). Process modeling and optimization of resistance welding for thermoplastic composites. *Journal of Composite Material* 36 (6): 721–744.

28 Hou, M., Ye, L., and Mai, Y.W. (1999). An experimental study of resistance welding of carbon fiber fabric reinforced polyetherimide (CF fabric PEI) composite material. *Applied Composite Material* 6: 35–49.

29 Hayes, S.A., Lafferty, A.D., Altinkurt, G. et al. (2015). Direct electrical cure of carbon fiber composites. *Advanced Manufacturing: Polymer and Composites Science* 1: 112–119.

30 Sancaktar, E., Ma, W., and Yurgartis, S.W. (1993). Electric resistive heat curing of the fiber-matrix interphase in graphite/epoxy composites. *Journal of Mechanical Design* 115: 53.

31 Boey, F., Gosling, I., and Lye, S.W. (1992). High-pressure microwave curing process for an epoxy-matrix glass fiber composite. *Journal of Material Processing Technology* 29: 311–319.

32 Boey, F. and Yue, C.Y. (1991). Interfacial strength of a microwave-cured epoxy glass composite. *Journal of Materials Science Letters* 10: 1333–1334.

33 Mijovic, J. and Wijaya, J. (1990). Review of cure of polymers and composites by microwave energy. *Polymer Composites* 11 (3): 184–191.

34 Singh, M.K. and Zafar, S. (2018). Influence of microwave power on mechanical properties of microwave-cured polyethylene/coir composites. *Journal of Natural Fiber* https://doi.org/10.1080/15440478.2018.1534192.

35 Hang, X., Li, Y., Hao, X. et al. (2015). Effects of temperature profiles of microwave curing processes on mechanical properties of carbon fiber – reinforced composites. *Journal of Engineering Manufacture* https://doi.org/10.1177/0954405415596142.

36 Ali, S., Bajpai, P.K., Singh, I., and Sharma, A.K. (2017). Curing of natural fibre-reinforced thermoplastic composites using microwave energy. *Journal of Reinforced Plastics* https://doi.org/10.1177/0731684414523326.

37 Yusoff, R., Aroura, M.K., Nesbitt, A., and Day, R.J. (2007). Curing of polymeric composites using microwave resin transfer moulding (RTM). *Journal of Engineering Science and Technology* 2 (2): 151–163.

38 Rao, R.M.V.G.K. and Rao, S. (2006). Studies on tensile and interlaminar shear strength properties of thermally cured and microwave cured glass–epoxy composites. *Journal of Reinforced Plastics and Composites* 25 (7): 785–795.

39 Balzer, B.B. and Nabb, J.M. (2008). Significant effect of microwave curing on tensile strength of carbon fiber composites. *Journal of Industrial Technology* 24 (3): 1–8.

40 Lionetto, F. and Maffezzoli, A. (2013). Monitoring the cure state of thermosetting resins by ultrasound. *Materials* https://doi.org/10.3390/ma6093783.

41 Krautkramer, J. and Krautkramer, H. (1977). *Ultrasonic Testing of Materials*. Berlin: Springer-Verlag.

42 Van Krevelen, D.W. (1990). *Properties of Polymers*. Amsterdam: Elsevier.

43 Beziers, D., Perilleux, P., and Grenie, Y. (1996). Composite structures obtained by ionization curing. *Radiation Physics and Chemistry* 4 (2): 171–177.

44 Saunders, C., Lopata, V., Barnard, J., and Stepanik, T. (2000). Electron beam curing – taking good ideas to the manufacturing floor. *Radiation Physics and Chemistry* 57: 441–445.

45 Berejka, A.J. and Eberle, C. (2002). Electron beam curing of composites in North America. *Radiation Physics and Chemistry* 63: 551–556.

46 Berejka, A.J., Montoney, D., Cleland, M.R., and Loiseau, L. (2010). Radiation curing: coatings and composites. *Nukleonika* 55 (1): 97–106.

47 Glauser, T. (1999). *Electron-Beam Curing of Thermoset Resins for Composites*. Stockholm: Department of Polymer Technology, Royal Institute of Technology. ISBN: 91-7170-443-4.

48 Guasti, F. and Rosi, E. (1997). Low energy electron beam curing for thick composite production. *Composite Part A: Applied Science and Manufacturing* 28: 965–969.

49 Zhao, X., Duan, Y., Li, D. et al. (2015). Investigation of curing characteristics of carbon fiber/epoxy composites cured with low-energy electron beam. *Polymer Composites* https://doi.org/10.1002/pc.23084.

50 Decker, C. (2002). Kinetic study and new applications of UV radiation curing. *Macromolecular Rapid Communication* 23: 1067–1093.

51 Decker, C. (1987). UV-curing chemistry: past, present and future. *Journal of Coating Technology* 59 (751): 97–106.

52 Endruweit, A., Johnson, M.S., and Long, A.C. (2006). *Polymer Composites* https://doi.org/10.1002/pc.20166.

53 Decker, C. (2001). UV-radiation curing chemistry. *Pigment and Resin Technology* 30 (5): 278–286.

54 Pogany, G. and Vancso-Szmercsanyi, I. (1979). Polymerization of glass-fibre reinforced unsaturated polyester resin by mends of ultraviolet radiation. *Plaste und Kautschuk* 26: 152.

55 Duan, Y., Wang, Y., Tang, Y. et al. (2010). Fabrication and mechanical properties of UV-curable glass fiber-reinforced polymer–matrix composite. *Journal of Composite Material* https://doi.org/10.1177/0021998310376107.

6

Friction and Wear Analysis of Reinforced Polymer Composites

Pawan Kumar Rakesh and Lalit Ranakoti

National Institute of Technology Uttarakhand, Department of Mechanical Engineering, Srinagar, Pauri Garhwal, 246174, India

6.1   Introduction

In everyday life, we come across many physical phenomena that are not possible without friction. Walking on the road to flying in the air, transmission of power in an automobile to any machining operation in industries are some occurrences that prove the existence of friction. For any process to happen one has to overcome the friction, which requires energy, and it cannot be restored but gets wasted in a different form of energy. These wasted energies can be in tangible or intangible forms. Intangible forms of energy are sound, light, heat, etc. and tangible form of energy is always associated with the material. Wear is the perfect example of loss of tangible energy in the form of material. Wear originates because of friction. Wear takes place in between the relative motion of two surfaces. It is a very important aspect to be considered in the field of engineering while designing a machine element that has to undergo sliding. Generally, wear may be reduced by adding some kind of lubricating materials. Wear deteriorates the material, at the microscopic or macroscopic level; but in both cases, the working performance of the material is degraded. Typically, wear does not cause any catastrophic failure but the cases mostly encountered show that the material weakens gradually [1]. This gradual failure of material due to wear causes vibration or misalignment. Sometimes, these vibrations lead to the complete failure of the material carrying out any process. The complete failure of material under running condition may lead to severe human or financial damages. These damages cannot be compensated but the time of failure can be increased [2]. In the present study, the mechanism of failure occurs due to wear in FRPs; the test method used to detect the failure of material due to wear is discussed.

6.1.1   Fiber Reinforced Plastics

Fiber (synthetic or natural) is combined with the matrix (polymer), resulting in polymer composites. The need for polymer composites is increasing day by day as they are proving to be a good alternative material to metals. Various applications such as aerospace, marine, automobile, and sports are fields where polymer composites are now being used extensively. These increasing applications are due to the favorable properties of polymer composites such as low weight, ease of manufacturing, low cost, and easy availability [3]. Although complete replacement of metals by the polymer composite from the industries is not yet possible, research is still going on for the design of ultrahigh strength polymer composites that may have strength to equal metal materials. The functioning of materials may vary according to the nature of work that a material undergoes to various kinds of loading. These loadings can be tensile load, compressive load, impact load, flexural load, wear, or a combination of any kind of loads. The hardness of materials that slide over one another decides the wear mechanism, wear rate, and the phenomena behind it. Loading parameters and sliding velocity are other parameters that somehow affect the wear phenomena but still the influence of hardness prevails [4]. The hardness of fiber reinforced polymer (FRP) composites depends on various factors such as fiber volume fraction, matrix selected, additives, fillers, and the method used for manufacturing.

6.1.2 Failure Mechanism of Fiber Reinforced Plastics

Fiber reinforced plastic composites are much softer and have low melting and operating temperatures than metals. These properties are not favorable for wear loading as the material deteriorates at a higher rate. Moreover, the mechanism of wear and friction in polymer composite is a complex phenomenon and is not fully understood. Friction takes place whenever two bodies slide over one another. For the required motion in sliding, there must be a contact between the bodies, which occurs at asperities of the high point. As shown in Figure 6.1, the wear is due to adhesion and deformation, as in the case of elastomer (a type of polymer). Figure 6.1 also shows the contact point of asperities (contact of asperities during adhesion and deformation).

Friction and wear depend on experimental parameters, which vary from material to material due to their difference in the chemical, mechanical, and physical properties. Upon loading a polymer composite, the close contact of surfaces

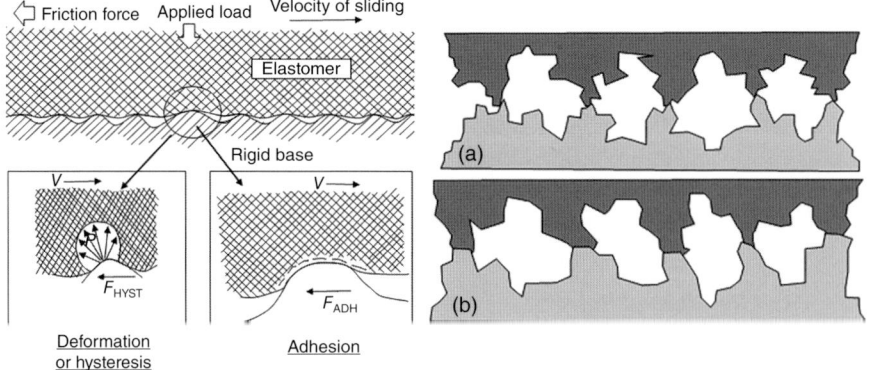

Figure 6.1 (a) Wear caused due to deformation and adhesion, (b) contact point of asperities.

encourages a force of adhesion. This adhesive force generated at the surface is due to the presence of van der Waals forces. These forces depend on the closeness of the mating surfaces. Particularly for the polymer, due to the presence of polarity, dipole interaction and hydrogen bonding elevate the adhesion. In polymer composites, the motion of molecules is higher even at ambient temperature, which causes the surface segments to lock into the asperities of the mating surface [5, 6]. Additional adhesion occurs due to the electrical charge set up at the contacting surface. During sliding of the polymer composite, a tangential force is needed to break down the formation of adhesive bonds. The formation and breakdown of bonds occur repeatedly, which shows that adhesion and de-adhesion are necessary for the body to slide. Adhesion also depends upon the temperature and rate of sliding due to the involvement of the viscoelastic effect. In some cases, the surface of soft polymer gets deformed by the hard particle asperities, which also contributes to the overall friction mechanism. Friction also intensifies because of hysteresis effect, which occurs due to the loss of energy in the front loading and recovery at the end. Deformation also plays a very important role in the friction mechanism for polymer composites. Whenever polymer composites are loaded in sliding, they undergo elastic deformation but not plastic deformation. The mechanism of elastic deformation for polymers is complex and the area of shear and elastic modulus have to be determined experimentally for the calculation of adhesion force due to deformation. The Hertz equation and Smith master curve of tensile strength are adopted for determining the force of adhesion. Mathematically, the force of adhesion can be calculated by the equation, $F = A \cdot S$, where A is the area of shear and S is the shear strength of the bond developed between the sliding surfaces [7].

Mainly, the mechanism of wear in sliding comprises adhesion, abrasion, surface fatigue, and corrosion as shown in Figure 6.2. Wear due to corrosion is the result of chemical or fretting corrosion, which includes chemical phenomena. Wear mechanism in the polymer is generally explained by the mechanism of adhesion and abrasion but practical considerations suggest that formation of wear debris occurs due to the significant effect of interplay, which takes place during polymeric sliding and hence results in a fracture in the substrate [8].

6.1.3 Adhesion Wear

The adhesive bond is formed at the contact points of asperities of two surfaces. Strong adhesive bonds are difficult to break but weak adhesive bonds rupture easily during sliding as high tangential force will be required to break the cohesive bond in the substrate. This implies that low wear breakage occurs at either an interface or at interfacial layers. On the other hand, strong adhesive bond increases the chances of failure of cohesive bonds. This results in a noticeable dissociation of material. The high reactivity of detached material is due to the failure of a cohesive bond with the material surface in contact. This bond formation can be stated as "film transfer formation," which administers the polymeric wear mechanism. Archard proposed a wear equation for metals:

$$\frac{V}{L} = K \cdot \frac{P}{H} \tag{6.1}$$

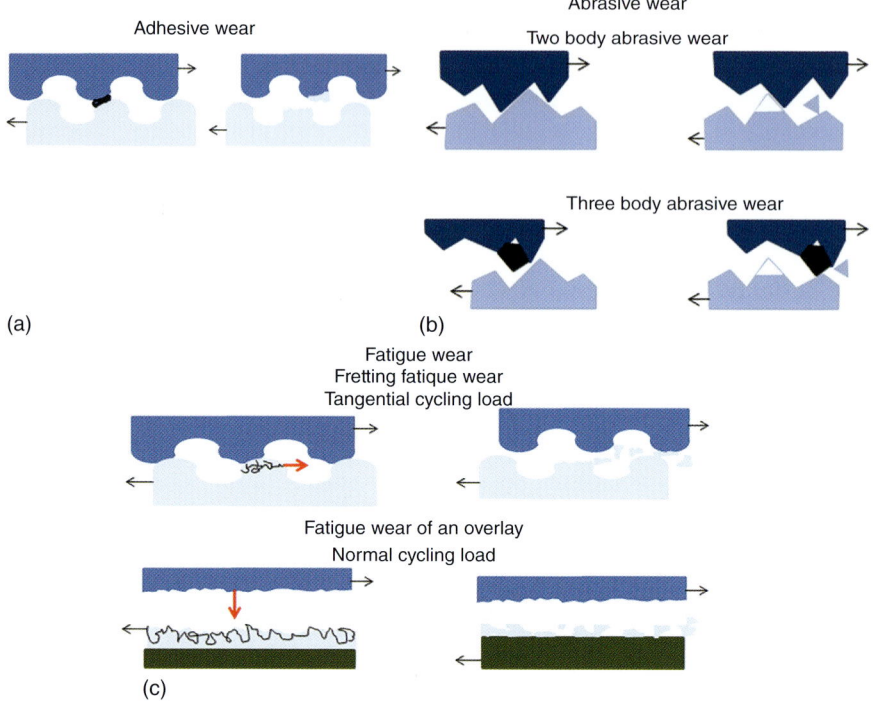

Figure 6.2 Wear due to (a) adhesion, (b) abrasion, and (c) fatigue.

where V is the volume of wear, P is the load, L is the distance moved during sliding, H is the hardness of the polymeric material, and K is the proportionality constant. Moreover, for polymeric material the same equation does not hold well due to the highly elastic deformation element.

Contact pressure, temperature, sliding velocity, material property, and roughness are some of the basic parameters that predominantly affect the adhesive wear. The correlation between wear rate and load remains intact until a significant change is noticed in other parameters such as surface texture and temperature. Consequently, change in sliding velocity also depends upon change in temperature. The polymer composite material when undergoing sliding forms a thin layer of segregated melted material at the interface. This occurs due to the softening and melting of material as the interface temperature reaches the melting point of the polymer due to the high pressure load. To avoid such a situation polymer composite materials are usually subjected to low pressure load. The effect of material crystallinity on wear rate is not found much in the literature but wear rate of polymer materials such as polyethylene (PE) and polytetrafluoroethylene (PTFE) increases and decreases respectively as crystallinity increases [9, 10]. On the other hand, it is expected that the increase in friction coefficient decreases the wear rate, which usually depends upon the molecular weight. Wear rate has little dependence on material properties such as tensile strength, hardness, percentage area reduction, and elastic modulus. Other phenomena such as the orientation of the surface, breaking off of spherulitic structure, and transfer of polymer particles further complicate the analysis of wear mechanism of the polymeric material.

6.1.4 Abrasive Wear

Removal of surface particles from a softer material by using a hard material is termed abrasive wear. Mathematically, abrasive wear can be expressed from a simple equation

$$\frac{V}{L} = K\frac{2}{\pi}P \tan \frac{\theta}{H} \tag{6.2}$$

Here, V/L = volume of wear per unit distance, θ = base angle subtended by the asperities, and K = the fraction of material displaced during sliding; by varying the intended angle greater than $30°$ the wear volume varies linearly with $\tan \theta$. It was found that at this angle, $12\,\mu m$ of roughness is achieved, which is generally not considered for engineering applications. In Eq. (6.2), other factors S and e are introduced, which denote breaking strength and breaking elongation, respectively, when analyzing the abrasion of abrasive paper and polymer. Equation (6.2) changes to

$$\frac{V}{L} = KP\mu\,(HSe) \tag{6.3}$$

Some polymers such as polyvinyl chloride (PVC) and polychlorotrifluoroethylene (PCTFE) have unusual wear mechanism. It was found that the rate of wear varies inversely with rupture energy and yield strain and directly varies with a modulus of flexure. Another parameter known as cohesive energy density plays an important role in the wear mechanism for thermoplastic polymers. A sign of inverse proportionality of wear rate with cohesive energy density is noticed for thermoplastic polymers. Low elastic modulus and elastic deformation for elastomers make them less susceptible to wear as large abrasiveness is found during sliding of elastomers with the materials having low subtended angles [11, 12]. To show the wear resistance, a graph has been drawn for different polymers. It can be seen from Figure 6.3 that polymers such as ultrahigh molecular weight polyethylene (UHMWPE) possess a maximum value of wear resistance.

6.1.5 Fatigue Wear

Wear due to fatigue was first noticed by Kragelskii [14]. During the service life of polymer materials, many kinds of friction bond evolve at the surface of contact. Various degradations such as chemical and chemical–mechanical were also observed along with fatigue wear. Mathematically, the equation of fatigue wear consists a power exponent "t" introduced by Wohler equation. The power exponent is expressed as $N \propto \left(\frac{S}{S_0}\right)^t$, where N denotes the number of cycles till failure, S denotes the stress applied, and S_0 denotes single stress cycle failure. In the wear mechanism of a rough surface with a smooth surface for a rubber material, the asperities are assumed to be deformed wholly but not individually. Modified factors are considered to update Eq. (6.1). These factors are the radius of curvature

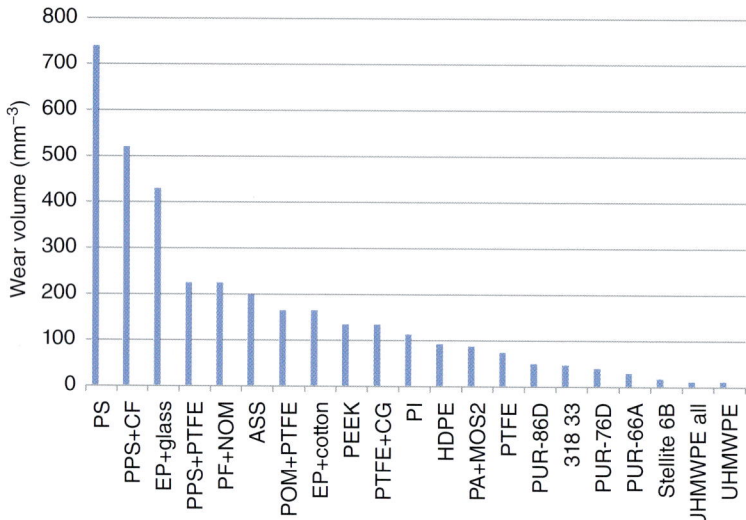

Figure 6.3 Wear rates of polymers that include ceramics for comparison at an angular speed $\omega = 20.9$ rad/s and constant load of 45 N. Source: Budinski 1997 [13]. Reproduced with permission of Elsevier. PS, polystyrene; EP, epoxy; PF, phenol-formaldehyde; POM, polyoxymethylene (Acetal); NOM, natural organic matter; POM, polyoxymethylene; PTFE, polytetrafluoroethylene; PEEK, polyether ether ketone; PI, polyimide; MOS_2, molybdenum disulfide; PU, polyurethane; UHMWPE, ultrahigh molecular weight polyethylene.

of asperities, the ratio of apparent area to real area, and the depth of penetration. Assuming plastic deformation at the contact asperities and ignoring the shear stress at the asperities, the fatigue wear equation developed for metals given by Halling [15] can be applied for polymers also. An additional factor called work hardening is introduced by Halling in the fatigue wear equation. The inclusion of factors such as distribution of asperities line and heights of asperities leads to the complex function K in Eq. (6.1). While considering elastic deformation for the hard asperities having spherical tips, Jain and Bahadur [16] introduced various factors such as Poisson's ratio and coefficient of friction and the equation modifies into $\frac{V}{L} = K_1 K_2 \frac{P \eta_L}{2 S_0} V_L$, where K_1 and K_2 are the constants, which includes Poisson's ratio, asperities density, elastic modulus, asperities heights, etc. In addition, fretting wear sometimes becomes prominent, which takes place during the oscillatory motion of contact surfaces in the tangential direction. The fretting wear does not anticipate much in wear mechanism due to the low amplitude of oscillation. It varies significantly at a higher amplitude. Sometimes, moisture plays a critical role in the area where humidity is prevalent [17]. The sudden collapse of a water bubble trapped between the contact surfaces leads to mechanical shock, which causes the removal of material from the surface. This wear is due to a cavitation that is is analogous to surface fatigue wear. The above discussions for fatigue wear are load dependent also as the initiation of crack propagation is influenced by normal and friction loads.

Apart from all the three mechanisms discussed earlier, an important aspect of transfer film formation also plays a very critical role in wear mechanism of polymer and polymer with metal materials. Transmission electron microscopy and electron diffraction confirm that a thin layer of polymer material emerges at the contact surface during sliding motion; in fact, the polymer material is sometimes transferred to the metal material in contact. It has been confirmed by ion field microscopy that when polymers such as PTFE come in contact with a glass or metal surface, a thin layer of melted polymer is maintained between the contact surfaces. A similar observation can be seen in the case of low density polyethylene (LDPE) when it slides with a glass surface. This film transfer is a function of time and temperature as confirmed by infrared spectroscopy. It is believed that the transient phenomena are transformed to steady state due to polymer film transfer. This layer also prevents the adhesive rubbing of hard asperities with soft asperities of a polymer. Polymers are generally self-lubricating, which leads to a reduction in the coefficient of friction [18]. This is because of the transfer of film to the metal surface, fracture at the substrate, and deformation of the material. Parameters that affect the film transfer mechanism are load, velocity, temperature, structure, etc. Topography, crystallinity, and roughness of the surface also affect the polymer film transfer mechanism. Thus, it is clear that wear of the polymer due to sliding comprises very complex phenomena that are not as simple as in metal material.

6.1.6 Wear Testing Method

Machines that are used for measuring wear and friction of material are generally called as tribo machines. The test setup mainly consists of two solid specimens that are kept in contact with each other accompanied by lubrication. The specimen used in the wear test is shaped like a disc or a cylinder, which leads to a line or point contact. Figure 6.4 helps understand the concept of contact used in the wear test. These contacts are a pin on the disc, pin on the cylinder (fixed or rotating), spherical pin on the cylinder, pin on the flat surface, etc. The specimen comes in contact having an initial dimension, which continuously changes under pre-defined parameters such as velocity, pressure, time, motion, and roughness. The rotating disc with which the specimen comes in contact transmits the mechanical energy through the surface area in contact, which leads to the generation of friction and wear to be measured. Generally, sliding, rolling, impact, and spin are the four major types of motion given to the test specimen. During testing, the specimen undergoes static or dynamic load either by using a dead weight spring mass system or electromagnetically. Normally, different types of static loading give the same value of normal forces while in case of dynamic loading the load and time behavior may vary according to the application of load. For the measurement of load, a strain gauge-based transducer is used. Sensors are used on the surface of the specimen, which relies on the pressure applied. In case of measuring the load with the help of electromechanical transducer, normal and frictional forces can be fed together to the electronic device, which directly gives the coefficient of friction. For polymers, increasing the load leads to a reduction in the coefficient of friction due to the increase in wear. The pressure applied changes during the testing as the surface of the specimen wears out, leading to change in the area. When the polymer material works at the glass transition temperature a change in

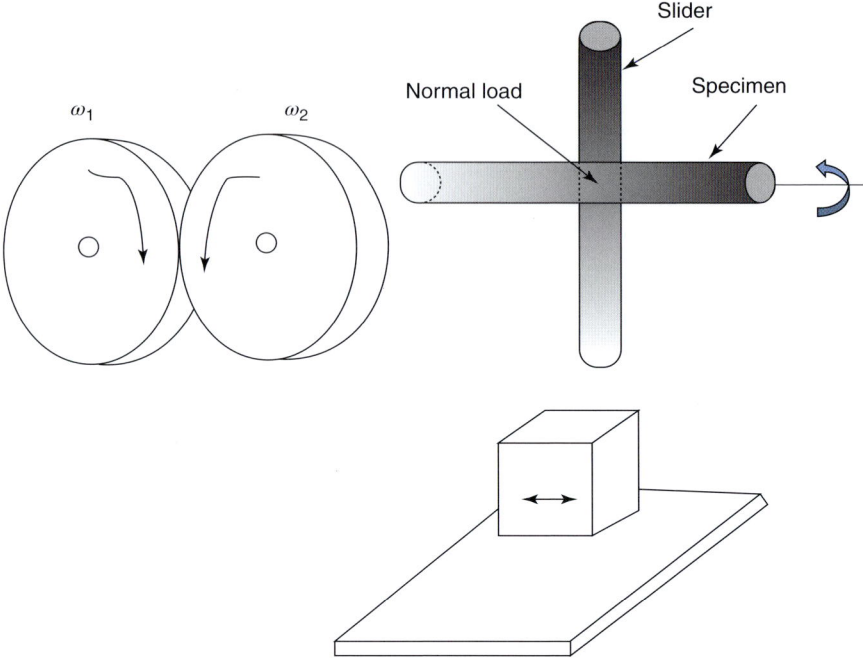

Figure 6.4 Types of contact in a tribological test.

specific wear is also observed. As load increases, the polymer material deforms elastically resulting in an increase in the real contact area. This increase in contact area may be attributed to the collapse of asperities at the junction. The heating produced by frictional resistance mostly depends on sliding speed. To control the sliding speed in the tribo machine, an electronic transducer is generally used.

A thermocouple or infrared pyrometer is used to control the temperature in the tribo machine. The effect of temperature on wear is prominent as the physical, mechanical, and damping characteristics of the contact surfaces change. On the other hand, the real contact area increases as the polymer undergoes thermal softening due to the increase in surface hardness. Time duration also plays a very important role in determining the wear rate of the test specimen as all the parameters such as speed, temperature, and load are time dependent [19]. The commonly used machines to measure wear are listed in Table 6.1.

Pin on the disc is a universally used machine for measuring friction and wear. Figure 6.5 shows the schematic diagram of a pin on disc. Hydraulic or pneumatic pressure is used to apply the load to test the specimen. In the pin on disc machine, the force transducer is used to measure the frictional force. The amount of wear and its rate are calculated by the loss in weight of the test specimen.

At first, a test specimen of the required dimension is prepared by means of mechanical work (shearing, shaping trimming, and cutting). The test specimen is weighed before bringing it to the abrasive disc. The specimen is then kept over a disc or drum (situated above the turntable) depending upon the machine to be used. Usually, discs are rotated for at least 5000 revolutions. To compute

Table 6.1 Wears testing of polymer materials.

S. No.	Machine	ASTM standard	Size of the specimen	Application
1	Dry sand rubber wheel	ASTM G65	Cubical, 70 mm × 20 mm × 7 mm	Tire treads, bushes, and rollers
2	Pin on drum	ASTM A514	Cylindrical (6–7 mm in diameter and 2–3 cm long)	Conveyor belts
3	Block on ring	ASTM G77, G137	Cubical, 10 mm × 20 mm × 50 mm	Pulley, camshaft, and bearing materials
4	Block on disc	ASTM G99	Cubical, 10 mm × 10 mm × 20 mm	Drawers and hinges
5	Pin on disc	ASTM G99	Cubical, 10 mm × 10 mm × 20 mm	Polymer composites

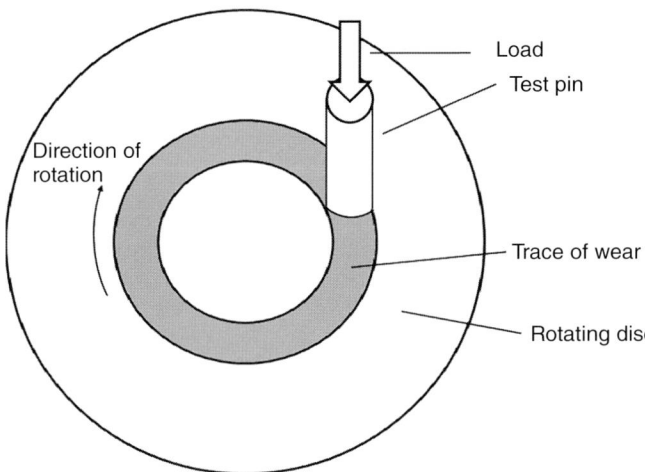

Figure 6.5 Schematic diagram of pin on disc.

the revolutions of the disc, an automatic record counter is used, connected to the turntable. This is followed by the measurement of weight loss, which subsequently describes the wear rate. Tabor abrader is also used to test the wear of wood composites with the variation of parameters such as the number of cycles and the load applied. A photoelectric photometer is used to measure wear by using the scattered light that falls on the abraded surface of the specimen. The test result is computed by taking the ratio of transmitted light to the diffused light. Sandpapers are used for calculating the wear of the flat specimen subjected to sliding on the abrasive particle.

Microscopic technique is used to estimate the irregularities present in the surface. These irregularities can be steps, scratches, or grooves present on the surface. Optical microscope can be used to analyze these irregularities. Scanning electron microscope (SEM) and atomic force microscope (AFM) are the updated techniques having very high magnification factor and are used to characterize the surface structure of a large variety of materials.

6.2 Results and Discussion

Polymer exhibits different tribological properties as compared to metals or ceramics. This difference occurs due to the low free surface energy and viscoelastic properties of the polymers. The introduction of fibers and fillers enhances the stiffness and strength of the polymer composites. The tribological property of polymer composite varies with respect to the polymer (thermoplastic or thermoset), fibers, and fillers used. Roughly, polymers are divided into two categories, thermoplastics and thermoset plastics. Both differ in their cooling characteristics due to which friction and wear analysis of both the polymers show different features. Polyethylene, polypropylene (PP), PVC, ethylene-vinyl acetate copolymer (EVA), polystyrene (PS), acrylonitrile butadiene styrene (ABS) terpolymer, polycarbonate (PC), polyamides (nylons or PA), polyesters, PTFE, and cellulose are among the common thermoplastics that are extensively subjected to the application of wear.

Polyethylene is one of the most common thermoplastics. It is further divided into two categories, high density polyethylene (HDPE) and LDPE. HDPE and LDPE have applications in bottles, bags, household ware, insulation of wire and cable, ultrathin films for wrapping purposes, etc. These products do not require high wear characteristics but UHMWPE and its composites have applications in orthopedic implants. Joints made of UHMWPE are being successfully implanted. These implants are hip and knee joints. These joints are subjected to friction and wear during movement. Even though UHMWPE is a suitable alternative in orthopedic implants, the problem of wear is still present. On combining UHMWPE with PP, it is found that wear rate of the coefficient of friction and wear reduced significantly. Further, a composite of kaolin (naturally occurring clay) with UHMWPE prepared by melt mixing resulted in enhancement in tribological properties. Ion implantation of UHMWPE by induction also resulted in up to 76% improvement in wear resistance [20–23]. Wear behavior of polymers is sometimes affected by the drawing ratio. Pins made of PP when sliding normal to the drawing direction lead to a reduction in wear loss of about 50%. Introduction of grease in dry sliding further reduces the wear by up to 10% for PP. Sliding in highly abrasive surfaces requires outstanding wear properties, which can be accomplished by adding mineral fillers to the PP matrix. The shape and size of the fillers also affect the wear performance of the PP composite [24, 25]. Second, the most common polymer used in industry is polyvinyl chloride (PVC). Fillers such as silicon carbide (SiC), aluminum oxide (Al_2O_3), boron carbide (B_4C), and calcium carbonate ($CaCO_3$), when added to the PVC, enhanced the wear properties of neat PVC remarkably [26, 27]. Being a good resistive material to atmospheric decay, EVA is more often used without any addition of fillers or fibers. Medical and surgical instruments are manufactured from EVA, which does not require special wear characteristics and is hence mostly applied in the neat form. ABS exhibits considerable abrasive property as it is formed by the combination of styrene–acrylonitrile and polybutadiene, which makes it both brittle and tough. Owing to its better abrasive resistance property, products such as casing boot shell and food containers are manufactured with ABS.

Experiments for wear analysis have been carried out for PC materials. PC materials have been tested with steel ball for sliding. Results show that the coefficient of friction can be reduced by the incorporation of siloxane segments to PC. This reduction of friction is attributed to the surface enrichment of the PC and also the molecular weight of siloxane. Addition of glass fiber to PC up to 40% by weight also enhanced its abrasive property [28–30]. Polyamides, also called nylon, are a special type of polymer having characteristics of high strength, toughness, and high melting point. Glass fiber reinforced nylon composite resulted in enhancement of stiffness. Some common applications of nylon include bearings, gears, sockets, and electric plug. These analyses include abrasion of PA with emery paper. The analysis shows that properties such as elongation, fracture energy, and toughness significantly affect the wear properties of PA. Various fibers and fillers have been mixed with PA to enhance its wear properties. Further addition of fillers such as copper and bronze helps in reduction in the coefficient of friction. It is to be noticed that copper is proved to be better than bronze in the enhancement of wear properties for PA polymers. Excellent wear resistance properties are also found in the gears made of carbon fiber reinforced polyamides [31–33]. Glass fiber, linen, and jute have been experimented with polyesters with different volume fractions. The results show that the volume fraction of glass fiber increases when the wear properties increase. In the case of linen reinforced polyesters, wear rate decreases up to 95% for the fiber volume fraction of 33% [34, 35]. Abrasive size particle and orientation of fiber also affect the wear rate of polyester composites.

In applications where low friction coefficient is required, PTFE is generally used. PTFE is a highly crystalline material that can be processed at a wide range of temperature. A large number of composites of PTFE have been fabricated and tested for friction and wear analysis. Most commonly used fillers/fibers with PTFE are graphite, carbon, glass, molybdenum disulfide (MoS_2), and poly-*p*-phenylene terephthalamide (PPTA). These fillers/fibers show enhancement in wear properties of PTFE. The volume percentage of fiber has a significant effect on the wear properties of PTFE. An optimum of 18% carbon + 7% graphite, 20% glass fibers + 5% MoS_2, and 10% PPTA fibers gives the best wear properties of PTFE composites. During wear testing of PTFE composites, it is also found that failure occurs due to the following reasons: (i) fiber pullout, (ii) embedment of fiber in the matrix, and (iii) three-body abrasion test. Surface characteristics of PTFE are improved by the addition of minimum concentration of 30% and a maximum of 40% of glass fiber. Sliding of PTFE composites for wear testing against hard steel surface has also been analyzed [36, 37]. It is found that under normal circumstances, the formation of film transfer plays a crucial role in the wear analysis of PTFE composites. The inclusion of graphite flakes in PTFE resulted in the enhancement of wear of composites. This is due to the film transfer at the interface during sliding. In some cases, decomposition of fillers in the composite takes place and discontinuity arises, which leads to a reduction in overall wear of the composite [38]. Wear test has been conducted for PTFE composite containing 10% bronze powder by volume. It was found that the wear rate decreases drastically as compared to the neat PTFE. The cause of reduction of wear may be attributed to the thermal conductivity of bronze [39]. Copper oxides (CuO) and copper sulfides (CuS) when added to polyether ether ketone (PEEK) and PA polymers lead to a reduction in wear drastically. Moreover, fillers such as zinc fluoride (ZnF_2), zinc sulfides (ZnS), lead sulfides (PBS), and silver sulfides (AgS_2) are those additives that help in reducing the wear rate of polymers such as PA and PPS [39, 40]. It may also be noticed that the size of fillers also affects the wear rate of the composite as nanofillers (size < 100 nm) generally give better wear property than micro-fillers. These fillers further improve the wear property when used in combination with fibers such as glass fiber and carbon fiber [3, 41].

6.3 Conclusions

The following conclusions are drawn from the present study:

(1) The wear resistance properties depend upon the hardness of FRP composites.
(2) The failure mechanisms of FRP composites are adhesive wear, abrasive wear, and fatigue wear.
(3) Among all the polymers, UHMWPE has the maximum wear resistance properties.

References

1 Nuruzzaman, D.M. and Chowdhury, M.A. (2012). Friction and wear of polymer and composites. In: *Composites and Their Properties* (ed. N. Hu). Intech.
2 Holmberg, K. and Erdemir, A. (2017). Influence of tribology on global energy consumption, costs, and emissions. *Friction* 5 (3): 263–284.
3 Friedrich, K. (ed.) (2012). *Friction and Wear of Polymer Composites*, vol. 1. Elsevier.
4 Bahadur, S. (2004). Friction and wear of polymers and composites. In: *Tribology of Mechanical Systems* (ed. J. Vizintin, M. Kalin, K. Dohda and S. Jahamir), 239–265. New York: ASME Press.
5 Shalwan, A. and Yousif, B.F. (2013). In state of art: mechanical and tribological behavior of polymeric composites based on natural fibers. *Materials and Design* 48: 14–24.
6 Santner, E. and Czichos, H. (1989). Tribology of polymers. *Tribology International* 22 (2): 103–109.
7 Friedrich, K. (2018). Polymer composites for tribological applications. *Advanced Industrial and Engineering Polymer Research* 1 (1): 3 39.
8 Chang, L. and Friedrich, K. (2010). Enhancement effect of nanoparticles on the sliding wear of short fiber-reinforced polymer composites: a critical discussion of wear mechanisms. *Tribology International* 43 (12): 2355–2364.

9 Jiang, Z., Gyurova, L.A., Schlarb, A.K. et al. (2008). Study on friction and wear behavior of polyphenylene sulfide composites reinforced by short carbon fibers and sub-micro TiO_2 particles. *Composites Science and Technology* 68 (3–4): 734–742.

10 Parikh, H.H. and Gohil, P.P. (2015). Tribology of fiber reinforced polymer matrix composites – a review. *Journal of Reinforced Plastics and Composites* 34 (16): 1340–1346.

11 Parikh, H.H. and Gohil, P.P. (2015). Composites as TRIBO materials in engineering systems: significance and applications. In: *Processing Techniques and Tribological Behavior of Composite Materials* (ed. R. Tyagi and J.P. Davim), 168–191. IGI Global.

12 Friedrich, K., Zhang, Z., and Schlarb, A.K. (2005). Effects of various fillers on the sliding wear of polymer composites. *Composites Science and Technology* 65 (15–16): 2329–2343.

13 Budinski, K.G. (1997). Resistance to particle abrasion of selected plastics. *Wear* 203: 302–309.

14 Kragelskii, I.V. (1965). *Friction and Wear*. London: Butterworths.

15 Halling, J. (1975). A contribution to the theory of mechanical wear. *Wear* 34 (3): 239–249.

16 Jain, V.K. and Bahadur, S. (1988). Tribological behavior of unfilled and filled poly(amide-imide) copolymer. *Wear* 123 (2): 143–154.

17 Bijwe, J., Indumathi, J., Rajesh, J.J., and Fahim, M. (2001). Friction and wear behavior of polyetherimide composites in various wear modes. *Wear* 249 (8): 715–726.

18 Bahadur, S. (2000). The development of transfer layers and their role in polymer tribology. *Wear* 245 (1–2): 92–99.

19 Nirmal, U., Hashim, J., and Lau, S.T.W. (2011). Testing methods in tribology of polymeric composites. *International Journal of Mechanical and Materials Engineering* 6 (3): 367–373.

20 Affatato, S., Bersaglia, G., Emiliani, D. et al. (2003). The performance of gamma-and EtO-sterilised UHMWPE acetabular cups tested under severe simulator conditions. Part 2: Wear particle characteristics with isolation protocols. *Biomaterials* 24 (22): 4045–4055.

21 Liu, G., Chen, Y., and Li, H. (2004). A study on sliding wear mechanism of ultrahigh molecular weight polyethylene/polypropylene blends. *Wear* 256 (11–12): 1088–1094.

22 Guofang, G., Huayong, Y., and Xin, F. (2004). Tribological properties of kaolin filled UHMWPE composites in unlubricated sliding. *Wear* 256 (1–2): 88–94.

23 Valenza, A., Visco, A.M., Torrisi, L., and Campo, N. (2004). Characterization of ultra-high-molecular-weight polyethylene (UHMWPE) modified by ion implantation. *Polymer* 45 (5): 1707–1715.

24 Bekhet, N.E. (1999). Tribological behavior of drawn polypropylene. *Wear* 236 (1–2): 55–61.

25 Sole, B.M. and Ball, A. (1996). On the abrasive wear behavior of mineral-filled polypropylene. *Tribology International* 29 (6): 457–465.

26 Yang, F. and Hlavacek, V. (1999). Improvement of PVC wearability by addition of additives. *Powder Technology* 103 (2): 182–188.

27 Baptista, A.P.M. and do Carmo Vaz, M. (1993). Comparative wear testing of flooring materials. *Wear* 162: 990–995.

28 Lee, J.H., Xu, G.H., and Liang, H. (2001). Experimental and numerical analysis of friction and wear behavior of polycarbonate. *Wear* 251 (1–12): 1541–1556.

29 Kim, Y.S., Yang, J., Wang, S. et al. (2002). Surface and wear behavior of bis-(4-hydroxyphenyl) cyclohexane (bis-Z) polycarbonate/polycarbonate–polydimethylsiloxane block copolymer alloys. *Polymer* 43 (25): 7207–7217.

30 Bergstrom, J., Thuvander, F., Devos, P., and Boher, C. (2001). Wear of die materials in full-scale plastic injection molding of glass fiber reinforced polycarbonate. *Wear* 251 (1–12): 1511–1521.

31 Rajesh, J.J., Bijwe, J., and Tewari, U.S. (2002). Abrasive wear performance of various polyamides. *Wear* 252 (9–10): 769–776.

32 Bijwe, J. and Indumathi, J. (2004). Influence of fibers and solid lubricants on low amplitude oscillating wear of polyetherimide composites. *Wear* 257 (5–6): 562–572.

33 Cayer-Barrioz, J., Mazuyer, D., Kapsa, P. et al. (2003). On the mechanisms of abrasive wear of polyamide fibers. *Wear* 255 (1–6): 751–757.

34 Pıhtılı, H. and Tosun, N. (2002). Investigation of the wear behavior of a glass-fiber-reinforced composite and plain polyester resin. *Composites Science and Technology* 62 (3): 367–370.

35 Chand, N., Naik, A., and Neogi, S. (2000). Three-body abrasive wear of short glass fiber polyester composite. *Wear* 242 (1–2): 38–46.

36 Khedkar, J., Negulescu, I., and Meletis, E.I. (2002). Sliding wear behavior of PTFE composites. *Wear* 252 (5–6): 361–369.

37 Tomescu, L., Ripa, M., Vasilescu, E., and Georgescu, C. (2003). Surface profiles of composites with PTFE matrix. *Journal of Materials Processing Technology* 143: 384–389.

38 Schwartz, C.J. and Bahadur, S. (2001). The role of filler deformability, filler–polymer bonding, and counterface material on the tribological behavior of polyphenylene sulfide (PPS). *Wear* 251 (1–12): 1532–1540.

39 Yu, L., Yang, S., Wang, H., and Xue, Q. (2000). An investigation of the friction and wear behaviors of micrometer copper particle- and nanometer copper particle-filled polyoxymethylene composites. *Journal of Applied Polymer Science* 77 (11): 2404–2410.

40 Xing, X.S. and Li, R.K.Y. (2004). Wear behavior of epoxy matrix composites filled with uniform sized sub-micron spherical silica particles. *Wear* 256 (1–2): 21–26.

41 Zhang, Z., Breidt, C., Chang, L. et al. (2004). Enhancement of the wear resistance of epoxy: short carbon fiber, graphite, PTFE and nano-TiO_2. *Composites Part A: Applied Science and Manufacturing* 35 (12): 1385–1392.

7

Characterization Techniques of Reinforced Polymer Composites

Manish K. Lila, Ujendra K. Komal, and Inderdeep Singh

Indian Institute of Technology, Department of Mechanical and Industrial Engineering, Roorkee 247667, India

7.1   Introduction

Nowadays, a variety of products are available based on our daily needs. These products are made up of various materials, i.e. metals, ceramics, polymers, semi-conductors, composites, etc. All materials possess certain distinct properties, which are also termed as their characteristics. Materials can be differentiated based on these characteristics, i.e. physical, chemical, mechanical, etc., which makes their characterization an unavoidable task, which can be done before as well as after fabrication of composites.

Characterization refers to an extensive and universal process by which a composite's structure, properties, and behavior are investigated. In broad terms, it is a vital process without which no scientific understanding of engineering materials can be established. Characterization of materials connects the entire product development process. The importance of characterization can be seen easily with the material science tetrahedron, as shown in Figure 7.1. It also helps in establishing the relationship between the structure of material and its properties/applications. The technical framework of materials characterization ranges from fundamental science to potential engineering applications, to quite practical "real-world" field tests, that can predict their performance and prevent component failure [1].

Characterization can be performed at different stages of the product design process to check the basic characteristic of the material. Various levels of characterization at the design stage are shown in Figure 7.2. The first four levels of characterization are mainly used at the material development stage, while macroscopic level characterization/examination is generally performed after fabrication of the product and the techniques depend on the intended application of the material/product.

Characterization leads to an advanced level of understanding required to resolve key issues at all the stages of the product, i.e. applied research,

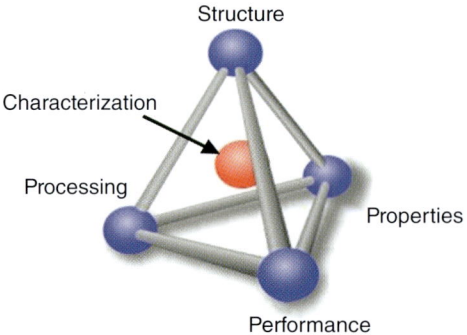

Figure 7.1 Material science tetrahedron illustrating the relationship among different phases.

Structure

Characterization

Processing

Properties

Performance

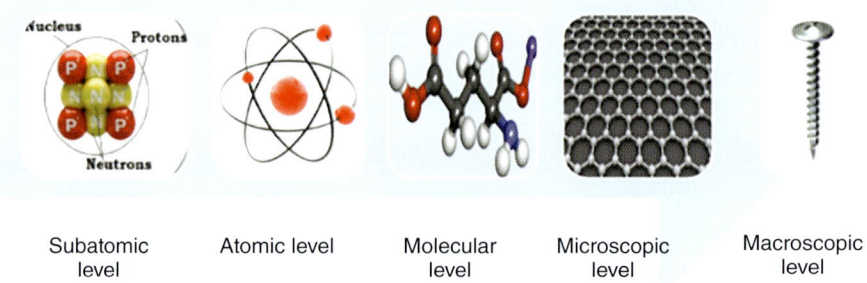

| Subatomic level | Atomic level | Molecular level | Microscopic level | Macroscopic level |

Figure 7.2 Levels of characterization.

design, development, and manufacturing, and subsequently their utilization, maintenance, and disposal also. It also helps in

- product development
- material comparison
- specialized method development
- failure analysis
- de-formulation and reverse engineering.

For fiber reinforced polymers (FRP), the major techniques used for characterization are of macroscopic level, as these are related with the physical and mechanical properties of the material after fabrication of fiber reinforced polymer composites.

7.2 Fiber Reinforced Polymers

FRPs originated during 1960s with an objective to produce lightweight materials with better performance. Nowadays, FRP composites are being used extensively in various high end applications, i.e. fiber glass boats, pressure vessels, and airplane panels. FRPs are unique composite materials in many respects, as these can be formulated to be corrosion, abrasion, and UV resistant, as well as smoke and fire retardant. FRPs are often a cost-effective choice in many industrial applications due to their long life cycles and durability in stringent environments with reduced maintenance costs. Modern composites are in their early phase of evolution and researchers are currently working on today's emerging technological challenges in characterization, fabrication, and applications, maneuvering the upcoming directions for research activities in composite materials.

Fiber reinforced polymers can also be termed as "composites," consisting of two phases, in which polymer is used as matrix and fibers as reinforcement. The polymer used in fabricating FRPs can be thermoset or thermoplastic in nature, based on the application requirement. In the same manner, the reinforcement can be of various types based on their origin (synthetic/natural fibers), length (long/short), structure (woven/non-woven), etc. The global characteristics of FRPs depend on various factors including individual properties of matrix as well as reinforcement, interfacial adhesion between both the constituents, fabrication process, and processing parameters. A variety of fabrication processes are used to fabricate FRP products and the selection of process mainly depends on the constituent materials, properties, and complexity of the product. The fabrication processes include hand lay-up, spray up, injection molding, extrusion, compression molding, etc. Primary processing of FRP deals with providing an initial or structural shape to the material. In the case of polymer matrix composites, manufacturing involves the conversion of resin to a soft state through heat and pressure and then to the condensed state by cooling. Secondary manufacturing of FRP deals with machining, drilling, and joining of the components fabricated by primary manufacturing. One of the two processes, either drilling or joining, is required as a secondary process in the case of intricate parts made of polymer matrix composites.

The number of polymers and reinforcing materials used in FRP applications have swiftly changed in the past two decades. During this period, the principles of testing have changed only a little but the standards, methods, and the interpretation of results have changed significantly. As materials and designs become progressively complex, the approaches and practices used to analyze and characterize them need to be increasingly state of the art.

7.3 Characterization of FRPs

Initial or prefabrication characterization of FRP deals with understanding the inherent behavior of the matrix and reinforcing material, which also helps in conceptualization and designing of a new product. Individual properties of matrix and reinforcement are characterized at the design stage from subatomic to microstructural level. This includes chemical and physical characterization of both the constituents, individually.

Further macroscopic characterization is performed to check the suitability of the fabrication process and process parameters along with the assessment of the final properties of the part. Post-fabrication characterization of FRP is used to analyze its mechanical performance, life cycle assessment (LCA), and material comparison for further optimization of materials or processes.

Physical characterization (surface roughness, hardness, etc.) is used to analyze other aspects of the product to check its suitability for various applications. Mechanical characterization is the most common method to check the mechanical performance (strength, modulus, etc.) of the developed FRPs, while environmental characterization is performed to analyze the performance and durability of FRP products in various work environments. Thermal characterization deals with thermal properties and burning behavior of FRPs. Information obtained from various characterization techniques can be used for the advanced level of analysis needed to resolve vital issues such as product failure modes and process related glitches, and also to make critical materials judgments. The characterization technique can be of destructive or nondestructive type based on specimen size and the test to be performed. Generally, mechanical properties (hardness, strength, etc.) are assessed on the basis of destructive type of characterization techniques.

7.4 Chemical Characterization

A good understanding of the type and quantity of the chemical constituents of a material is the first step to ensure product quality and is normally employed in the material development process. In broad terms, chemical characterization is the utilization of analytical chemistry to identify and quantify the number of chemicals/constituents in a material and an evaluation of the toxicological risk associated with the exposure level. Along with this, it is associated with the reactivity and disposal of the material. Chemical characterization is required in many areas, i.e. quality assurance, research and development of product, workplace, and environmental monitoring, which also includes gravimetric and volumetric determination methods.

Chemical characterization methods are used to analyze organic and inorganic substances as well as their compounds using standardized methods. It consists of a variety of analytical or mathematical techniques to identify and quantify materials that may have migrated to/from the product [2]. The general techniques used for chemical characterization of material are wet chemical method, chromatography, and spectroscopy.

Wet chemical method comprises mathematical calculation for elemental or functional group analysis using different chemicals present in the FRPs by using precipitation, extraction, and distillation techniques. Yet, the results of wet chemical methods are not reliable due to the accuracy and the reaction environment, which may affect the reactions as well as the results. For chromatography and spectroscopy, highly sophisticated instruments are used and the complexity of the technique depends on the level/information required about the material.

Chromatography is an analytical technique that originated from the Greek words "*Chromo* – color" and "*gram* – band." The working principle is based on separation of compounds into different bands (color graphs) and, subsequently, identification of those bands. It is used for separating chemical substances into their individual components to analyze them thoroughly. The techniques used for chromatography are paper chromatography (PC), thin layer chromatography (TLC), gas chromatography (GC), liquid column chromatography (LCC), etc.

Paper chromatography (PC): It is an inexpensive and powerful technique used for separating dissolved chemical substances by considering different reactions and their rates across sheets of paper.

Thin layer chromatography (TLC): It is an analytical method for identifying the components in a mixture by separating its compounds. TLC can be used to determine the number of components, identity of compounds, as well as the purity of a compound. The progress of a reaction can also be monitored by observing the appearance of a product or the disappearance of a reactant.

Gas chromatography (GC): Similar to paper chromatography, if the analysis is performed by a gas chromatograph, it is called gas chromatography. In gas chromatography, the sample is dissolved in a solvent and subsequently vaporized to isolate the analytes by assigning the sample between two phases: a stationary phase and a mobile phase. The mobile phase consists of a chemically inert gas, which is used to bring the molecules of the analyte through the heated column. The stationary phase is either a solid adsorbent, termed gas–solid chromatography (GSC), or a liquid on an inert support, termed gas–liquid chromatography (GLC).

Liquid column chromatography (LCC): In this, the sample is dissolved in a liquid and subsequently distributed between the stationary and mobile phases. The mobile phase used in LCC is in liquid form, which slowly filters down through the solid stationary phase, bringing the separated components with it. A variety of stationary/mobile phase combinations can be employed for the separation of a mixture depending on the physical state of the phases.

In FRP, the molecular structure is highly dependent on the curing/cooling conditions as the resin/melt solidifies, which makes their characterization an inevitable task. Spectroscopy is most commonly employed to study the molecular behavior before and after fabrication of composites. Spectroscopy is a family of analytical techniques, using an array of frequencies (light, UV, IR, X-rays, etc.) on materials for obtaining their spectral features/spectra in the form of a graph/image. This spectrum is exhibited due to the unique interaction of material components with certain frequencies of electromagnetic waves. The nature of interaction can be in form of absorption, reflection, emission, refraction, resonance, impedance, or (elastic/inelastic) scattering (Figure 7.3a). The equipment used for spectroscopic analysis varies in terms of radiation source, interaction, level, and material and can be of increasing complexity.

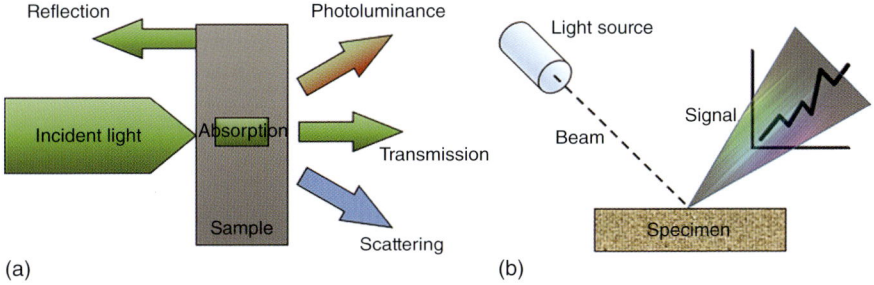

Figure 7.3 (a) Modes of interaction of light with sample, (b) schematic of spectroscopy.

Modes of interaction and a schematic of reflective spectroscopy are shown in Figure 7.3b.

Some of the spectroscopic techniques based on difference in radiation source or material interaction and commonly used for analysis, are Fourier transmission infrared (FTIR) spectroscopy, γ-ray spectroscopy, laser spectroscopy, mass spectrometry, and nuclear magnetic resonance (NMR). X-ray diffraction (XRD) spectroscopy is used to analyze the crystal structure, elemental analysis, as well as the surface crystallinity of FRPs. Some common techniques used for characterization of FRP are mentioned below.

FTIR: This is the most preferred method for spectroscopy, in which infrared radiation is passed through a sample. A fraction of infrared rays is absorbed by the sample and the remaining passes through (transmitted) it. The transmitted signal is detected by the detector and presented in the form of spectrum/graph, demonstrating a molecular "fingerprint" of the sample. Dissimilar chemical structures (molecules) produce unlike spectral fingerprints, which are further studied for analysis of any molecular change in FRPs. In natural fiber reinforced composites, FTIR is used to analyze the hydroxyl ($-OH$), carboxyl ($-C=O$), and other groups.

Raman spectroscopy: Raman spectroscopy involves laser light to study the molecular fingerprints. In FTIR, infrared bands appear from a change in dipole moment of the molecule due to the interaction of light, while in Raman spectroscopy, these bands appear due to change in the polarizability of the molecule due to the same interaction. High energy photons interact with the molecules, which results in inelastic scattering due to change in rotational, vibrational, or electronic energy of the molecule. This inelastic scattering is termed Raman effect.

Mass spectrometry: This technique is commonly used for characterizing bio-composites as it is a powerful analytical technique, which is used to detect unknown compounds within a sample as well as to reveal the structure and chemical properties of different molecules.

NMR: This is an analytical chemistry technique to be used for quality control and research for defining the constituents and purity of a sample as well as its molecular structure. The intramolecular magnetic field surrounding an atom in a molecule changes at the resonating frequency and provides details of the electronic structure of the molecule and its individual functional groups. In FRPs, this technique is used to study the interfacial behavior or change in molecular structure with respect to time [3].

Spectroscopic methods, combined with other analytical techniques, offer multiple advantages. For example, scanning electron microscopy with energy dispersive spectroscopy (SEM/EDS) can be used to analyze molecular structure as well as chemical composition with very high resolution. Similarly, gas chromatography associated with mass spectrometry (GC–MS) can be performed to identify the weakest aroma [4].

Apart from chemical characterization, some physicochemical characterization is also performed on FRPs, using X-rays. XRD analysis is widely used for the same. XRD analysis or X-Ray Powder Diffraction (XRPD) analysis is a unique method to determine crystallinity of a compound as well as elemental analysis. It is based on constructive interference of monochromatic X-rays and a sample in which X-rays are generated. Subsequently, these X-rays are filtered to produce monochromatic radiation, parallelized and directed toward the sample. The interaction of the X rays with the sample results in a constructive interference due to diffraction. The diffraction of X-rays is at a certain angle, and the diffraction conditions satisfy Bragg's law ($n\lambda = 2d \sin\theta$). This law relates the wavelength of electromagnetic radiation to the diffraction angle and the lattice spacing in a crystalline sample. Based on the diffraction spectrum obtained, the crystallinity of the sample can be calculated using either Segal's empirical equation (Eq. (7.1)) or Ruland–Vonk method (Eq. (7.2)). Both the methods use the intensity profile after diffraction from the specimen. The diffraction pattern generated provides a unique "fingerprint" of the crystals present in the sample. When properly inferred, by comparing with standard reference patterns and measurements, these fingerprints allow identification of the crystalline form. Along with crystallinity analysis, X-ray diffraction analysis is used to perform the elemental analysis by comparing the obtained profile with JCPDS (Joint Committee on Powder Diffraction Standards) data. JCPDS database is a standard database of XRD reference patterns for various materials.

$$\mathrm{CrI} = \frac{I_{002} - I_{\mathrm{Am}}}{I_{002}} \tag{7.1}$$

$$\%\mathrm{Cry} = \frac{A_{\mathrm{T}} - A_{\mathrm{Am}}}{A_{\mathrm{T}}} \tag{7.2}$$

where CrI is the crystallinity index of the specimen, I_{002} is the intensity of diffracted rays at angle of Bragg plane (002), and I_{Am} is the intensity, above the baseline, between the angles of Bragg planes (020) and (110). % Cry is the percentage crystallinity of the specimen, A_{T} is the total area under the XRD curve (above the background), and A_{Am} is the amorphous area (Figure 7.4).

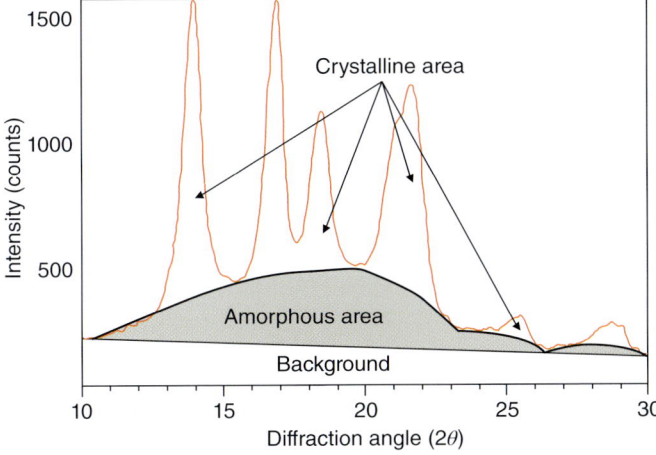

Figure 7.4 Typical XRD curve for polypropylene.

Suitable software, i.e. "X'pert High Score," "Origin," etc. can be used to properly analyze the XRD curve obtained for FRPs.

7.5 Physical Characterization

Physical characterization of FRP includes determination of various physical properties (i.e. density, hardness, surface roughness) and functional aspects (i.e. mechanical, optical, electrical, thermal properties, etc.) by qualitative or quantitative methods. It is a decisive technique to evaluate whether a product is big/small, heavy/light, smooth/rough, porous/tough, strong/weak, or hard/soft, etc. Physical characterization is generally performed after fabrication of FRPs to check the final properties for the intended use. Some of the techniques used for characterization of FRP composites are described in Section 7.5.1.

7.5.1 Microscopic Characterization

One of the most basic characterization techniques for the analysis of solid materials, optical microscopy is the simplest one as comparatively less effort is required for sample preparation before characterization. Two modes of optical microscopy that are used generally are transmission and reflection, based on the measurement of light from a transparent or opaque material, respectively (Figure 7.5).

Most of the FRPs are opaque or translucent in nature; therefore, reflection type microscopy is usually performed on these. For transmission microscopy, a very thin layer (in microns) can be used, but the preparation of this thin layer is quite

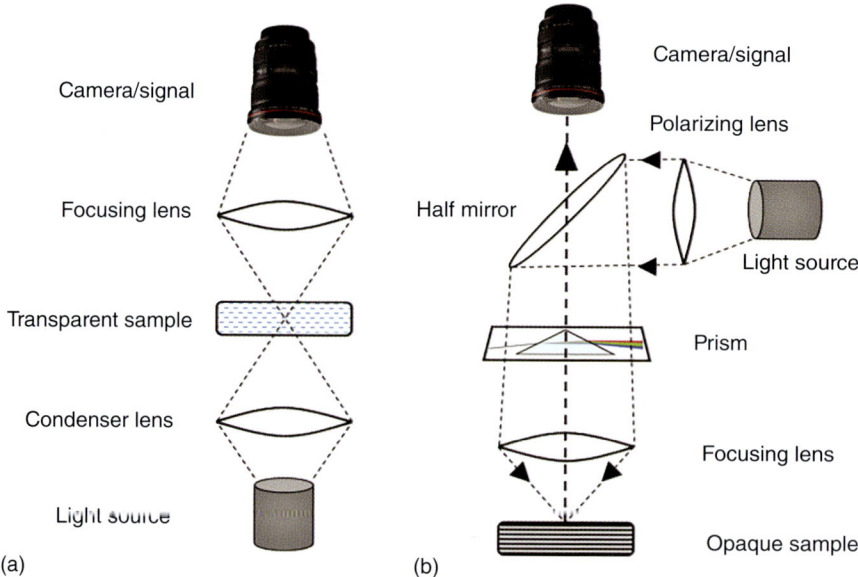

Figure 7.5 Schematics of microscope: (a) Transmission, (b) reflection.

a tedious task; therefore, reflective microscopy is useful for imaging of most of the reflective samples, which include minerals, metals, semiconductors, glasses, polymers, and composites. Surface asperities such as depressions and particulates lead to path differences of reflected beams and the interference between these beams results in the formation of the final image.

The closest spacing of two points that can be seen through the microscope as separate entities is referred as resolution of the microscope, and is often expressed in nanometers (nm). Theoretical resolution (R) of an optical system can be calculated using Abbe's equation (Eq. (7.3)):

$$R = \frac{0.61\lambda}{\eta(\sin\theta)} \tag{7.3}$$

where R is the resolution of the microscope, λ is the wavelength of the light, η is the refractive index, and θ is the incident angle. Generally, the denominator is termed as "numerical aperture" of the objective lens. It can be observed from the equation that lower wavelength may result in better resolution of the image, which can be achieved by increasing the "numerical aperture" as well as by reducing the wavelength of incident light. Therefore, low wavelength lights (i.e. electron beams) are used for better resolution of the images, and hence the characterization is further termed "electron microscopy." With the electron beam under reflective mode and transmission mode, SEM and transmission electron microscopy (TEM), respectively, are widely used for optical characterization of FRPs. Instead of lens, electromagnetic coils are used for focusing/condensing the high energy electron beam. Precise focusing of the energetic electrons results in an unprecedented spatial resolution compared to optical based microscopic techniques as shown in Figure 7.6 [5].

SEM is the most common method for microscopic/morphological examination of FRPs. It is used to study the fractural behavior in fiber reinforced plastic. SEM enabled with the latest features also interprets the fiber volume fraction, fiber orientation, as well as their aspect ratio, in the case of short fiber reinforcement. The resolution of normal SEM instrument is to distinguish particles up to 500 nm, which can be enhanced by replacing the thermionic beam gun with an electromagnetic one (FE-SEM). To study the nanocomposites, field emission scanning electron microscopy (FESEM) is preferred due to its high resolution at different magnification characteristics.

Transmission electron microscope comprises an accelerated beam of electrons, which passes through a very thin specimen to exhibit the structure and

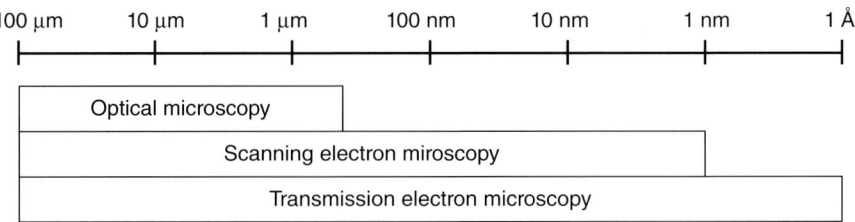

Figure 7.6 Characterization size regime for different microscopic techniques.

morphology of the FRP specimen. TEM is used to observe microstructure at a comparatively higher magnification and resolution than a light microscope. The structure, crystallization, morphology, and stress can also be studied with TEM, while SEM can provide information about the morphology of a specimen only. The main drawback of TEM is that it requires very thin specimens (1–10 nm) that are semitransparent to electrons; therefore, sample preparation takes a longer time.

Apart from imaging, electron microscopy can be employed for measuring surface roughness on a fine scale, even down to the level of molecules, i.e. scanning probe microscopy (SPM). It is also a branch of microscopy that outlines the image of surfaces using a physical *probe* to scan the specimen. The principle of the same is explained in Section 7.5.4.

7.5.2 Density

Density is considered as an important aspect as FRPs are considered as lightweight and possess high strength to weight ratio as compared to ceramics and metals. Therefore, these are widely used in lightweight applications. Their light weight also lends them well to logistics as these are easier to ship and install. The density of FRP mainly depends on the individual density of the matrix and reinforcement, volume fraction of each constituent, as well as the void content. The void content is a function of the fabrication process and indirectly an indicator of porosity and fiber wetting in the FRP. The bulk density of material can be defined as the ratio of its weight to its volume, which can be expressed as Eq. (7.4).

$$\text{Density } (\rho) = \text{Mass } (m)/\text{Volume } (v) \tag{7.4}$$

where the mass and volume can be measured directly using weighing scale and measuring instruments. Archimedes' principle can be used as another method for determination of density of FRP. Archimedes' principle states that the apparent weight of an object, when immersed in a liquid, decreases by an amount that is equal to the weight of the displaced liquid. Therefore, density can be measured by using Eq. (7.5), by knowing the mass of material in air and immersion liquid.

$$\rho_a = \frac{m_a}{[m_a - m_l]} \rho_l \tag{7.5}$$

where m_a is the mass of material in air, m_l is the mass in immersion liquid, and ρ_l is the density of the immersion liquid. Equation (7.5) can be used to measure the bulk density of FRP, but is not preferred to calculate the exact density of FRPs, as FRP is considered a two-phase material, contains two materials of different density, with different volume fraction. Therefore, the equation can be written as Eq. (7.6) (in terms of volume fraction) and as Eq. (7.7) (in terms of weight fractions) of the constituents, respectively.

$$\rho_c = \rho_f v_f + \rho_m v_m \tag{7.6}$$

$$\rho_c = \frac{1}{\left[\frac{m_f}{\rho_f} + \frac{m_m}{\rho_m}\right]} \tag{7.7}$$

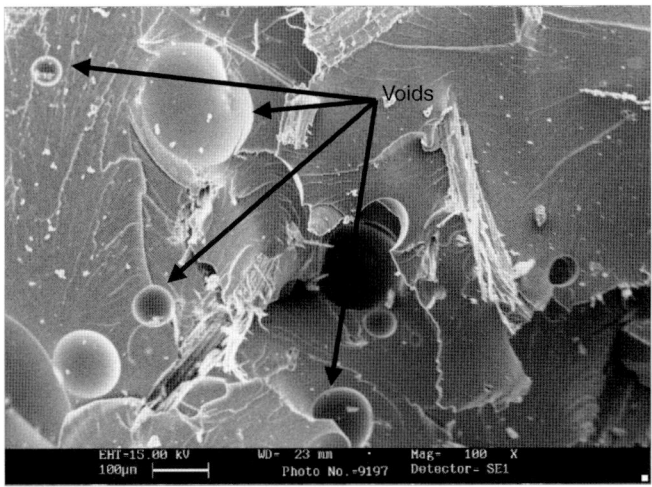

Figure 7.7 Presence of voids in FRP structure.

7.5.3 Void Fraction

While fabricating FRP, during the incorporation of fibers into the matrix, air or other volatiles may be trapped, in the form of microvoids (Figure 7.7), which may further significantly affect its properties. Higher void content may result in lower fatigue resistance, greater susceptibility to water diffusion, and increased variation (scatter) in mechanical properties.

Therefore, it is necessary to analyze the void volume fraction in FRP, which may help in the necessary modification in the process. The void content in FRPs can be calculated by comparing the bulk density with its actual density, using Eq. (7.8).

$$\%v_{\text{void}} = \left[\frac{\rho_{\text{bulk}} - \rho_{\text{actual}}}{\rho_{\text{bulk}}} \right] \times 100 \tag{7.8}$$

Apart from the mathematical techniques, some instrumental techniques such as Brunauer–Emmett–Teller (BET) are also used to measure the surface area as well as void volume/size in FRPs. BET surface area analyzer is commonly used to measure the same using the principle of gas adsorption. Adsorption occurs when a solid surface is exposed to some fluid. FRPs are porous solids, which contain voids, cavities, or channels. These pores/voids are classified according to their diameter as micropores (<20 Å), mesopores ($200–500$ Å), and macropores (>500 Å). An inert gas (nitrogen) is used due to its weak chemical interaction with the FRP specimen, which is most suitable for adsorption phenomenon. The specimen is placed in the work chamber and vacuum is applied to remove unwanted gases. Subsequently, nitrogen is released in the chamber and at different pressure, the amount of nitrogen absorbed over the surface is measured. Based on the results obtained, surface area, void size, and volume can be calculated using empirical relations.

7.5.4 Surface Hardness

Hardness is the property of a material that enables it to resist plastic or localized deformation, usually by penetration or by indentation. In other words, this is the property of a material that gives it the ability to resist being permanently deformed under loaded conditions. This is also mentioned as stiffness or resistance to bending, scratch, abrasion, or cutting. The greater the hardness, the greater the resistance it has to deformation. In metals, ceramics, and most of the polymers, the deformation is considered as plastic deformation of the surface. Indentation methods are used mainly to measure the hardness of a material, which measure the depth/area of the indent at a particular load and provide a unique number, which is considered as the hardness number of the material. A greater number indicates higher hardness of the material. Common methods used for measuring hardness of metals, ceramics, and high-end FRP materials are Brinell, Rockwell, and Vicker hardness (Figure 7.8). These three methods are different in the load applied and the shape of the indenter used. Ball, pyramidal, and conical diamond with round tip indenters are used for Brinell, Rockwell, and Vicker hardness tests, respectively. Knoop hardness (HK) is an alternate of Vickers test and lies in the microhardness testing range (applied load < 10 N) to analyze the hardness of thin layers. The indenter is an asymmetrical pyramidal diamond, and the indentation is measured using optical device along the long diagonal. Typical FRP composites are comparatively softer and of less thickness; therefore, traditional methods are not used to check their hardness. For softer materials and FRPs with thermoplastic matrix, Shore hardness is the most common method for measuring the hardness (Figure 7.9). Shore hardness is measured using a durometer gauge, which uses a spring-loaded steel rod to compress the surface of the FRP sample and provide a particular reading, showing its Shore hardness number.

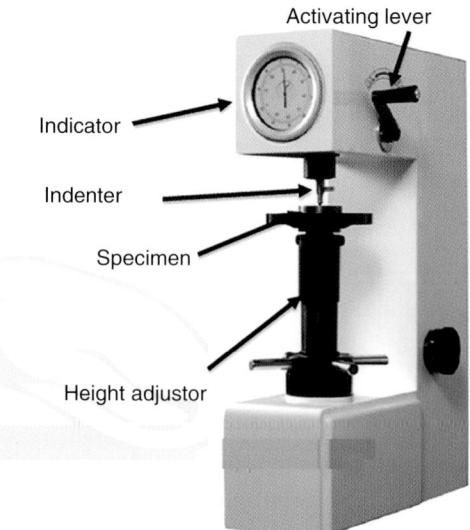

Activating lever

Indicator

Indenter

Specimen

Height adjustor

Figure 7.8 Typical structure of hardness tester.

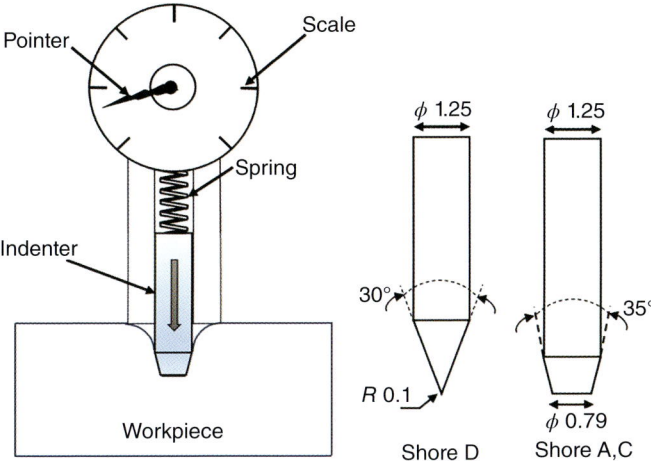

Figure 7.9 Shore hardness tester (durometer).

Mainly two types of durometers (Shore A and Shore D) are used to measure the Shore hardness of FRP composites. Both have the scale marked 1–100, but are used for different types of FRPs. Shore A is used for measuring the hardness of rubbery materials, i.e. shoe sole, tires, erasers, etc. while Shore D scale is used for harder FRPs based on thermoplastic matrices such as polypropylene, polyethylene, and PLA. The working of the durometer and geometries of the indenter used are shown in Figure 7.9.

7.5.5 Surface Roughness

Surface roughness of FRP is also an important aspect, when intended to use for some specific application (paint, wear, etc.) or secondary processing (drilling, joining, etc.). Surface roughness is a quantitative calculation of the relative average surface roughness (R_a) on a line or an area. This also includes surface shape, surface finish, surface profile roughness or in-surface area roughness (S_a), surface texture, asperities, and structural characterization. FRPs made by closed mold processes (injection molding, flow compression molding, etc.) have better surface finish as compared to open mold processes. Higher surface roughness may lead to poor mechanical properties under loading condition due to stress concentration. Rough surfaces also possess higher surface area, which is prone to higher moisture absorption and other reactions, when subjected to specific environments. In secondary processing of FRP, high surface roughness is required over the joining area for a better application of adhesives, while in the case of drilling, lower surface roughness is required over the peripheral surface of the hole to reduce stress concentration. Corrective measures, i.e. application of gel coat, machining, etc. can be taken to achieve the desired surface roughness of the FRP. Both contact type and noncontact type methods can be used for measurement of surface roughness of FRPs. The contact type of instrument consists of a stylus, which tracks across the surface under inspection. This signal is then

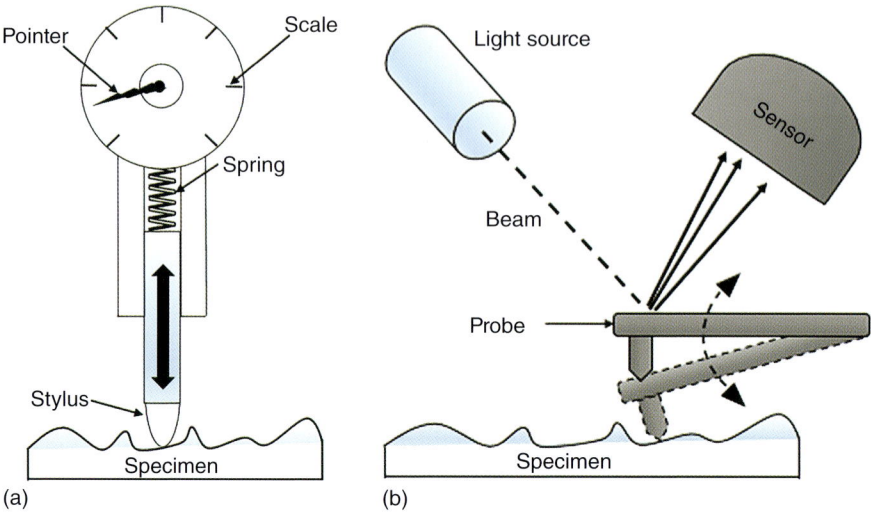

Figure 7.10 Modes of roughness measurement (a) contact type, (b) noncontact type.

measured through analog or digital signals, which can be further processed using software to get a value that represents the surface roughness (Figure 7.10a).

Stylus load also introduces error in the measurement in the case of contact type measurement. A sharp stylus under low loads may result in local elastic deformation of the surface being measured as well as the stylus tip at the area of contact. Therefore, nowadays, piezoelectric transducers are used in contact type surface measurement techniques. Optical methods (scanning probe microscope) are also used for noncontact type surface measurement, in which a light beam is incident on the surface, where it is reflected or scattered. This reflection/scattering is measured through optical sensors to get surface roughness values of the FRP surface. Atomic force microscopy (AFM) is commonly used as a noncontact type surface measurement method for measuring the surface roughness of a specimen at the nanoscale level. In AFM, a very sharp tip is used as probe (to map the morphology of a surface) at the end of a micro-fabricated cantilever with a low spring constant, which helps measure the tip–sample force as the tip presses against the sample [6]. Deflection/bending of the cantilever occurs due to the forces between the tip and sample, which is measured, and further exhibits the surface topography of the sample (Figure 7.10b). In AFM, there is no requirement for the sample to be conductive, nor is it necessary to measure a current between the tip and sample to produce an image.

7.6 Mechanical Characterization

FRPs are anisotropic materials because their properties vary as a function of direction. The mechanical properties of FRPs depend on composite design parameters, i.e. aspect ratio of reinforcement, reinforcement orientation, volume fraction, thickness, and number of layers of reinforcement (in case of composite laminates). Quasi-isotropic FRP laminates will demonstrate consistent mechanical performance along different axes, while for non-isotropic laminates, properties vary adequately as a function of the direction of applied load. Rule of mixture is generally used to make an estimate of the properties of FRPs at the starting stage. By using the rule of mixture the properties as well as strength, modulus of elasticity, density, and electrical and thermal conductivity can be predicted easily. Typical tests used to measure the mechanical properties (strength, modulus, etc.) of FRPs include tensile, flexural, impact, compression, and shear test. These tests mainly differ in the method of load application and are specifically used according to the application of the product.

7.6.1 Tensile test

Tensile strength is the maximum stress a material can bear without fracture when stretched. It is an intensive property of the FRP, and therefore does not change with the size of the specimen. Tensile strength can be calculated by dividing the maximum load by the cross-sectional area of the FRP specimen. Universal testing machine (UTM) is used to measure the tensile strength as well as modulus of FRPs. The tensile strength of FRP composites depends on various factors including the inherent properties of fiber and matrix, aspect ratio of fibers, fabrication process, processing parameters, and most importantly interfacial adhesion between the fibers and the matrix. In tensile testing, the test specimen is gripped (gauge length) between two jaws and both the jaws are stretched in the opposite direction at a predefined speed according to the standards used (Figure 7.11).

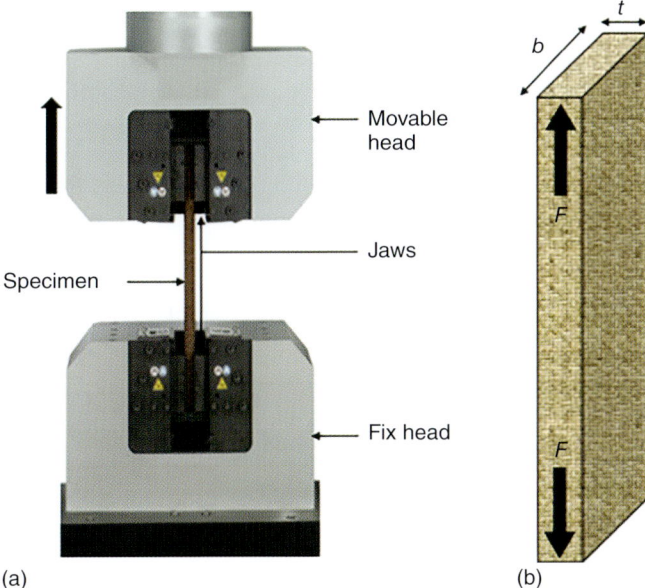

(a)   (b)

Figure 7.11 Tensile testing of FRP composites.

Maximum load and the elongation are measured before failure of the specimen, and based on the reading the tensile strength and modulus are calculated by Eqs. (7.9) and (7.10), respectively. A stress–strain curve can be drawn by measuring the values of load and extension data at various points of time, which can be further used to calculate chord modulus, secant modulus, etc. of the FRPs.

$$\sigma = \frac{F}{A} \tag{7.9}$$

$$E = \frac{\sigma}{\varepsilon} \tag{7.10}$$

where σ is the tensile strength, ε is the strain, E is the modulus, F is the maximum load, and A is the cross-sectional area of the specimen. Different units can be used for tensile strength of FRPs and some common units are newton per square meter (N/m^2), kilogram (force) per square centimeter (kg/cm^2), pounds per square inch (psi), or megapascal (MPa).

7.6.2 Compression Test

Compressive strength or compression strength is the capability of a material or structure to withstand gradual loads, which tends to reduce its size, normally under compressive load (Figure 7.12). Compressive properties are dependent on factors such as fiber properties, matrix properties, fiber/matrix interfacial bond strength, and fiber misalignment.

Composites, when loaded in compression, may experience a number of failure modes, including fiber crushing, shear splitting, and elastic buckling. FRPs based on thermoset polymers exhibit better compressive strength as compared to thermoplastic-based FRPs. When compressive load is applied to FRP, it deforms elastically up to a certain point, and then plastic deformation occurs. Further increase in the load may completely flatten a test piece without any definite fracture occurring. Compressive strength is also measured using UTM, on which a definite size of test specimen is placed between two plates and compressive load is applied gradually as per the selected test standard. Maximum load and the corresponding deformation are recorded and compressive properties are calculated using the same equations as in tensile testing.

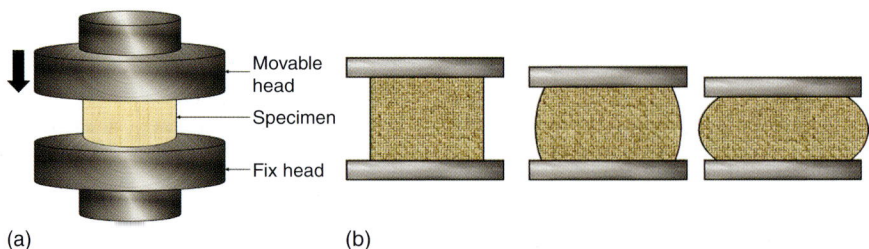

(a)   (b)

Figure 7.12 Compressive testing of FRP composites.

7.6.3 Flexural Test

Flexural strength, or bending strength, or transverse rupture strength is a material property, defined as the stress in a material just before it yields when subjected to bending load. Flexural strength is typically governed by tensile and compressive properties and the lamina stacking sequence in FRP laminates. In homogeneous material, generally flexural strength is equal to the tensile strength. Flexural test is also performed on UTM in either three point bending or four point bending mode (Figure 7.13).

Three point flexural test is advantageous due to the ease of specimen preparation and testing but has a disadvantage of higher sensitivity to loading geometry and strain rate. The sample is placed on two supporting pins set a distance apart, i.e. simply supported beam, and the load is applied at the middle point of the FRP specimen. Four point bend test is normally performed on highly brittle materials. A schematic of both modes has been shown in Figure 7.13. Applied load and the corresponding deflection are measured at various points of time; stress–strain curve is drawn and the strength is calculated by using the following equations:

$$\sigma_f = \frac{3FL}{4bd^2} \text{ (for four point bend test, loading span is half of the support span)}$$

$$(7.11)$$

$$\sigma_f = \frac{3FL}{bd^2} \text{ (for four point bend test, loading span is } 1/3 \text{ of the support span)}$$

$$(7.12)$$

$$\sigma_f = \frac{3FL}{2bd^2} \text{ (for three point bend test)} \tag{7.13}$$

The flexural strain can be calculated by using Eq. (7.14).

$$\varepsilon_f = \frac{6Dd}{L^2} \tag{7.14}$$

where σ_f is the flexural stress, F is the maximum load, L is the support span, D is the deflection, and b and d are the width and thickness of the test specimen. The most common standard used for flexural test of FRPs is ASTM D-790 and ASTM D-7264.

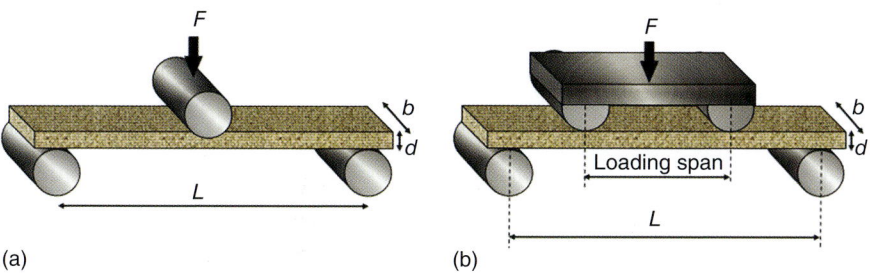

(a)   (b)

Figure 7.13 Flexural testing of FRP composites (a) three point bending, (b) four point bending.

7.6.4 Impact Test

Impact test is a method for evaluating the toughness and notch sensitivity of FRPs. In other words, it is a method for determining shock loading behavior of FRPs. Impact energy is measured as the energy absorbed in breaking the specimen in a single blow. In some applications (paint quality assessment) multiple blows of increasing intensity are exercised (drop ball impact). The most common test modes for impact testing of FRPs are Izod and Charpy test either on (V-type/U type) notched or unnotched specimen. Notched tests are usually performed to analyze the notch sensitivity of the FRP specimen. Both the tests (Izod and Charpy) are different only in the arrangement of the specimen during testing. In Izod test, the specimen is held in cantilever configuration, while it is a simply supported beam configuration in the case of Charpy impact test. The specimen is fixed and stroked by a hammer pendulum of specific weight from a specified height. When the pendulum is in a raised position, it possesses certain energy and on releasing, it strikes the specimen; a part of the energy is consumed to break the specimen, which is measured. This energy is considered as the impact energy/strength of the specimen. The weight and height of the pendulum depend on the material and the test standard used. Typical standards used for impact test are DIN 50115, EN 10045-1, ASTM E23, BS 131-1, and ASTM E-604. A schematic diagram of an impact tester and testing are shown in Figure 7.14.

7.6.5 Shear Test

Shear strength is the strength of FRPs against the type of yield or structural failure, where it fails in shear. Shear force tends to produce a sliding failure along the planes, parallel to the direction of force (Figure 7.15).

In other words, it is the ability of FRPs to resist forces that can cause the internal structure of the material to slide against itself. Shear strain can be

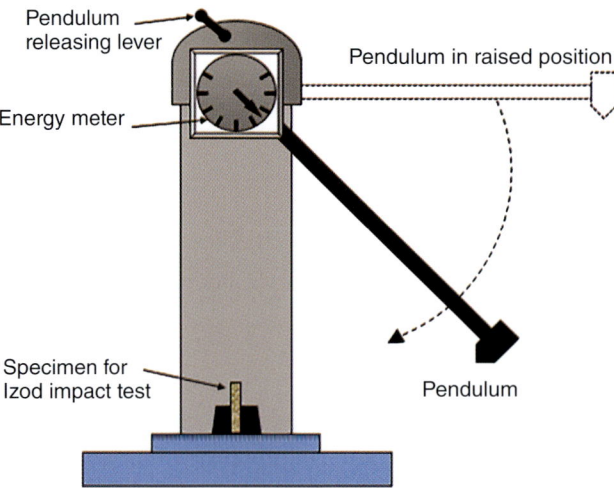

Figure 7.14 Impact testing of FRP composite specimen.

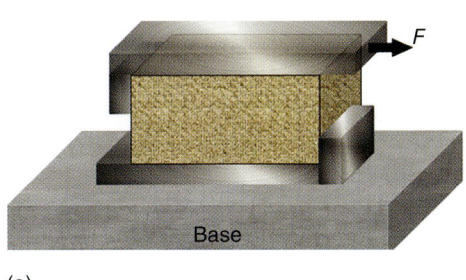
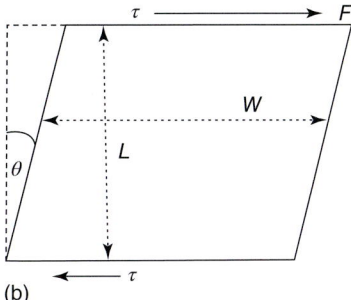

(a)   (b)

Figure 7.15 Shear testing of FRP composites.

defined as the tangent of the sliding angle and is equal to the length of defor-mation at its maximum, subsequently divided by the perpendicular length in the direction of force. The ratio of shear stress (τ) to shear strain (γ) is defined as the shear modulus of FRP. Shear stresses may be described as in plane, transverse to the plane (through thickness), or between individual laminas. This test is generally employed on FRP composite laminates to check the shear behavior between different plies. In-plane shear properties are controlled by fiber orientation and laminate stacking sequence. Transverse, or punching, shear strength is highly dependent on reinforcement type and volume. Interlaminar shear strength is primarily dependent on the matrix and matrix/fiber interfacial properties.

7.7 Thermal Characterization

FRPs, when used in structural applications, are subjected to various thermal conditions. The physical properties of FRPs vary with the change in temperature, which necessitates the thermal characterization of FRPs. Usually, FRPs exhibits better modulus and stiffness properties at lower temperatures as compared to elevated temperatures. With the increase in temperatures, a gradual decrease in modulus is exhibited up to a point where it transits from glassy to a rubbery state. This point is termed as glass transition temperature (T_g). Similar to mechanical properties, thermal performance of FRP composites also depends on the matrix, fillers, fiber type, and solidification (curing) process. Thermal characterization techniques deal in the analysis of change in physical properties of FRP as a function of temperature. Usually, FRPs are subjected to a controlled temperature environment during analysis. The range of temperature may vary from normal (ambient) to extreme, i.e. cryogenic temperature to the burning point of the FRPs. Thermal characterization of FRPs includes measurement of thermal properties, thermogravimetric analysis (TGA), differential thermal analysis (DTA), differential scanning calorimetry (DSC), and dynamic mechan-ical analysis (DMA). Thermal properties include thermal conductivity (k), thermal diffusivity (α), specific heat capacity (C_p), and coefficient of thermal expansion.

7.7.1 Thermal Properties

Almost all the polymers possess low thermal conductivity, which means that FRPs are good heat insulators. This may be due to the low density of the matrix and the high strength of the embedded fibers. Thermal properties of FRPs are mainly dependent on individual properties of fiber and matrix as well as the aspect ratio and the interfacial adhesion. Rule of mixture and other statistical methods can be applied to calculate the theoretical thermal properties of FRPs, but certain advanced methods are also available to measure them practically. The advanced methods include transient plate source (TPS), heat flow meter (HFM), transient line source (TLS), laser flash apparatus (LFA), monotonic heating (MMH), and transient hot wire (THW) methods. DSC is sometimes used to calculate the specific heat capacity of FRPs by analytical method. All the methods are different in the measurement range, phase of specimen (solid/liquid), and mode of applications. TPS and TLS methods are widely used to measure the thermal properties of FRPs. The transient plane source method utilizes a flat sensor with electronic accessories and special mathematical model describing the heat conductivity, considering the material as isotropic. This technique typically employs two samples halves, and the sensor is placed between these two. The flat sensor consists of a continuous double spiral of electrically conducting nickel (Ni) metal, etched out of a thin foil. The spiral is located between two thin layers of Kapton (polyimide film), which provides insulation and mechanical stability to the sensor. A constant electric current is passed through the conducting spiral, which increases the temperature of the coil; heat is generated and it dissipates into the sample at a rate depending on the thermal transport properties of the material. The temperature and time response are measured and then thermal properties are calculated with help of mathematical model. TLS method employs a needle-type sensor in place of a disc type sensor.

FRPs undergo dimensional fluctuation in response to variations in temperature, a phenomenon that has critical implications for their structural applications. Therefore, FRPs must be designed to accommodate this fluctuation for safe application. The coefficient of linear thermal expansion of FRPs is the increase in length (linear dimension) per unit increase in temperature, which can be defined at a definite temperature or a temperature range. The change in length with temperature for a solid material can be expressed as

$$\alpha_1 = \frac{l_f - l_i}{t_f - t_i} \tag{7.15}$$

where α_1 is the linear coefficient of thermal expansion, and l_i and l_f are the initial and final length of the specimen over the initial (t_i) and final (t_f) temperatures. Volumetric coefficient (α_v) of thermal expansion can also be measured by measuring the volumetric change over the change in temperature.

$$\alpha_v = \frac{v_f - v_i}{t_f - t_i} \tag{7.16}$$

The technique used to measure the coefficient of thermal expansion are based on interferometry and dilatometry and the normal temperature range may vary from $-253\,°C$ (20 K) to $1100\,°C$ (1373 K), which further depends on the type of

FRP. In interferometry, optical interference techniques are used to measure the displacement of the specimen ends in terms of the number of wavelengths of monochromatic light with respect to change in temperature. In dilatometry, the specimen is heated in a furnace while the displacement of the specimen ends is transmitted to a sensor by means of push rods. Interferometry is a noncontact type measurement technique and is preferred over dilatometry due to high precision.

7.7.2 Thermogravimetric Analysis

TGA is performed to check the change in physical/chemical properties of FRP as a function of gradually increasing temperature or as a function of time in isothermal environment. The variation is mainly due to the structural transformation or degradation of either the matrix or the reinforcement in FRP. The analysis provides information about physical changes, such as phase transitions, absorption, and desorption, as well as chemical changes including chemisorption, thermal decomposition, and solid–gas reactions. A variety of test environment can be employed, while testing FRPs, i.e. argon, nitrogen, air, or vacuum.

A thermogravimetric instrument consists of a precise weight balance with a pan inside a heater/furnace with a programmable control temperature. The specimen of certain weight (1–150 mg) is placed in the pan and the temperature is increased at constant rate to start a thermal reaction. A reference pan is also used for comparison between the specimen and the reference material. Alumina is used as the reference pan due to its higher thermal stability. The change in mass of the specimen is recorded over time and temperature and subsequently compiled into a plot of mass or percentage of initial mass to temperature/time, which is termed as a TGA curve. The first derivative of this TGA curve is termed as derivative thermogravimetric (DTG) curve, which is further used for in-depth analysis as well as DTA.

7.7.3 Differential Thermal Analysis (DTA) and Differential Scanning Calorimetry (DSC)

In DTA, the difference in temperature/energy between a substance and a reference material is measured as a function of temperature/time under controlled temperature. DSC is a similar technique but it utilizes the difference in energy of the specimen and reference material/system. Both the techniques are used to analyze the glass transition temperature (T_g), melting point (T_m), crystallinity, and burning behavior of FRPs. Generally, the temperature program for a DSC analysis is designed such that the sample holder temperature increases/decreases linearly as a function of time. The change in enthalpy in case of DSC is measured and the crystallinity of the FRP is calculated using Eq. (7.17).

$$\text{Crystallinity} = \frac{\Delta H_f - \Delta H_s}{\Delta H_r} \times 100 \qquad (7.17)$$

where ΔH_r is the change in enthalpy of the reference material (100% crystalline), and ΔH_f and ΔH_s are the change in enthalpy of specimen during fusion (melting) and solidification, respectively. Measurement of crystallinity is generally performed on FRPs, based on thermoplastic matrix only, as it is a function of fusion and solidification only.

7.7.4 Dynamic Mechanical Analysis (DMA)

DMA is a versatile and flexible technique that is widely used to characterize the physicomechanical properties of FRPs as a function of temperature, time, frequency, stress, atmosphere, or a combination of these parameters. It can be used on solids and polymer melts for viscoelastic study, modulus, and damping, and is programmable to measure force, stress, strain, frequency, and temperature. When DMA results are combined with temperature response, the results also exhibit the rheology of solids termed as dynamic mechanical thermal analysis (DMTA). Initially, thermomechanical analysis (TMA) was used on FRPs to study the change in mechanical properties with respect to temperature only.

In DMA, an oscillating force (stress) in sinusoidal mode is applied on the specimen, which is held in single cantilever or double cantilever configuration, and the oscillating sample response is recorded. While performing the test on FRP, the frequency is kept constant and the temperature is gradually increased. Modulus and damping are calculated from the elastic and viscous response, respectively.

7.8 Durability Characterization

FRPs possess better strength to weight ratio over metallic and ceramic materials and are also suitable for many applications. Because of the nature of the applications, it is necessary to check their short-term as well as long-term durability as the properties of FRPs deteriorate over a period of time due to load application, environmental conditions, etc. Therefore, it is critical to know whether the FRPs used in structural application will continue to perform satisfactorily over its lifespan. This includes the FRPs' ability to endure a load as well as resist further deformation. Similarly, if the material is subjected to harsh environmental conditions, i.e. fire, chemicals, moisture, etc., the characterization becomes more critical due to the uncertainty in the actual conditions. Therefore, it is mandatory to check their durability by characterizing them in such conditions. Durability characterization mainly includes the analysis of FRPs subjected to continuous loading conditions (creep, fatigue, and wear), uncontrolled temperature (fire, thermal shock), and chemical environment (moisture uptake, degradability, corrosion). Durability characterization also assists in predicting the life of a component. As it is almost impossible to check the characteristics in the real environment, specimens are subjected to the specified conditions over a period of time. The obtained results are analyzed using mathematical models or simulation techniques to assess the properties, when subjected to the specific conditions for a prolonged time [7].

7.8.1 Creep Testing

FRPs are increasingly used in many structural applications, particularly in the building and automotive industries, which necessitates the prediction of their load bearing integrity. A significant contributor to FRP structural integrity is the viscoelastic behavior of composite materials based on thermoplastic matrices. Depending on the magnitude, creep and fatigue behavior could cause a significant time-dependent deformation in FRP structures in the long run [8].

Creep is defined as a test when a sample is subjected to a constant stress and the strain is monitored as a function of time. The deformation rate of the material at a constant temperature is known as the creep rate, which is the slope of the creep and time curve. Creep generally occurs at elevated temperatures; therefore, the testing is performed in an environmental chamber for precise heating/cooling control. Creep generally occurs in three stages: primary (stage I), secondary (stage II), and tertiary (stage III). Primary creep occurs at the beginning of the tests (Figure 7.16); at this stage creep is mostly transient. Subsequently, resistance to creep increases until the secondary stage is reached, where the rate of creep becomes roughly steady. This stage is often referred to as steady-state creep. In the tertiary stage, the creep rate begins to accelerate due to the decrease in cross-sectional area of the specimen. The fracture in material will occur mostly at this stage.

7.8.2 Fatigue Testing

Fatigue testing can be defined as the process of progressive localized permanent structural change or deformation in a material, when it is subjected to fluctuating stresses and strains. This leads to cracks or complete fracture in the specimen after a sufficient number of cycles. Similar to creep, fracture also occurs in three stages, as crack nucleation or initiation (stage 1) followed by crack propagation (stage 2) and subsequent failure (stage 3). The basic difference between creep

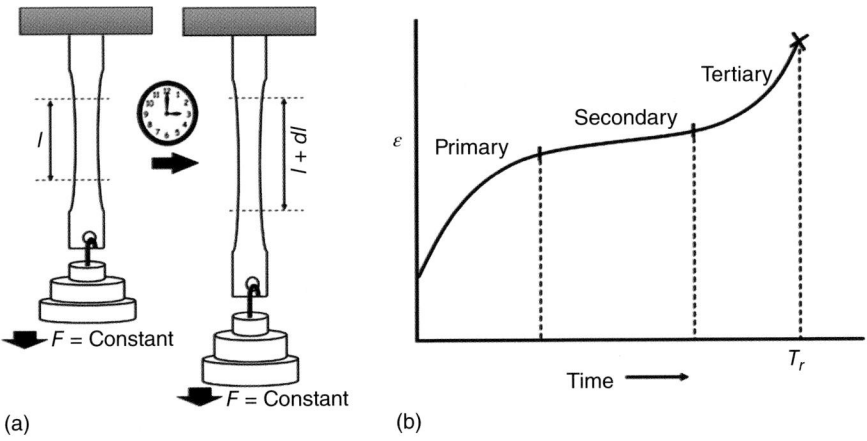

Figure 7.16 Creep testing of FRP composites.

and fatigue testing is the load application, where it is static in nature during creep testing and dynamic in fatigue testing. Loads can be applied axially (axial fatigue), in torsion (cantilever fatigue), or flexurally (three point/four point) during fatigue testing. Depending on the amplitude of the mean and cyclic load, the net stress developed in the specimen may be in one direction or reverse directions through the loading cycle. The data obtained from fatigue testing are generally exhibited as an *S–N* curve, which is a graph between the number of cycles and the cyclical stress developed. The cyclical stress can be represented as stress amplitude, maximum stress, or minimum stress. ASTM D-671 standard deals with the fatigue testing of plastics in flexural mode.

7.8.3 Wear Testing

Thermoset-based polymer composites are now gaining attention in various industrial applications in sliding/rolling/transmission components, such as, bearings, seals, gears, cams, belts, grinding mills, and clutches, due to their self-lubricating properties, to a certain extent. Also these are used in tribological applications for their elasticity, accommodation to shock loading, low friction, and wear resistance properties. Wherever the contact between sliding pairs is present, there is the problem of friction and wear, which necessitate the understanding of wear and wear mechanism of FRPs. Friction and wear behavior of glass fiber reinforced composites show that friction and wear are highly influenced by the process parameters, i.e. applied load, sliding speed, sliding distance, as well as fiber orientation.

The wear in FRPs can be subdivided into three main groups: adhesion, abrasion, and surface fatigue. Each of these is governed by its own laws and limitations; furthermore, on many occasions, it may act in such a way as to affect the others. Abrasive wear is the result of hard asperities on the counterface, which penetrates into the contact surface of the polymer and eradicates material, which further results in micromachining of the surface leading to wear grooves, tearing, etc. This type of wear is the most common form of dry wear and it is substantial in the FRP, when it slides repeatedly over the same track, on a smooth metal counterface [9]. Surface fatigue wear occurs when friction contact undergoes cyclic stress at rolling and reciprocal sliding. Apart from these, delamination wear, fretting wear, and corrosive wear are the other forms that can be seen in FRPs [10].

A variety of techniques are available to check the tribological characteristics of FRPs. Pin on disc method is usually adopted by the researchers due to its simplicity and time-saving features. In all the techniques, the sample of specific dimension (flat/pin/cylinder) is moved (rolled/slide/spin) over a specific geometrical part (disc/cylinder, etc.) continuously with a defined set of parameters. The parameters are applied load, sliding speed, time, sliding distance, and applied pressure. Coefficient of friction, material removal rate (MRR), and specific wear rate (SWR) in terms of volume and mass along with the increasing in temperature are measured at a given set of parameters. Friction and wear characteristics of FRPs are system specific, as different test methods may exhibit different values for the same material and the same set of parameters. Thus, it is difficult to rely on any method for exact results but at the same time, these methods can assist in determining the mechanisms as well as tribological ranking of the materials.

7.8.4 Fire Testing

Usage of FRPs in architectural applications can be observed nowadays due to their better mechanical properties, lightweight, and tailored shape. Their low thermal transmissibility makes FRP panels suitable for facade applications, also under controlled temperature. Almost all the matrices in FRPs are hydrocarbon in nature; therefore, these burn vigorously in case of fire and produce heavy, dark, and dense smoke. To prevent fire or to slow down the burning of polymer in FRPs, fire retarding agents are used, which helps in reducing the propagation of flames, fume generation, and heat generation [11].

Previously, the flammability behavior of FRPs was determined entirely through testing, which was costly and time consuming, but with the evolution of technology and mathematical models, it is now easy to predict the fire behavior of FRPs. Traditional procedure consists of subjecting the FRP member to standard firing conditions in a specialized furnace. Rather than performing laboratory tests at full scale, it is now possible to conduct detailed numerical analysis to obtain the flammability characteristics of a variety of FRPs by conducting small-scale experiments, i.e. calorimetry [11, 12]. A variety of standards are available to measure the flammability characteristics of FRPs based on their applications, i.e. ASTM E-2058, D-7309, E-108, E-84, etc.

7.8.5 Environmental Testing

Long-term durability, weathering resistance, and exceptional mechanical properties have recently proposed the adoption of FRPs for building facade systems in an increasing number of buildings worldwide. However, some challenges related to the environmental and thermal aspects still exist. Therefore, LCA is used to assess the environmental impact of FRPs. LCA is considered an appropriate and rigorous method for evaluating the environmental impact. This is a complicated assessment that is valid only for a specific material system in a specific project [10].

In terms of environmental testing, a critical factor that is directly associated with application and end use of FRPs is its behavior during service conditions. This behavior can be analyzed by conducting proper aging and biodegradability studies on FRPs. Aging is considered as a phenomenon of variation in properties of FRPs with respect to time, due to various factors, i.e. temperature, pressure, etc. Actual study of environmental behavior may involve exposing the FRPs for a longer time span, which makes the actual experimentation practically difficult; therefore, accelerated studies are conducted [13]. Two factors are most important for analyzing the environmental behavior of FRPs, namely, exposure conditions and time duration. Generally, to reduce one, the other must be increased. Accelerated aging is exposing the material to conditions harsher than it encounters in reality during service.

During service conditions, FRPs are open to a multiplicity of environmental situations such as moisture, solvents, oil, heat, loads, and radiation. Generally, natural fiber-based FRPs absorb moisture in humid environments and when dipped in water. This absorption leads to the degradation of fiber–matrix interface. This degradation of interface leads to poor stress transfer capabilities, which further results in a variation in mechanical as well as dimensional properties. The absorption rate depends on many variables, such as fiber constituent and structure, matrix, temperature, water distribution within the composite, and reactivity of the medium [14].

Nonavailability of specific standards to check environmental effect has led to various studies on FRPs in different environments based on their intended use. Earlier studies on FRPs used a weathering chamber, but the logic behind the selection of environmental cycle remained arbitrary, although it was closely related to the application area of composites as the final product.

7.9 Conclusion

Application of FRPs in structural and nonstructural applications is increasing day by day. Various characterization techniques have been discussed in the chapter based on their chemical, physical, mechanical, and thermal aspects. Recyclability and secondary processing (joining/machining) of FRP are a major concern for the manufacturer. Also due to ecological and environmental concerns, natural fiber-based bio-composites and bioplastics have gained the attention of researchers. Therefore, the authors believe that still a lot of research work is required in the domain of characterization of FRPs in order to fully realize their application spectrum.

References

1 Cross, J.O., Opila, R.L., Boyd, I.W., and Kaufmann, E.N. (2015). Materials characterization and the evolution of materials. *MRS Bulletin* 40 (12): 1019–1033.

2 Gad, S.C. and Gad-McDonald, S. *Biomaterials, Medical Devices, and Combination Products: Biocompatibility Testing and Safety Assessment*, 1e, ISBN 9781482248371. Raton: CRC Press Boca.

3 Hoh, K., Ishida, H., and Koenig, J.L. (1990). Multi-nuclear NMR spectroscopic and proton NMR imaging studies on the effect of water on the silane coupling agent/matrix resin interface in glass fiber-reinforced composites. *Polymer Composites* 11 (3): 192–199.

4 Hill, A.D., Lehman, A.H., and Parr, M.L. (2007). Using scanning electron microscopy with energy dispersive X-ray spectroscopy to analyze archaeological materials. Introducing scientific concepts and scientific literacy to students from all disciplines. *Journal of Chemical Education* 84 (5): 810–813.

5 Fahlman, B. (2011, ISBN: 978-94-007-0693-4). *Material Chemistry*. Dordrecht, Netherlands: Springer.

6 Sharma, S.K. (2018). "*Handbook of Materials Characterization*", ISBN: 978-3-319-92955-2. Springer International Publishing AG.

7 (2016). *Guidelines and Recommended Practices for Fiber-Reinforced-Polymer (FRP) Architectural Products.* Arlington: American Composites Manufacturers Association.

8 Sullivan, J. (1991). Measurement of composite creep. *Experimental Techniques* 15: 32–37.

9 Abdelbary, A. (ed.) (2015). Polymer tribology. In: *Wear of Polymers and Composites"* , ISBN:978-1-78242-177-1. Woodhead Publishing.

10 Bajpai, P.K., Singh, I., and Madaan, J. (2013). Tribological behavior of natural fiber reinforced PLA composites. *Wear* 297 (1–2): 829–840.

11 Berardi, U. and Dembsey, N. (2015). Thermal and fire characteristics of FRP composites for architectural applications. *Polymers* 7: 2276–2289.

12 Lau, D., Qiu, Q., Zhou, A., and Chow, C.L. (2016). Long term performance and fire safety aspect of FRP composites used in building structures. *Construction and Building Materials* 126: 573–585.

13 Lila, M.K., Shukla, K., Komal, U.K., and Singh, I. (2019). Accelerated thermal ageing behaviour of bagasse fibers reinforced poly (lactic acid) based biocomposites. *Composites Part B: Engineering* 156: 121–127.

14 Lila, M.K., Singh, B., Pabla, B.S., and Singh, I. (2018). Effect of environmental conditioning on natural fiber reinforced epoxy composites. *Materials Today: Proceedings* 5 (9–1): 17006–17011.

8

Detection of Delamination in Fiber Metal Laminates Based on Local Defect Resonance

Tanmoy Bose, Subhankar Roy, and Kishore Debnath

Department of Mechanical Engineering, National Institute of Technology Meghalaya, Bijni Complex, Laitumkhrah, Shillong 793 003, Meghalaya, India

8.1 Introduction

Composite materials are defined as the combination of two or more materials in different percentages by weight. The properties achieved with a composite material are mostly better than the properties possessed by the original materials. Properties such as high strength, low weight, low thermal expansion, and resistance to corrosion can be achieved in composite materials. Polymer matrix composite (PMC) is the most widely used composite material. PMC possesses properties such as high strength to weight ratio, toughness, stiffness, and resistance to chemical reactions, which make them quite suitable for applications in automotive, construction, sports, marine, home appliances, electrical industries, and many more [1–3]. Different types of PMC are available based on their reinforcing material, such as glass fiber, carbon fiber, and Kevlar fiber reinforced PMC. Glass fiber reinforced polymer (GFRP) and carbon fiber reinforced polymer (CFRP) composites play a major role in the application of PMCs. Recently, another category of composite materials known as fiber metal laminates (FMLs) is finding wide applications in the aerospace industry. Glass reinforced aluminum (GLARE) and aramid reinforced aluminum laminates (ARALL) are the most common types of FMLs used in the aerospace industry. GLARE is fabricated from a stack of aluminum sheets that are bonded with unidirectional glass fibers reinforced epoxy prepregs [4]. Other FMLs such as carbon reinforced aluminum alloy (CARALL, Carbon Reinforced Aluminium Laminate) are under development and testing.

In recent times, study of wave spectroscopy has become very significant in the characterization of defects in any material. Damages such as delamination and cracks can be detected using conventional methods of structural health monitoring (SHM) and nondestructive evaluation (NDE). The defects are detected due to the scattering and reflection of incident vibration waves when the incident wave passes over discontinuities and irregularities at the damage location [5, 6]. Conventional SHM techniques cannot address the case of barely visible

Reinforced Polymer Composites: Processing, Characterization and Post Life Cycle Assessment,
First Edition. Edited by Pramendra K. Bajpai and Inderdeep Singh.
© 2020 Wiley-VCH Verlag GmbH & Co. KGaA. Published 2020 by Wiley-VCH Verlag GmbH & Co. KGaA.

damages (BVD) such as cracks and delaminations due to minimal effect of the BVD on scattering and reflection of incident waves. This difficulty in health monitoring can be overcome by implementing nonlinear wave spectroscopy (NWS) technique [6]. The detection of early delamination and cracks can be effectively performed using NWS technique, which helps in studying the nonlinear behavior of the material as well as the effect of damages on the propagating incident wave. The presence of BVD leads to a rise in nonlinearity, which can be well detected by higher harmonic frequencies of the damage frequency. The higher harmonic of damage frequencies can be classified using the clapping phenomena of the delamination or crack faces. The clapping phenomenon may also lead to friction generation at the free surface, which is one of the causes for generation of higher harmonics [7, 8].

If the frequency of incident ultrasonic wave matches with the local frequency of a defect present in composite structures, local defect resonance (LDR) occurs. The phenomenon of LDR is one of the recent developments in detection of defects and has been studied by many researchers. The excitation of a damage using LDR frequency may lead to an increase in the temperature gradient of the defect area as investigated by Solodov et al. [9]. The technique of thermographic imaging can be used in order to detect the rise in temperature at the defect area, thus locating the defect. Transmission of energy to the defect location by using wave propagation is another type of NWS, known as resonant ultrasonic spectroscopy (RUS). Solodov et al. [10] reported that defect in the form of delamination can be considered analogous to a flat bottom hole (FBH) in order to understand the concept of LDR frequency. An increase in the normalized residual thickness of the FBH may lead to a change in boundary conditions of the defect. The simply supported boundary condition of an FBH with more depth converts to a case of clamped boundary condition when there is a decrease in the depth [10]. This change in boundary conditions helps in differentiating different defects with different depths in material. Many other studies based on LDR frequency have been carried out for the detection of delaminations using thermosonics [9, 11], shearosonics [11], laser vibrometry [12, 13], and fully acoustic [14] techniques. Solodov et al. [11, 14] used loudspeaker and shearosonic imaging of defects as well as fully acoustic based noncontact delamination detection over large areas. The LDR frequencies of an FBH in aluminum plate have been detected using the laser Doppler vibrometer [13]. Rahammer and Kreutzbruck [15] used a periodic chirp signal in order to determine the LDR frequency from thermal signatures. Most of the articles in the literature were found to discuss the LDR-based defect imaging of a composite material. Some authors [12, 16, 17] detected the LDR frequencies using fast Fourier transform (FFT) plots by applying narrow sweep excitation (NSE) to the composite structure. However, the identification of LDR frequencies from the FFT plot is a tedious job due to the presence of multiple peaks at the vicinity of LDR frequency.

Another technique for detection of LDR frequency of any defect can be carried out using an NWS technique based on bispectral properties. The normalized parameter of the bispectrum known as the bicoherence can be employed to determine the defect resonance frequencies related to any damage. There is much significant research carried out on the detection of damages using bispectral analysis and bicoherence estimation. Rivola and White [18] investigated fatigue cracks in beams due to excitation with white noise using bicoherence estimation. It was observed that the bicoherence technique is able to identify the defect although the bicoherence depiction of the signal is challenging to describe. On the basis of a non-Gaussian autoregressive model, a parameter dependent bispectrum technique is recommended by Raghuveer and Nikias [19]. It was observed that the technique is very useful for obtaining bispectral estimates as compared to the traditional estimates for revealing the phase couplings due to sinusoids. The method provides significantly better determining ability of the LDR in the detection of any damage. Kim and Powers [20] showed that the bicoherence spectrum may be used to segregate between nonlinear coupled waves and excited waves in order to determine the power induced from the quadratic coupling in a self-excited variation spectrum. Collis et al. [21] protracted two traditional methods of bicoherence and skewness function into their fourth order counterparts, i.e. tricoherence and kurtosis functions. A new process of pre-whitening a signal is proposed where the abovementioned normalization technique fails for narrowband signals. Nichols and Olson [22] derived the expressions for bispectrum and bicoherence functions for quadratic nonlinearities in the case of a multi-degree of freedom spring mass system. The probability of nonlinearity detection has been shown to improve using this expression. The bicoherence analysis discussed in the literature can be further utilized in the detection of LDR frequencies, which can be used for defect imaging, and is discussed in the subsequent sections.

8.2 Local Defect Resonance Based Nondestructive Evaluation

8.2.1 Concept of Local Defect Resonance

The stiffness and the mass of any structure get reduced at the damage location due to the presence of free surfaces. The natural frequency of the damage due to the effective mass and stiffness can be interpreted as LDR frequency (f_{LDR}) of that defect. The concept of LDR frequency of any defect is shown by presenting the defects in the form of FBH. FBH simulates vibrations of typical defects such as that of a spherical cavity or a disc-like crack with elliptical cross section as a delamination in composites. The LDR frequency can be written as [10]

$$f_{LDR} = \frac{1}{2\pi} \sqrt{\frac{K_{eff}}{M_{eff}}} \tag{8.1}$$

where K_{eff} = effective stiffness and M_{eff} = effective mass of the defect area.

The potential energy relation in the case of FBH with depth "h" and the total plate thickness "H" is utilized for obtaining the effective stiffness (K_{eff}) as shown in Figure 8.1. The remaining part of the material in the thickness direction is known as the residual thickness of the defect (t). If the depth of the FBH (h) is nearly equal to the total thickness of the plate (H), the boundary condition for

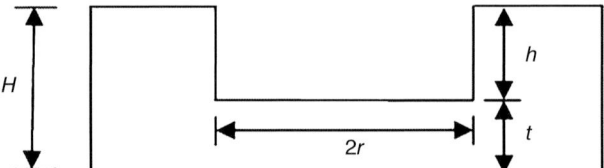

Figure 8.1 Schematic of a defect in the form of flat bottom hole.

the residual thickness of the defect becomes clamped. Otherwise, the boundary condition is considered to be simply supported for the defect area. The following expression is used for determining the potential energy of vibrations for a circular plate [10].

$$\text{P.E.} = \frac{K_{\text{eff}} \cdot U_{\text{eff}}^2}{2} = \frac{32\pi D u_0^2}{3r^2} \tag{8.2}$$

where U_{eff} = effective displacement of vibration of the FBH, D = bending stiffness of the plate, r = radius of the FBH, and u_0 = amplitude of vibration at the center of the FBH. From Eq. (8.2), substituting the term for effective displacement of vibration, the effective stiffness is found to be given by the relation

$$K_{\text{eff}} = \frac{192\pi D}{r^2} \tag{8.3}$$

Similarly, the effective mass of the defect is determined from the relation of kinetic energy of the FBH given by the following equation [10]:

$$\text{K.E.} = \frac{M_{\text{eff}} \cdot U_{\text{eff}}^2}{2} = \frac{m u_0^2}{10} \tag{8.4}$$

where m = mass of the plate below the FBH.

From Eq. (8.4), the value of effective mass for the flat bottom hole is found to be around 1.8 times of m

The effective stiffness (K_{eff}) and the effective mass (M_{eff}) are then combined to give the general relation for determining the LDR frequency. The relation for LDR frequency is then modified as in Eq. (8.5) [10].

$$f_{\text{LDR}} = \frac{1.6 \ t}{r^2} \sqrt{\frac{E}{12\rho(1 - v^2)}} \tag{8.5}$$

where E = modulus of elasticity, ρ = mass density of the material, and v = Poisson's ratio.

8.2.2 Modeling of GLARE-Fiber Metal Laminate

In this section, a study for the detection of LDR frequency of a circular delamination in a fiber metal laminate (FML) GLARE has been carried out. The GLARE plate is modeled and analyzed using the ABAQUS software. The GLARE is considered to have a 3/2 lay-up, i.e. three layers of aluminum alloy and two layers of glass fiber/epoxy prepreg. The specification of the GLARE FML with delamination is listed in Table 8.1. The dimension of the GLARE plate modeled

Table 8.1 Specification of the GLARE model.

Layer	Material type	Number of layers	Thickness of each layer (mm)
Aluminum	Al 2024-T3	3	0.3
Prepreg	S2-glass fiber + FM94 prepreg	2	0.35

is $180 \times 180 \times 1.6\,\mathrm{mm}^3$ having circular delamination between the second and third layers from the top. The position of the circular delamination has been changed for generating two different models of GLARE plate, viz. one with circular delamination at the center and the other with eccentric delamination at a co-ordinate of $x = 135\,\mathrm{mm}$ and $y = 45\,\mathrm{mm}$ (considering the origin at the left bottom corner of the plate). The radius of the delamination (r) is considered to be 12 mm and the height of delamination (h) is generally very small and taken as 0.02 mm. Thus, the residual thickness of the defect area (t) is found to be 1.58 mm. Figure 8.2 shows the schematic of the GLARE plate for both the configurations of circular delamination.

The material properties considered for modeling the GLARE plate are shown in Table 8.2. The material property of the aluminum 2024-T3 alloy and S2-glass fiber/FM94 epoxy prepreg are used in order to obtain the material property for the whole GLARE plate using the rule of mixtures for composite materials. The

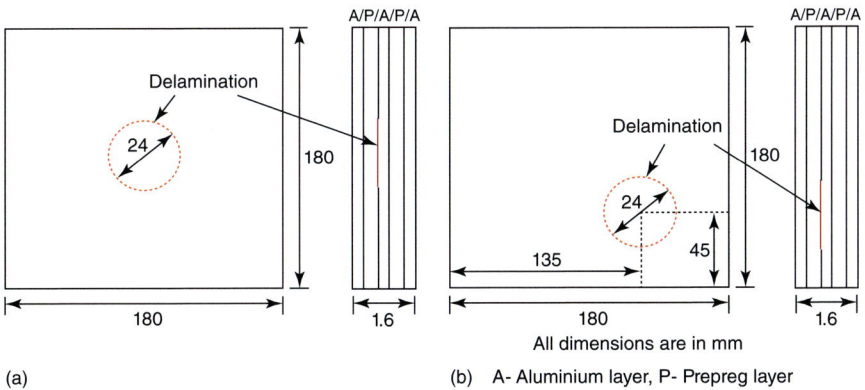

Figure 8.2 Schematic of GLARE plate with circular delamination at the (a) center and (b) $x = 135\,\mathrm{mm}$ and $y = 45\,\mathrm{mm}$.

Table 8.2 Material properties of the GLARE model.

Material	Mass density (kg/m³)	Modulus of rigidity, E (GPa)	Poisson ratio, v
Al 2024-T3	2780	73.1	0.33
S2/FM94	1980	52.0	0.25
GLARE	2362	62.1	0.29

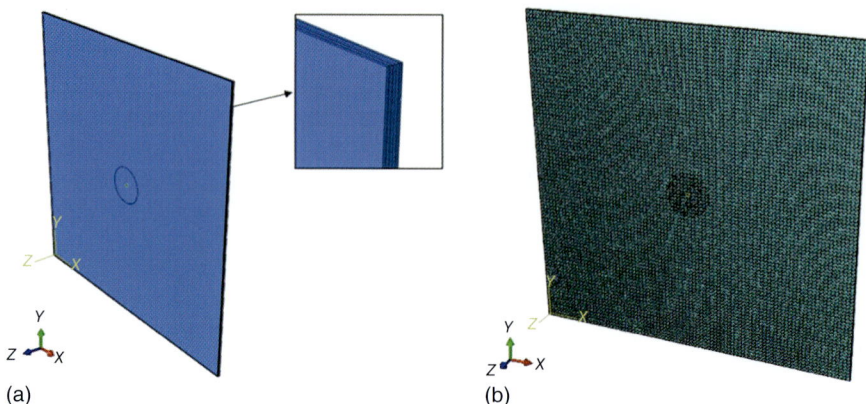

(a)   (b)

Figure 8.3 (a) 3D model and (b) meshing of the GLARE plate consisting of delamination generated in ABAQUS.

orientation of the S2-glass fiber is considered to be $0°$ throughout the model of the GLARE plate.

The model of the GLARE plate generated in ABAQUS and an enlarged view of its cross section are shown in Figure 8.3a. The elements used on the plate for analysis purpose are quadratic tetrahedral C3D10 and C3D10M with 10 nodes for steady state dynamic analysis and explicit dynamic analysis, respectively. The meshing of the GLARE plate model is illustrated in Figure 8.3b.

8.2.3 Determination of LDR Frequency from Steady State Analysis

The theoretical model of the LDR described in the previous section is used to determine the LDR frequency of the circular delamination in the GLARE plate. The fundamental LDR frequency (f_{LDR}) is obtained at 27.12 kHz from the analytical relation. In order to verify the analytical solution with the numerical model generated using the ABAQUS software, a steady state dynamic analysis is performed for both the GLARE plate models with different positions of circular delamination.

Three sets of points are selected in the GLARE model, namely transmitter, receiver, and delamination (at the center of the defect area), which act as the point of reference for extracting the results. The forcing frequency has been applied with 10 N force on the transmitter point in negative z-direction, while the results are extracted at the receiver and delamination points. Both the models gave the same peak of interest for the LDR frequency at around 26.86 kHz with displacement of 2.65×10^{-4} m for the central defect and 2.85×10^{-3} m for the eccentric defect. Another peak is observed before the LDR frequency, which corresponds to a natural frequency of GLARE plate at 26.08 kHz. The plot for frequency vs. displacement of the GLARE plate obtained from the steady state dynamic analysis is shown in Figure 8.4. The deviation of the fundamental LDR frequency obtained from the analysis is calculated to be 0.96% from that of the theoretical value. Thus, the model is validated and the precision of the theoretical model in predicting the LDR frequency is observed.

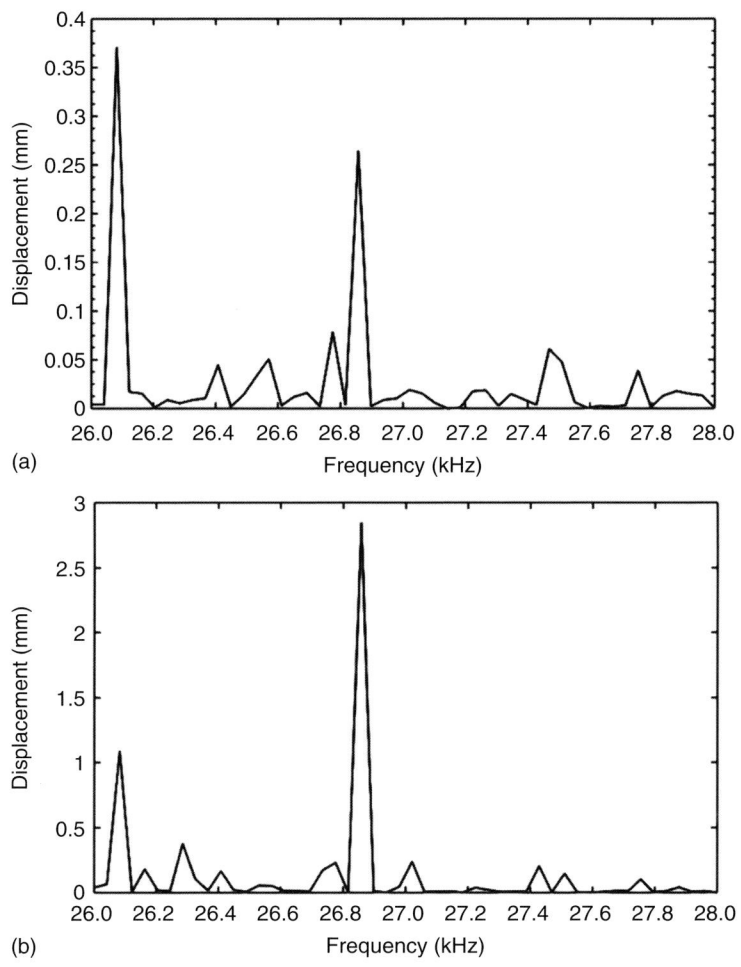

Figure 8.4 Frequency vs. displacement of the GLARE plate with delamination at (a) the center and (b) $x = 135$ mm and $y = 45$ mm.

8.3 Super-Harmonic and Subharmonic Excitation in Fiber Metal Laminates

The FML (GLARE) with delamination is further investigated under different super-harmonic and subharmonic LDR frequencies in order to study the non-linear effects arising due to intermodulation frequencies. Four different forcing frequencies (sinusoidal waveforms) $f=f_{LDR}/3, f=f_{LDR}/2, f=f_{LDR}$, and $f= 2f_{LDR}$ are used in order to observe the nonlinear effects arising in the GLARE model. The nonlinear effects due to the intermodulation and interaction terms are observed by performing the explicit dynamic analysis of the model in ABAQUS. The results for the explicit dynamic analysis are extracted at the receiver point, which is far away from the damage location. Figure 8.5 shows the nonlinear

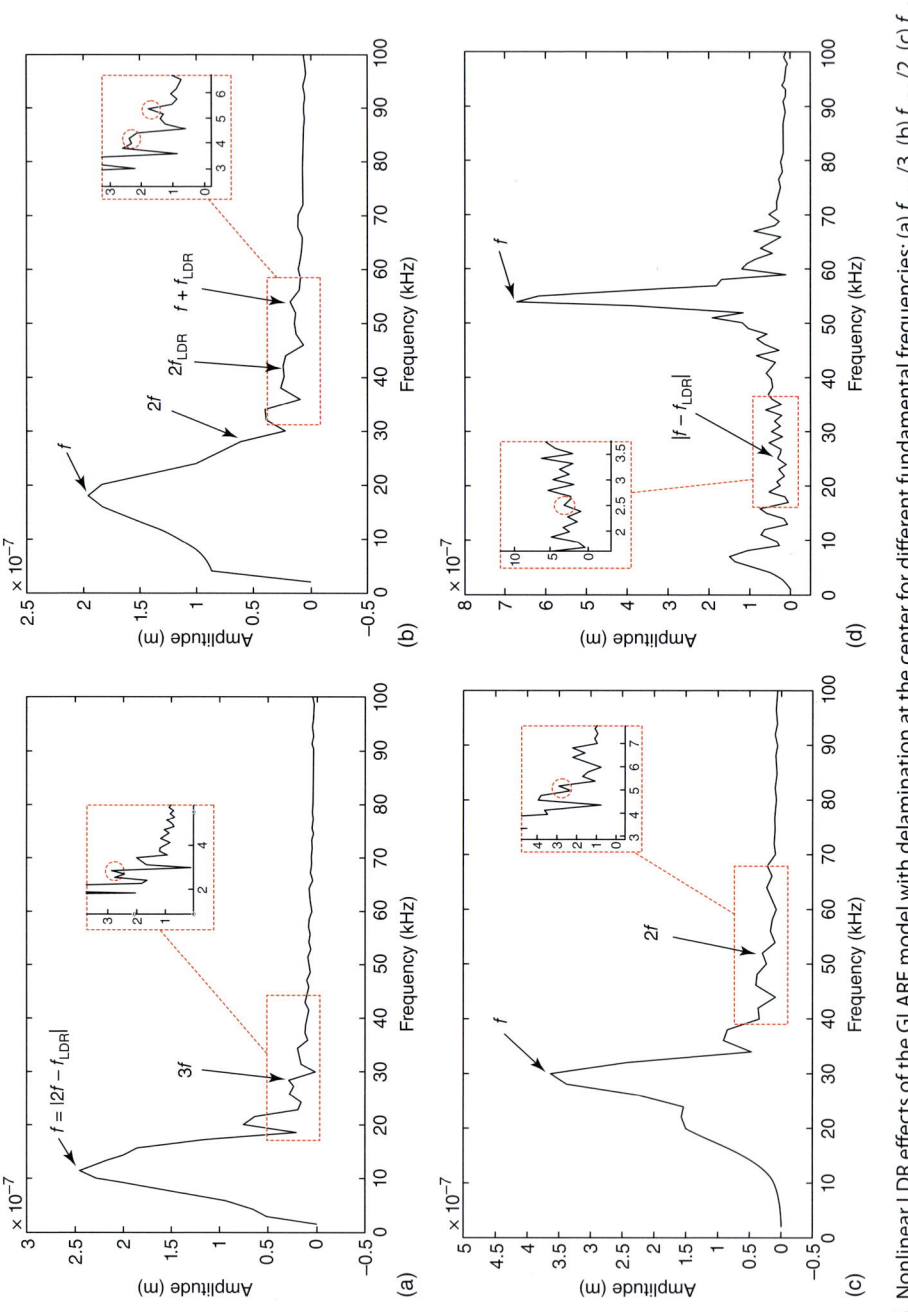

Figure 8.5 Nonlinear LDR effects of the GLARE model with delamination at the center for different fundamental frequencies: (a) $f_{LDR}/3$, (b) $f_{LDR}/2$, (c) f_{LDR}, and (d) $2f_{LDR}$.

LDR effect of the GLARE model with circular delamination at the center for the different forcing frequencies chosen. Figure 8.5a shows the amplitude of vibration corresponding to the driving frequency $f_{LDR}/3$ (56 234.5 rad/s) revealing the third order harmonic term $3f$. For the second forcing frequency $f_{LDR}/2$ (84 383.2 rad/s), the second order harmonic term $2f$, the super-harmonic term $2f_{LDR}$, and the intermodulation responses $f + f_{LDR}$, $|f - f_{LDR}|$ are obtained, as shown in Figure 8.5b. In case of the forcing frequency $f = f_{LDR}$, only the second order harmonic term $2f = 2f_{LDR}$ (168 766.4 rad/s) is obtained, as illustrated in Figure 8.5c. Finally, the forcing frequency corresponding to the second harmonic frequency of the obtained LDR, $2f_{LDR}$ (337 532.7 rad/s), is applied for the explicit dynamic analysis. It was observed that only the intermodulation response term $|f - f_{LDR}|$ is obtained from the analysis (Figure 8.5d). Thus, it is clear that by applying forcing frequency at LDR in super-harmonic and subharmonic range, we can observe the nonlinear effects arising due to intermodulation and interaction responses.

Similarly, the nonlinear effects of the other model with delamination at the center and at $x = 135$ mm and $y = 45$ mm are determined, by repeating the analysis for the four different forcing frequencies. A similar trend of plots for the amplitude vs. frequency is observed in the case of the second model, as illustrated in Figure 8.6. The peaks for the higher harmonics and the intermodulation terms can be better interpreted if the frequency resolution is decreased to around 10–100 Hz. This may lead to an increase in the computation time by more than 10 times and so may require a high configuration computational workstation for carrying out the analysis.

Thus, the analysis performed was found to be efficient in predicting the nonlinear elastic effects and the nonlinear defect resonance intermodulation for a nonlinear LDR response. It is evident that by using continuous periodic excitations, nonlinear effects of the defect can be observed from locations other than that of the defect region. However, the excitation of GLARE plate using super-harmonic and subharmonic LDR frequencies shows a maximum peak at the forcing frequency only. The peak corresponding to the actual LDR frequency and its second order harmonic is not so distinct in Figures 8.5 and 8.6, respectively. This difficulty in LDR frequency detection can be overcome by implementing bicoherence analysis of the receiver signal. The technique of bicoherence analysis for LDR frequency detection is discussed in Section 8.4.

8.4 Detection of LDR Frequency Using Bicoherence Analysis

8.4.1 Theory of Bicoherence Estimation

The bicoherence estimate is a normalized bispectrum, which is sensitive to the nonlinearity arising due to interaction between the incident wave frequency and the LDR frequency. Bispectrum estimation consists of two different frequency indicators that are perpendicular to each other while the bicoherence value forms the third axis perpendicular to both the indicator axes. The bicoherence value can

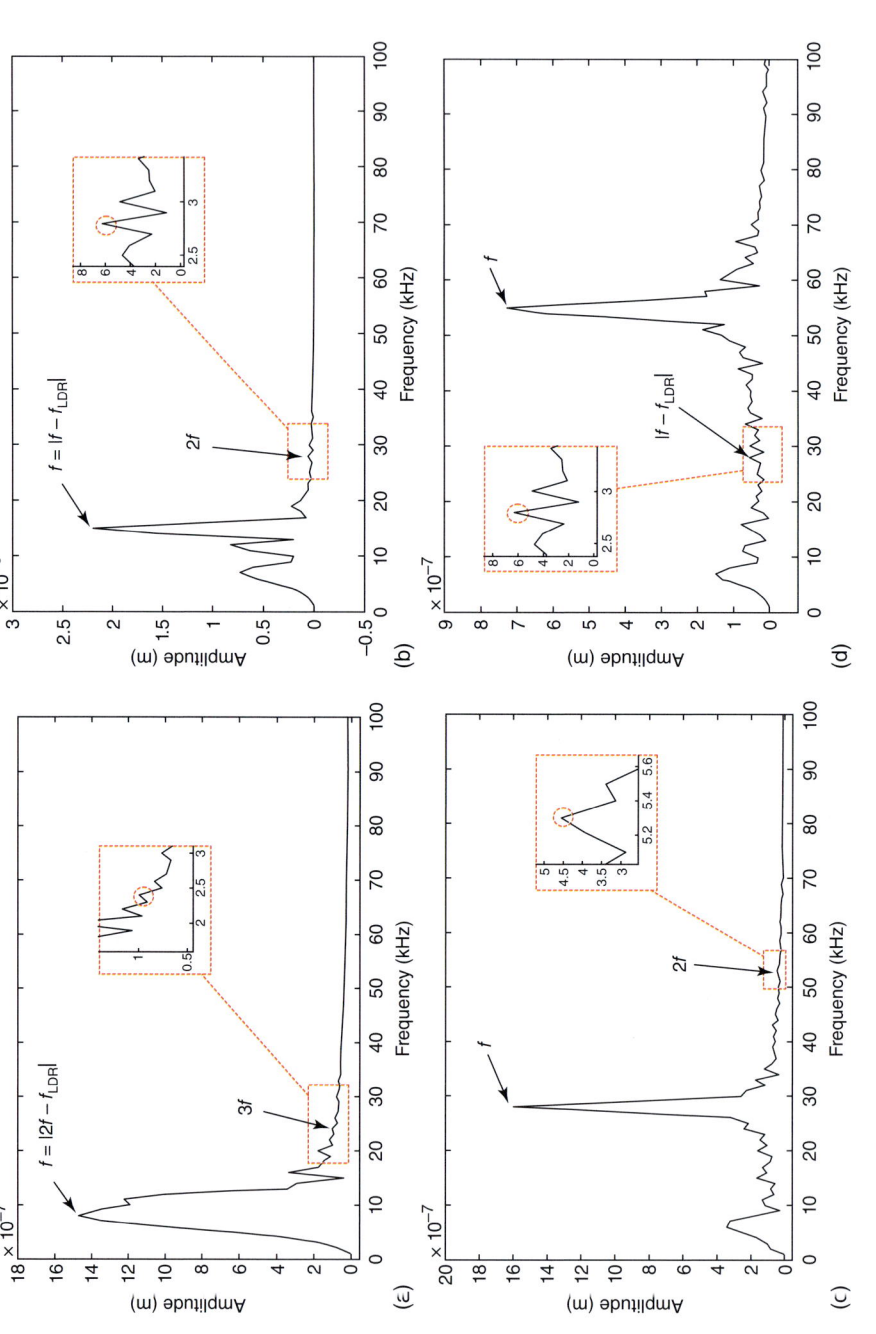

Figure 8.6 Nonlinear LDR effects of the GLARE model with delamination at $x = 135$ mm, $y = 45$ mm for different fundamental frequencies: (a) $f_{LDR}/3$, (b) $f_{LDR}/2$, (c) f_{LDR}, and (d) $2f_{LDR}$.

range from 0 to 1 according to the amount of quadratic phase coupling (QPC) present in the system. The bicoherence estimator is derived from the power spectrum relations. The power spectrum with time series $p(t)$ is signified by performing the discrete Fourier transformation (DFT) as [23]

$$S_{xx}(f) = E[P(f)P^*(f)] \tag{8.6}$$

where E is the expectation operator and x is the specific frequency value.

The DFT of this signal is known as the bispectrum, which is given by [23]

$$B_x(f_1,f_2) = D[E(f_1)E(f_2)E^*(f_1 + f_2)] \tag{8.7}$$

The variance of the bispectral estimate is dependent on the energy of the signal, which causes complications in the estimate. These complications are solved by normalizing the bispectrum to get the skewness function, which is not dependent on the energy of the signal. The skewness function can be written as

$$s^2(f_1,f_2) = \frac{|D[B(f_1,f_2)]|^2}{D[V(f_1)]D[V(f_2)]D[V(f_1 + f_2)]} \tag{8.8}$$

where $V(f_n)$ is the variance of the bispectral estimator.

One of the demerits of this normalization is that the magnitude of skewness function is not confined and carries undesired variance properties. It can be resolved by dividing the bispectrum with the power spectrum. This technique of normalization is known as the bicoherence estimation and is denoted by b^2. The bicoherence estimation of the signal is obtained from the relation given by Eq. (8.9) [23].

$$b^2(f_1,f_2) = \frac{|D[E(f_1)E(f_2)E^*(f_1 + f_2)]|^2}{D[|E(f_1)E(f_2)|^2]D[|E(f_1 + f_2)|^2]} \tag{8.9}$$

The bispectral plane obtained during the analysis has 12 symmetric planes [20]. The bispectral information from these symmetrical planes is redundant and any one of the planes is enough to study the nonlinearity arising due to QPC. The other 11 regions in the bicoherence plane are repetitive in nature and can be neglected while performing the bicoherence analysis. Each symmetric plane is triangular in shape and is known as the primary domain of the bicoherence estimation (Figure 8.7). The primary domain is further divided into two regions: inner domain (ID) and outer domain (OD) [23]. All the points in the bicoherence plane fall under the Nyquist frequency range, which is half of the sampling frequency. The QPC due to second order nonlinearity arising because of interaction between two frequencies can be detected using bicoherence estimate. Bicoherence provides the information of both magnitude and phase of measured signals [7].

8.4.2 Case Study of a Flat Bottom Hole

8.4.2.1 Modeling of a Flat Bottom Hole

As discussed in Section 8.2.1, delamination can be considered analogous to FBH in order to determine the LDR frequency of the defect. In this section, a defect

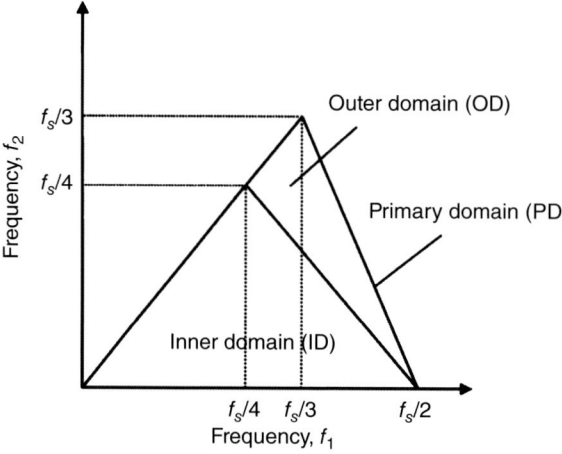

Figure 8.7 Non-redundant zone of the bispectral plane known as the primary domain.

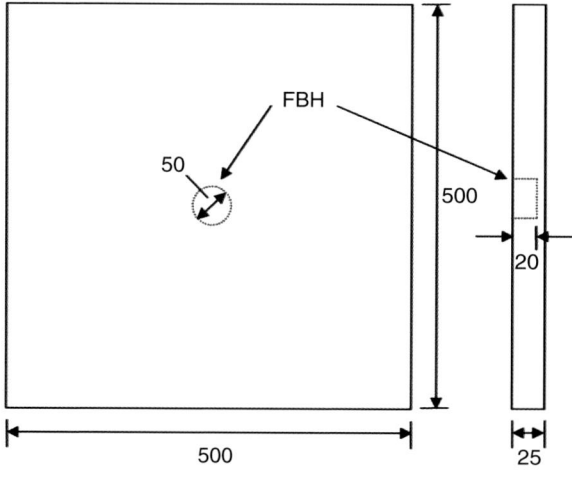

Figure 8.8 Schematic of the circular FBH model.

All dimensions are in mm

in the form of FBH has been modeled using the ABAQUS software. The dimension of the plate considered for modeling the FBH is $500 \times 500 \times 25$ mm^3 having a circular FBH. The position of the circular FBH is considered to be at the center (Figure 8.8). The radius of the FBH (r) is considered to be 25 mm and the depth of FBH (h) is 20 mm. Thus, the residual thickness of the defect area (t) is found to be 5 mm. The material properties considered for modeling the FBH are shown in Table 8.3.

Table 8.3 Material properties.

Mass density (kg/m³)	Modulus of elasticity, E (GPa)	Poisson ratio, v
2580	72	0.33

8.4.2.2 Determination of LDR Frequency Using Fast Fourier Transform (FFT) Plot

At first, the LDR frequency has been calculated analytically according to the relation given in Eq. (8.5). The LDR frequency estimated by the analytical model is found to be 20.68 kHz. This helps in determining the range of sweep excitation to be provided to the model in order to determine the LDR frequency. In this study, a chirp signal is used to excite the FBH model at a sweeping range of 10–40 kHz while performing the explicit dynamic analysis. The sampling frequency (f_s) used during the analysis is 5 MHz with a time increment of 2×10^{-7} s. The output signal is extracted from a receiver node away from the FBH location. The output signal obtained from the explicit dynamic analysis in ABAQUS is then processed using MATLAB. The frequency spectrum of the output signal is obtained by performing the FFT of the signal, as shown in Figure 8.9. From the figure, it is observed that the maximum amplitude of vibration is 28.11 kHz, which corresponds to the LDR frequency of the FBH. The FFT plots show multiple peaks at the vicinity of LDR frequency, which makes the identification of LDR frequency a tedious task. Therefore, an automated technique of bicoherence analysis is performed on the receiver signal to obtain the second order harmonic of LDR frequency.

8.4.2.3 Determination of LDR Frequency Using Bicoherence Analysis

The bicoherence estimation technique determines the second order harmonic of LDR frequencies corresponding to the FBH. The peaks obtained in the bicoherence analysis are dependent on the DFT and segment length. The variation of normalized frequency with respect to DFT length has been illustrated in Figure 8.10. In the figure, F_L is the exact LDR frequency and F_0 is the frequency obtained from the bicoherence analysis at a specific set of DFT and segment length. The DFT length and segment length used in order to obtain a normalized frequency

Figure 8.9 Frequency spectrum of the output signal showing maximum amplitude of vibration at the LDR frequency.

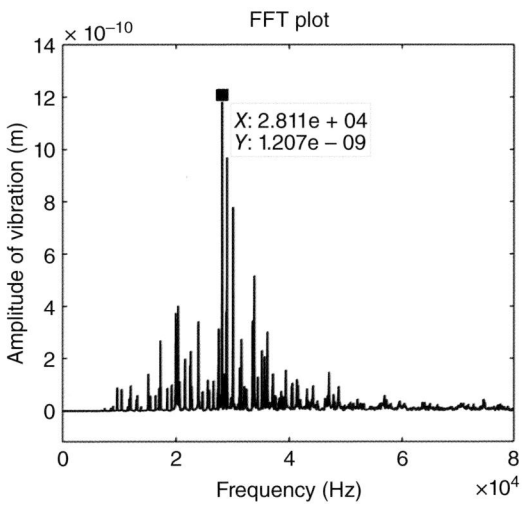

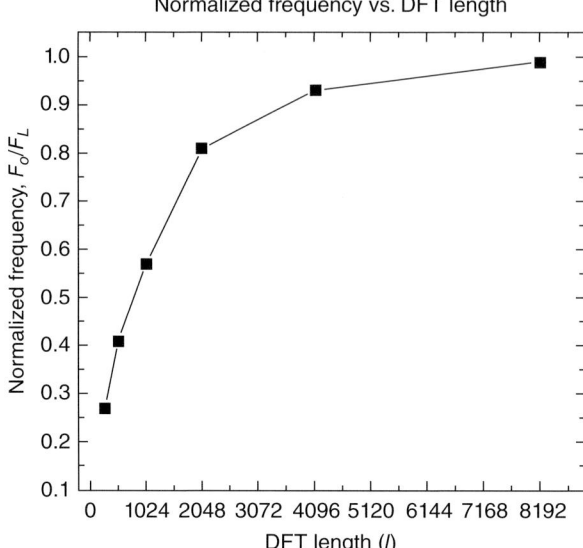

Figure 8.10 Variation of normalized frequency with DFT length in the case of numerical simulation.

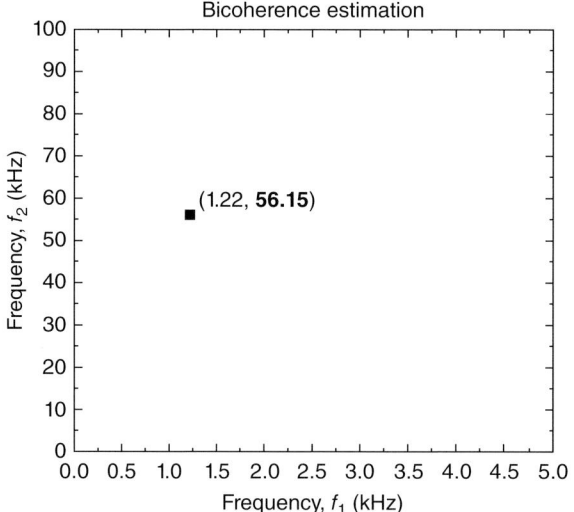

Figure 8.11 Bicoherence plot of the receiver signal showing second order harmonic of the LDR frequency at 56.15 kHz.

equal to unity are 4096 and 2048, respectively. The bicoherence value gets saturated after a DFT length of 4096 and there is negligible change in the normalized frequency of the model.

The bicoherence plot corresponding to chirp signal excitation in the case of an FBH is shown in Figure 8.11. Here, the second order harmonic LDR corresponding to FBH is found to be 56.15 kHz, which is approximately twice the LDR frequency obtained from the FFT plot. This shows the advantage of bicoherence analysis compared to FFT plots, in which the second order harmonic of the defect

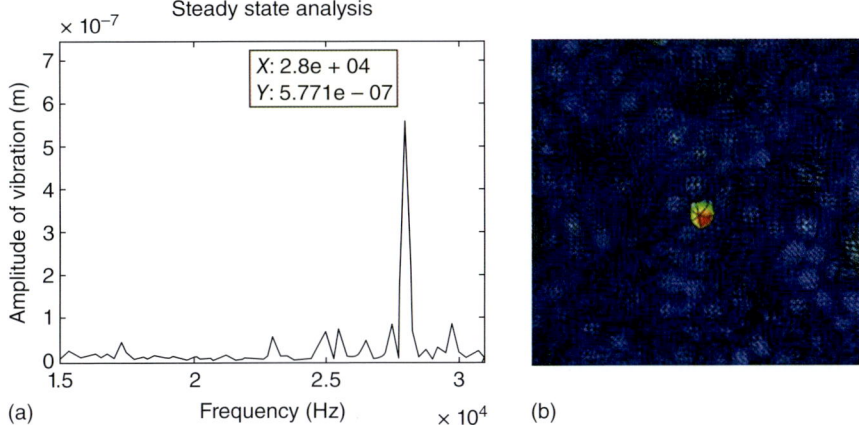

Figure 8.12 In the case of circular FBH at the center (a) steady state analysis showing LDR frequency at 28 kHz and (b) mode shape of the plate at the LDR frequency of 28 kHz.

frequency is effectively determined. Thus, the LDR frequency estimation process can be automated easily using the bicoherence analysis.

8.4.2.4 Validation of the LDR Frequency Using Steady State Analysis

The LDR frequencies obtained from the FFT plots as well as the bicoherence estimation plot in the previous sections are validated using the steady state analysis. A steady state analysis is performed using a sweep frequency of 15–35 kHz in order to determine the LDR frequency of the FBH. Figure 8.12 shows the LDR frequency corresponding to the FBH along with its mode shape at that particular frequency.

The LDR frequency for the circular FBH obtained from the steady state analysis is found to be 28 kHz, which is almost similar to the results obtained from the FFT plot. Twice the LDR frequency obtained in steady state plot is 56 kHz, which is also in close agreement with the results obtained from the bicoherence analysis. The mode shape corresponding to the LDR frequency is shown in Figure 8.12b. The mode shape shows maximum amplitude of vibration only at the FBH location when subjected to LDR frequency. The mode shapes corresponding to other peaks in the steady state plot are corresponding to different natural frequencies of the plate and are not related to the defect. Thus, the detection of LDR frequency using FFT and bicoherence plots has been verified and an automated technique of LDR frequency detection is established.

8.5 Concluding Remarks

In this chapter, a brief description of the theoretical model for LDR frequency and the bicoherence estimation technique are discussed. The LDR frequency is estimated using the analytical relation that is used to decide a range for the frequency sweep of the chirp signal on the model. A study based on super-harmonic and subharmonic LDR frequency excitation on an FML has been described. Two different configurations of circular delamination are investigated on a GLARE plate. The nonlinear effects due to intermodulation and interaction responses are observed at four different forcing frequencies in the form of continuous sinusoidal periodic excitation. It is observed that the nonlinear effects arising due to the presence of the defect in a composite material can be extracted from any location of the plate, which is far away from the defect area. However, the excitation of GLARE plate using super-harmonic and subharmonic LDR frequencies does not show distinct peaks corresponding to the actual LDR frequency and its second order harmonic. Moreover, the output signal obtained by performing explicit dynamic analysis is used to obtain the exact LDR frequency from the FFT plot. In the case of the FFT plot, multiple peaks in the vicinity of the LDR frequency are observed. The identification of LDR frequency in such a case is a cumbersome task and must be done manually. In order to overcome this difficulty in LDR frequency detection, an automated technique based on bicoherence analysis of the receiver signal is implemented. The bicoherence estimation is found to be very sensitive to quadratic nonlinearity and shows the second order harmonic frequency of the LDR. Thus, it can be concluded that the bicoherence estimation technique can be used for an automated detection of LDR frequency in any material. The technique of bicoherence estimation can be further used for location of defects in the case of composites and FML. The concept of LDR frequency and its detection using bicoherence analysis has a vast scope in post life cycle assessment of a composite material.

References

1 Satishkumar, T.P., Satheeshkumar, S., and Naveen, J. (2014). Glass fiber-reinforced polymer composites-a review. *Journal of Reinforced Plastics and Composites* 33 (13): 1258–1275.

2 Debnath, K. and Singh, I. (2017). Low-frequency modulation-assisted drilling of carbon-epoxy composite laminates. *Journal of Manufacturing Processes* 25: 262–273.

3 Vinay, H.B., Govindaraju, H.K., and Banakar, P. (2015). Processing and characterization of glass fiber and carbon fiber reinforced vinyl ester based composites. *International Journal of Research in Engineering and Technology* 4 (5): 401–406.

4 Guocai, W. and Yang, J.M. (2005). The mechanical behavior of GLARE laminates for aircraft structures. *The Journal of the Minerals, Metals and Materials Society* 57 (1): 72–79.

5 Lima, W.J.N.D. and Hamilton, M.F. (2003). Finite-amplitude waves in isotropic elastic plates. *Journal of Sound and Vibration* 265: 819–839.

6 Ciampa, F., Scarselli, G., and Meo, M. (2017). On the generation of nonlinear damage resonance intermodulation for elastic wave spectroscopy. *Journal of the Acoustical Society of America* 141: 2364–2374.

7 Ciampa, F., Pickering, S., Scarselli, G., and Meo, M. (2016). Nonlinear imaging of damage in composite structures using sparse ultrasonic sensor arrays. *Structural Control and Health Monitoring* 24: 1–13.

8 Meo, M. and Zumpano, G. (2005). Nonlinear elastic wave spectroscopy identification of impact damage on sandwich plate. *Composite Structures* 71: 469–474.

9 Solodov, I., Bai, J., Bekgulyan, S., and Busse, G. (2011). A local defect resonance to enhance acoustic wave-defect interaction in ultrasonic non-destructive evaluation. *Applied Physics Letters* 99: 211911.

10 Solodov, I., Bai, J., and Busse, G. (2013). Resonant ultrasound spectroscopy of defects: case study of flat-bottomed holes. *Journal of Applied Physics* 113: 223512.

11 Solodov, I., Rahammer, M., Gulnizkij, N., and Kreutzbruck, M. (2016). Noncontact sonic NDE and defect imaging via local defect resonance. *Journal of Nondestructive Evaluation* 35 (48): 1–8.

12 Fierro, G.P.M., Calla, D., Ginzburg, D. et al. (2017). Nonlinear ultrasonic stimulated thermography for damage assessment in isotropic fatigued structures. *Journal of Sound and Vibration* 404: 102–115.

13 Hettler, J., Tabatabaeipour, M., Delrue, S., and Aveele, K.V.D. (2017). Detection and characterization of local defect resonances arising from delaminations and flat bottom holes. *Journal of Nondestructive Evaluation* 36 (2): 1–10.

14 Solodov, I., Dillenz, A., and Kreutzbruck, M. (2017). A new mode of acoustic NDT via resonant air-coupled emission. *Journal of Applied Physics* 121: 245101.

15 Rahammer, M. and Kreutzbruck, M. (2017). Fourier-transform vibrothermography with frequency sweep excitation utilizing local defect resonances. *NDT and E International* 86: 83–88.

16 Dionysopoulos, D., Fierro, G.P.M., Meo, M., and Ciampa, F. (2018). Imaging of barely visible impact damage on a composite panel using nonlinear wave modulation thermography. *NDT and E International* 95: 9–16.

17 Dyrwal, A., Meo, M., and Ciampa, F. (2018). Nonlinear air-coupled thermosonics for fatigue micro-damage detection and localisation. *NDT and E International* 97: 59–67.

18 Rivola, A. and White, P.R. (1998). Bispectral analysis of the bilinear oscillator with application to the detection of fatigue cracks. *Journal of Sound and Vibration* 216 (5): 778–809.

19 Raghuveer, M.R. and Nikias, C.L. (1985). Bispectrum estimation: a parametric approach. *IEEE Transactions on Acoustics, Speech, and Signal Processing* 4: 869–891.

20 Kim, Y.C. and Powers, E.J. (1979). Digital bispectral analysis and its applications to nonlinear wave interactions. *IEEE Transactions on Plasma Science* 7 (2): 120–131.

21 Collis, W.B., White, P.R., and Hammond, J.K. (1998). Higher-order spectra: the bispectrum and trispectrum. *Mechanical Systems and Signal Processing* 12 (3): 264–283.

22 Nichols, J.M. and Olson, C.C. (2010). Optimal bispectral detection of weak, quadratic nonlinearities in structural systems. *Journal of Sound and Vibration* 329: 1165–1176.

23 Fackrell, J.W.A., White, P.R., Hammond, J.K. et al. (1995). The interpretation of the bispectra of vibration signals-I. Theory. *Mechanical Systems and Signal Processing* 9 (3): 146–155.

9

Secondary Processing of Reinforced Polymer Composites by Conventional and Nonconventional Manufacturing Processes

Manpreet Singh[1], Sarbjit Singh[1], and Parvesh Antil[2]

[1] Punjab Engineering College (Deemed to be University), Department of Mechanical Engineering, Chandigarh, 160012, India
[2] College of Agricultural Engineering & Technology, Chaudhary Charan Singh Haryana Agricultural University, Department of Basic Engineering, Hisar, 125004, Haryana, India

9.1 Introduction

In the past few decades, reinforced polymer matrix composites (PMCs) materials have becomes popular in the fields of aircraft, marine, automobiles, and sports equipment due to their better mechanical properties such as high strength to weight and stiffness to weight ratios. The PMCs mainly comprise of carbon fiber reinforced polymer (CFRP) [1], glass fiber reinforced polymer (GFRP) [2], and fiber metal laminates (FMLs) composite [3]. Fiber reinforced PMCs materials are formed by combination of reinforcing polymer and polymer matrix. The superior characteristics of fibers include light weight, stiffness, weight reduction, and high strength of composites. The orientation of fiber reinforcement in composites is unidirectional or bidirectional [4]. Composites formed by unidirectional fiber orientation have high strength and stiffness in fiber direction, whereas the direction perpendicular to fiber orientation exhibits low strength and stiffness, i.e. anisotropic nature. On the contrary, the composites formed by bidirectional fibers produce uniform strength and stiffness in every direction.

The polymer matrix used in composites provides binding and environmental protection to fibers. Basically, the polymer matrix is of two types, i.e. thermoplastic and thermosetting. Thermoplastics can be easily shaped under application of heat, whereas thermosets require cross-linking curing. Apart from CFRP and GFRP, researchers developed advanced composites with superior mechanical properties, i.e. FMLs composite. The FMLs are formed by binding metal thin alloy sheet to plies of reinforced PMC. FMLs include glass reinforced aluminum laminate, carbon reinforced aluminum laminate, carbon reinforced titanium laminate, etc. [3, 5].

Secondary processing of composites deals with machining of composites by conventional and nonconventional machining processes to get the required geometrical products with better tolerance. Conventional machining processes such as drilling, milling, and turning are successively attempted by different researchers for secondary processing of composites. However, composite characteristics such as anisotropy, hardness of fibers, fiber arrangement, and orientation make them difficult to machine and cause surface damage, fiber pullouts, and delamination [6, 7]. Several undesirable damages to machining products reduce the long-term performance of the composites. Among the conventional machining processes, drilling became popular owing to its ability to produce delamination-free products. However, nonconventional machining processes such as electrical discharge machining (EDM), electrochemical machining (ECM), water jet machining (WJM), abrasive jet machining (AJM), and electrochemical discharge machining (ECDM) have been successively attempted by researchers for the secondary processing of composites. This chapter addresses the secondary processing of composites with different domains such as conventional machining and nonconventional machining including machined surface characteristics such as delamination, cracks, and surface damages.

9.2 Secondary Processing of Reinforced Polymer Matrix Composites by Conventional Machining

The manufacturing of composites is grouped into two processes, primary and secondary processing. Primary processing deals with the conversion of raw material into simple shapes, whereas secondary processing involves machining of composites to achieve the desired shape with tolerance. Primary processing is carried out by different techniques such as hand lay-up, compression molding, and pultrusion [8]. On the other hand, secondary processing involves machining of composites by conventional and nonconventional machining processes. In conventional machining processes, the drilling process remains popular because a small aircraft requires 100 000 holes for fastening [9, 10]. The following section presents a comprehensive state-of-the-art review on secondary processing of composites by drilling, milling, and turning.

9.2.1 Conventional Drilling of Reinforced Polymer Matrix Composites

The drilling behavior of composites materials is quite different than that of metallic materials. While machining, the drill has to be passed through alternative layers of fiber (reinforcement) and polymer (matrix). Hence, the machining mechanism becomes complex due to different chemical and physical properties of reinforcement and matrix. The machining mechanism in drilling of composites is due to a number of fractures caused by irregular load sharing between the matrix and reinforcement [11]. This tends to generate an uneven and damaged surface around a drilled hole. The uneven damaged surface around the drilled hole suppresses the performance characteristics of the composite product in the respective applications. The uneven surface damage to drilled holes are characterized by different researchers as delamination (failure of inter-laminar bond), fiber pullout (failure of matrix reinforcement bonding), cracks, burning, spalling (uncut fibers), etc. [12–16]. Apart from different damages, the

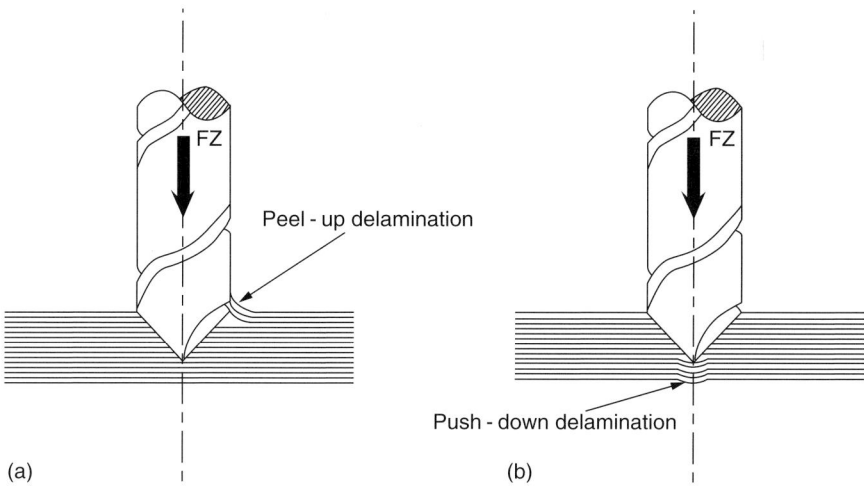

Figure 9.1 (a) Peel up delamination and (b) push down delamination. Source: Singh et al. 2013 [19]. Reproduced with permission of Elsevier.

delamination of a drilled hole is of great interest to researchers since nearly 60% of the rejection of parts in aircraft assembly relates to delamination [17]. The delamination mechanism in drilling of reinforced PMCs is referred as peel up (at drill entrance) and push down (drill exit) [12, 13, 18, 19] as shown in Figure 9.1. While drilling, the laminated plies of composites are abraded due to contact with the cutting edge of the drill bit. With further advancement of the drill bit, the laminated plies are peeled up along flutes of drill as shown in Figure 9.1a. On the other hand, as the drill bit approaches the exit side the uncut thickness of the composite becomes very low. This tends to generate push down delamination caused by failure of interlamina bonding as shown in Figure 9.1b.

9.2.2 Grinding Assisted Drilling of Reinforced Polymer Matrix Composites

The delamination of drilled hole by conventional drilling is addressed by different researchers using a core drill bit bonded with PCD particle during the machining operation [20]. Grinding assisted drilling process uses a hollow drill bit coated with diamond particles as shown in Figure 9.2. Park et al. [20] revealed that the thrust force is lowered in drilling with the use of hollow grinding drill bit as compared to the conventional drill bit. Similarly, Tsao and Hocheng [21] and Jain and Yang [22] observed reduction in thrust force by drilling composites using core drill.

The delamination factor in drilling of composites is highly affected by excessive thrust force, i.e. force higher than the critical value. The excessive thrust force in drilling of composites causes high delamination [23]. Therefore, the delamination factor and uneven damage to the hole are least below the critical thrust force. Similarly, the drilling of CFRP composites by core drill produces better surface quality holes with minimum delamination [24, 25].

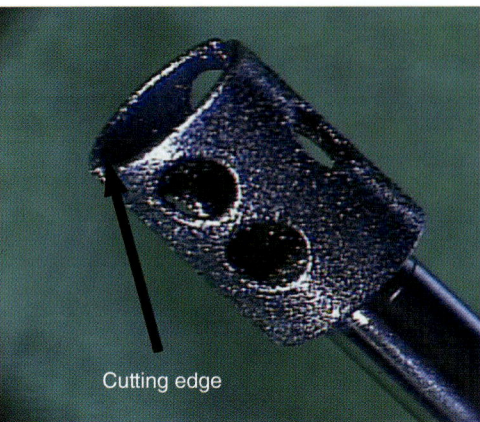

Figure 9.2 Core drill bit for drilling of composites. Source: Tsao and Hocheng 2007 [21]. Reproduced with permission of Elsevier.

9.2.3 Vibration Assisted Drilling (VAD) of Reinforced Polymer Matrix Composites

The challenges in delamination factor and uneven surface damage while machining fiber reinforced PMCs have been addressed by different researchers incorporating low amplitude vibrations to the drill bit [26, 27] and the work piece [28], known as vibration assisted drilling (VAD).The VAD process is intermittent in nature, whereas conventional drilling is a continuous machining process. The researchers concluded that thrust force is reduced to 20–30% in VAD of fiber reinforced PMCs as compared to conventional drilling [28, 29]. Therefore, the delamination factor and uneven damages to machining surface can be reduced using VAD.

9.2.4 High Speed Drilling (HSD) of Reinforced Polymer Matrix Composites

High speed drilling (HSD) has been attempted by different researchers for reducing delamination and surface damage in machining [30]. Rawat and Attia [30] revealed that combination of high speed and low feed rate in drilling of CFRP reduces delamination rate due to reduction in thrust force. Similarly, in various other studies researchers revealed that the delamination factor is minimized in drilling of CFRP with high speed, low feed rate, and low point angle [31–33].

9.2.5 Drilling of Reinforced Polymer Matrix Composites with Drill Bit Geometry of Different Materials

Drill bit geometry is an important aspect that highly influences the delamination factor and quality of the machined hole [34]. In the literature, researchers attempted drilling of fiber reinforced PMCs with different drill geometries such as twist drill, step drill, slot drill, brad drill, straight flute drill, Jo drill, trepanning tool, and U shape. Piquet et al. [35] studied drilling of carbon epoxy composites with twist drill geometry (4.8 diameter, twist angle 25°, rake and clearance

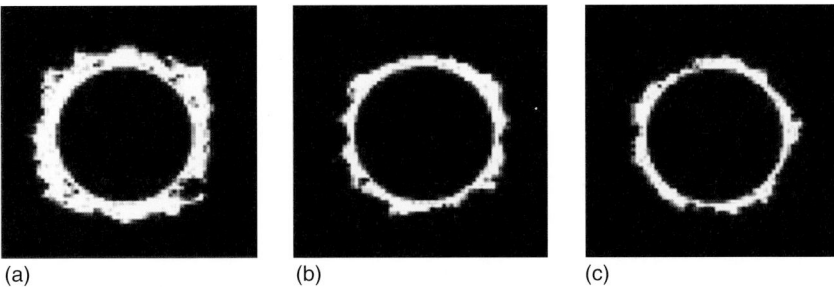

(a) (b) (c)

Figure 9.3 Delamination of hole drilled with (a) Twist drill, (b) candle stick drill, and (c) saw drill. Source: Tsao and Hocheng 2004 [24]. Reproduced with permission of Elsevier.

angle 6°) and special geometry drill (4.8 diameter, twist and rake angle 0°, clearance angle 6°, and three cutting edges). They concluded that holes machined with special geometry drill have less delamination. This is because of low contact between the drill bit and the composite laminate. Mathew et al. [36] investigated the effect of trepanning tool in the drilling of GPRP composite. The authors revealed that thrust force is reduced to 50% as compared to the conventional twist drill. The reduction in thrust force by trepanning tool causes low delamination. El-Sonbaty et al. [37] analyzed the effect of twist drill diameter on thrust force and delamination in drilling of GPRP composites. The increase in drill diameter results in high thrust force due to increase in shear area. Tsao and Hocheng [24] reported drilling of CFRP composites by using twist drill, candle stick drill, and saw drill. The ratio of delamination factor to holes machined with twist drill is high as compared to candle stick drill and saw drill as shown in Figure 9.3a–c. This is because of the difference in the cutting edges of drills.

Davim and Reis [15] studied the effects of drill geometries such as helical flute drill (high speed steel, HSS), helical flute K10 drill (cemented tungsten carbide), and four flute K10 drill (cemented tungsten carbide). They concluded that helical flute carbide drill gives superior performance in terms of delamination and tool wear. Therefore, this can be used as an optimum geometry for drilling of CFRP composites. Additionally, they observed cutting speed as the dominant factor for delamination in drilling of fiber reinforced PMCs as shown in Figure 9.4.

Debnath and Singh [38] evaluated the effect of carbide drill geometries such as four faceted drill, parabolic, and step drill (Figure 9.5) in drilling of CFRP composites. The four faceted drill geometries have minimum surface damage as compared to other geometries in drilling of CFRP. Therefore, the four faceted drill is highly recommended for drilling of CFRP composites. Likewise, in other studies Jo drill and eight faceted drill geometries gave superior machining performance due to their low thrust force [39].

Rakesh et al. [40] reported machining of GFRP by using modified drill geometries such as Jo drill, trepanning tool, U shape, and twist drill. They revealed that drilling induced delamination is highly dependent on drilling force, i.e. drill point geometry. The hole machined with twist drill geometry exhibits high delamination. On the other hand, the hole machined with U shape exhibits minimum surface damage, i.e. delamination at same parametric conditions.

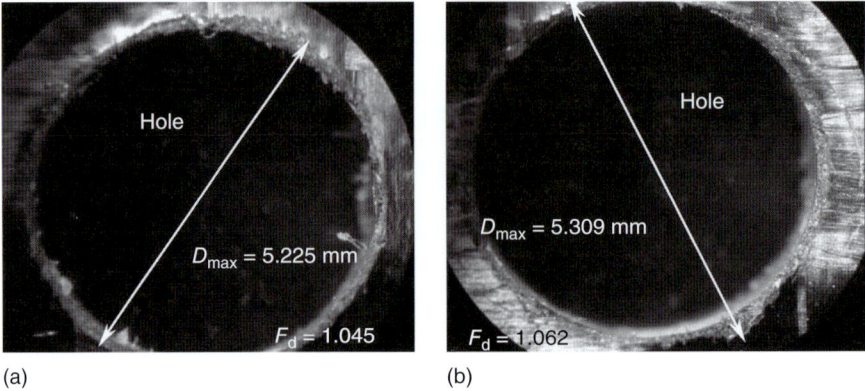

(a)   (b)

Figure 9.4 Delamination factor with variation in cutting speed: (a) $V = 16$ m/min and (b) $V = 32$ m/min. Source: Davim and Reis 2003 [15]. Reproduced with permission of Elsevier.

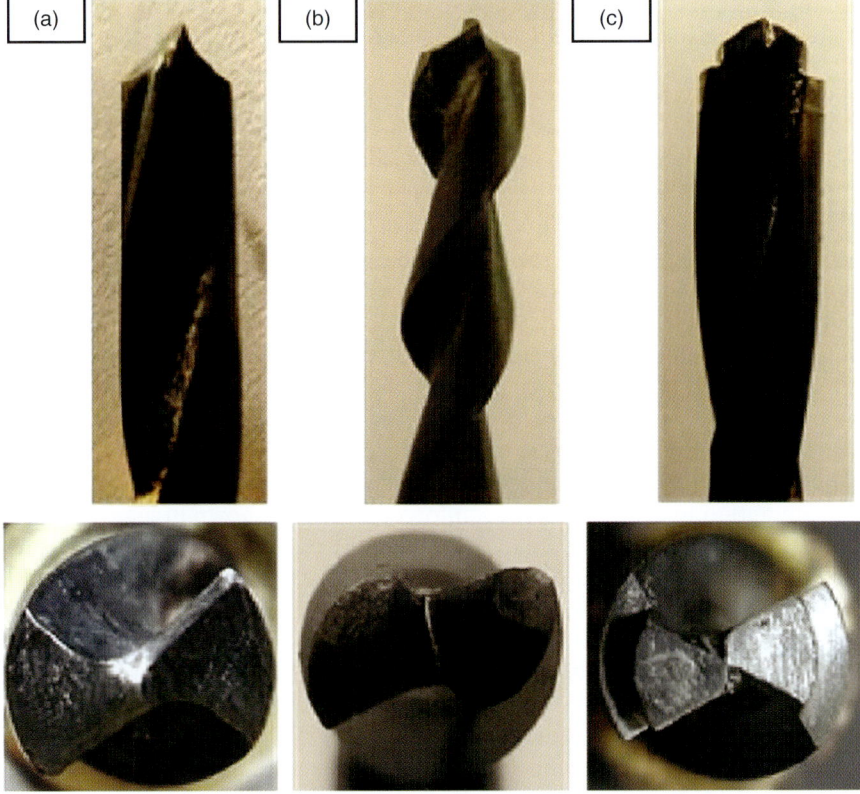

Figure 9.5 Different drill geometries for machining (a) four facet, (b) parabolic, and (c) step Source: Debnath and Singh 2017 [38]. Reproduced with permission of Elsevier.

9.2.6 Milling of Reinforced Polymer Matrix Composites

Milling process is highly recommended for trimming and cutting of composites to get the desired shape and tolerance. The milling of CFRP composite was initially attempted by Hocheng et al. [41], who addressed some experimental aspects of milling. The chips with different characteristics such as powder ribbon and brush-like shapes were produced during the cutting of composites. Additionally, having the cutting direction in parallel to fiber orientation is an optimal method for milling of composites. However, cutting at 45° and 90° to fiber orientation exhibits poor surface roughness owing to high cutting forces. Rahman et al. [42] analyzed the surface finish mechanism in milling of carbon/PEEK composite. They concluded that machined surface quality of composites at critical speed is far better than that of metallic materials. The cutting temperature at critical cutting speed is higher than the glass transition temperature of material to be cut. Therefore, the better surface finish is due to a rubbery regime in the machining zone.

Davim et al. [43] reported the effect of feed rate and cutting velocity on milling of GFRP composites. The delamination rate is significantly increased with increase in feed rate and cutting velocity in milling. However, the feed rate has a dominant effect on delamination and surface damage by milling. Likewise, in another study Davim and Reis [44] studied the milling of CFRP composite by two flute and six flute end mill. The authors concluded that surface characteristics of composite machined with two flute end mill are far better than six flute end mill due to low delamination. Inoue et al. [45] studied the behavior of end mill materials such as HSS, polycrystalline diamond (PCD), and tungsten carbide in the milling of CFRP. The PCD mill tool exhibited more tool life as compared to HSS and tungsten carbide mill tool. Azmi et al. [46] reported the cutting mechanism in milling of GFRP using input process parameters such as feed rate, depth of cut, and spindle speed. They concluded that fibers reinforcement is subjected to buckling and bending failure as the tool moves toward the fiber direction. Therefore, the surface below tool edge is fractured due to fiber and matrix failure. This finding is justified by scanning electron microscope (SEM) micrograph as shown in Figure 9.6. The brittle fractures of fibers are shown in Figure 9.6a,b, whereas Figure 9.6c,d exhibits fiber pullouts from the cutting surface due to fiber debonding and failure of matrix. Likewise, previous research work has shown that feed rate has a dominant effect on surface roughness in milling.

9.2.7 Turning of Reinforced Polymer Matrix Composites

The machining behavior of fiber reinforced composites in turning is quite different from that of metallic materials due to anisotropic nature. With growing need of composites in different applications various researchers explored machining of PMCs by turning process. Kim et al. [47] reported the effect of cutting speed and feed rate on surface roughness in the turning of CFRP composite. They concluded that the feed rate and fiber winding angle have a dominant effect on the surface roughness of the machined composite. In another study, Santhanakrishnan et al. [48] successfully attempted the turning of CFRP composite by using

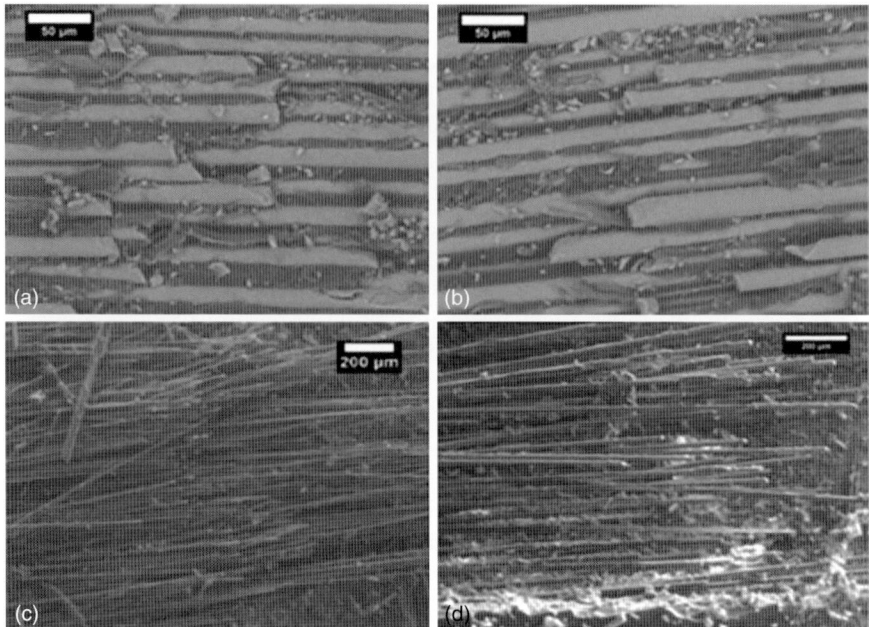

Figure 9.6 SEM micrographs of GFRP machined by milling. Source: Azmi et al. 2013 [46]. Reproduced with permission of Springer Nature.

carbide insert tool. The machined surface had uniform texture and less fiber pull-out caused due to crushing and sharp fracture of fibers.

Davim and Mata [49] reported surface roughness behavior in turning of GFRP composites, confirming the authors' previous research work [47, 48], which shows that surface roughness increases with increase in feed rate. Palanikumar et al. [50, 51] observed the surface roughness mechanism in the turning of GFRP. They concluded that surface roughness is better at 30° fiber orientation angle as compared to 90° angle. This is because of the generation of compressive strain in composites at a higher orientation angle. Additionally, the surface roughness is better at low feed rate and high cutting speed.

The following findings have been drawn from a comprehensive analysis of the secondary processing of fiber reinforced PMCs using conventional machining processes:

- The delamination factor in conventional drilling of fiber reinforced PMCs has been successfully addressed by using VAD, HSD, and variation in drill bit geometry.
- The VAD process gives superior performance in terms of minimum surface damage, delamination, and tool wear rate (TWR) due to reduction in thrust force up to 20–30%.
- Similarly to VAD, the HSD process is a promising method for secondary processing of PMCs. Moreover, the combination of low feed rate and drill point angle in HSD exhibits better quality machined surface with minimum delamination.

- The incorporation of special drill geometries such as Jo drill, eight faceted drill, four faceted drill, and parabolic drill are highly recommended for secondary processing of fiber reinforced PMCs due to minimum surface damage area.
- Cutting at 45° and 90° to fiber orientation exhibits poor surface roughness in the milling process owing to high cutting forces. Additionally, the feed rate and cutting velocity have a dominant effect on the surface quality of the machined surface.
- The use of two flute and six flute end mill in milling operation exhibits better quality machined surface due to minimum delamination.
- In turning of fiber reinforced PMCs, cutting speed and feed rate have dominant effects on surface roughness. Additionally, machining quality is better with fiber orientation at 30° as compared to 90° angle.

9.3 Secondary Processing of Reinforced Polymer Matrix Composites by Nonconventional Machining

The secondary processing of composites by conventional machining is a critical task due to tool failure and the anisotropic nature of material. Due to these factors, there is a requirement of manufacturing processes that overcomes these difficulties. Nonconventional manufacturing processes such as laser machining, ultrasonic machining (USM), abrasive water jet machining (AWJM), EDM, and ECDM are highly suitable for secondary processing of fiber reinforced PMCs. Some of the important aspects relevant in these processes have been discussed below.

9.3.1 Laser Beam Machining of Reinforced Polymer Matrix Composites

Laser beam machining (LBM) process is highly recommended for machining of fiber reinforced PMCs due to the thermal nature of the cutting process. The difficulties encountered in conventional machining process such as TWR and cutting forces are well addressed by incorporation of nonconventional machining processes. In LBM process, the focusing of a high energy beam on a small spot causes melting, vaporization, and chemical degradation of the polymer matrix material. Despite several advantages, the use of LBM process in industries is limited due to heat affected zones (HAZ), matrix charring, delamination, and fiber damage in machining [52]. Machining of GFRP by CO_2 laser process was first attempted by Tagliaferri et al. [53]. They revealed that the quality of the machined surface is highly dependent on cutting speed and the thermal properties of fibers. Later on, researchers compared machining of fiber reinforced PMCs using excimer laser (maximum power 95 W) and CO_2 laser (maximum power 1800 W) [54]. They found that a surface machined with excimer laser exhibits better surface characteristics such as low fiber damage and uniform matrix owing to the low thermal energy in machining regime. Uneven damages to a surface machined by gas CO_2 laser are well addressed by using solid state Nd:YAG laser (power less than 1000 W) [55]. The incorporation of Nd:YAG laser in machining of CFRP composites yields low HAZ and fiber damages caused by low thermal action.

Despite low productivity, the researchers compromised with the use of Nd:YAG laser due to better surface characteristics of the machined surface [56].

The generation of HAZ in machining of fiber reinforced PMCs by laser process leads to high rejection rate of products. The HAZ is caused by excess thermal energy in machining and variation in thermal properties of fiber and matrix material. The challenges to HAZ in machining of fiber reinforced PMCs were addressed by different researchers. Pan and Hocheng [57] studied the effects of anisotropic thermal properties of fibers on HAZ. They concluded that input laser energy has a dominant effect on the size of HAZ. Likewise, Uhlmann et al. [58] analyzed the effect of fiber orientation on HAZ in laser cutting. They revealed that the damaged area is maximum in the cutting direction perpendicular to fiber orientation. Fürst et al. [59, 60] reported machining of CFRP composite by using CO_2 and fiber laser. They revealed that simultaneous use of CO_2 and fiber laser source in CFRP machining exhibits better cutting rate and HAZ due to low interaction time. Apart from these, different researchers reduced HAZ in machining of CFRP by incorporating Nd–YAG laser [55, 61, 62]. Lau et al. [55] observed minimum HAZ in machining by Nd–YAG laser due to better focusing and high beam intensity. Mathew et al. [61] investigated the effects of process parameters in Nd–YAG laser cutting of CFRP. They revealed that pulse frequency and cutting speed have a dominant effect on HAZ. Moreover, at high pulse frequency, laser acts continuously and thus slows down the cooling process in machining. Likewise, in another study the pulse duration and cutting speed show a significant effect on HAZ [62]. Staehr et al. [63] concluded that HAZ in laser cutting is directly related to scanning speed. Scanning speed in turn affects the number of passes and the elapsed time for the laser to pass a point on the cutting path again. The elapsed time is dependent on scanning length and scanning speed. Long scanning length in laser cutting reduces HAZ owing to increase in the time period between two successive passes.

The challenges in HAZ for laser cutting of fiber reinforced PMCs are limited by the use of ultrashort pulsed laser sources [64–66]. These sources release power for a minimum time in machining, which overcomes HAZ because of minimum heat transfer to the machining zone. Leone et al. [65] successfully attempted machining of 0.5 mm thickness CFRP by using ultrashort pulsed laser (less than 100 W) with the multiple scanning technique. On the other hand, high power laser (more than 1 kW) can cut 3 mm thickness CFRP in 30 passes [67].

9.3.2 Ultrasonic Machining of Reinforced Polymer Matrix Composites

The USM is an optimum method for machining of hard and brittle materials. In this process, material removal occurs as a result of hammering actions caused by abrasive particles, i.e. imparting high frequency vibrations to the tool electrode [68]. In recent years, the machining of fiber reinforced PMCs has been explored by different researchers using different techniques such as ultrasonic drilling (UD) [69], rotary ultrasonic machining (RUM) [70–73], rotary ultrasonic elliptical machining (RUEM) [74], ultrasonic vibration cutting (UVC) [75], and rotary mode ultrasonic drilling (RMUD) [76]. Hocheng and Hsu [69] initially attempted the ultrasonic drilling of fiber reinforced carbon epoxy composites.

They revealed that the surface quality of a hole machined with UD is far better than that with conventional drilling with minimum delamination at hole entrance and exit. Cong et al. [70–72] concluded that RUM is a hybrid process that combines the process characteristics of diamond grinding and USM. A comparative analysis of RUM with ultrasonic drilling of CFRP exhibited low thrust force, surface roughness, and delamination. Additionally, the RUM tool can produce 1400 holes as compared to five holes produced by a conventional drilling tool before being worn out [70]. In other studies, researchers observed the effects of cutting fluid, cold air as coolant, and power consumption in RUM of CFRP [71, 72]. They concluded that surface roughness of RUM with cold air is higher as compared to that with cutting fluid. Additionally, cold air as coolant in RUM causes the machined surface to burn. Ding et al. [73] reported a comparative analysis of holes drilled with conventional drilling and RUM. They concluded that better quality holes were machined with RUM method due to reduction in drilling forces and torque while machining as shown in Figure 9.7.

Apart from these studies, Liu et al. [74] attempted RUEM of CFRP composites using core drill and elliptical tool vibrations. RUEM can produce holes of better surface quality with minimum delamination as compared to conventional drilling. This is because of the ultrasonic cleaning action of RUEM, which washes chips efficiently during machining. Likewise, Kim and Lee [75] observed the effect of ultrasonic vibration in cutting of CFRP as compared to ultrasonic cutting. They revealed that UVC at a speed below the critical value produced better surface characteristics. Debnath et al. [76] presented a new approach by rotating the work piece in ultrasonic drilling, named as RMUD for machining of GFRP composites. They revealed that hole wall machined with RMUD showed minimum exposed fibers as compared to conventional drilling. The RMUD process emerged as a proficient method for drilling of fiber reinforced PMCs with better surface characteristics.

9.3.3 Abrasive Water Jet Machining of Reinforced Polymer Matrix Composites

Among various methods, AWJM is a viable machining process for secondary processing of fiber reinforced PMCs [77]. The machining mechanism of AWJM involves thermally free operation, which shears off fibers with better surface characteristics. However, the machining of fiber reinforced PMCs leads to delamination and moisture absorption [78]. Delamination in the AWJM process is caused by shock waves arising due to high energy abrasive particles impinging on the machined surface [79]. The delamination factor in AWJM is characterized by different researchers using different hypothesis and techniques. Mayuet et al. [80] studied the delamination mechanism in AWJM of CFRP composites by varying different process parameters. They revealed that abrasive particles size and flow rate have a dominant effect on delamination. Delamination rate increases with increase in abrasive flow rate and abrasive size due to high energy shock waves. Shanmugam et al. [81] concluded that shock waves in AWJM initiate cracks on the machined surface. Furthermore, the impact of abrasive particles on cracks induces delamination on composites. In another study,

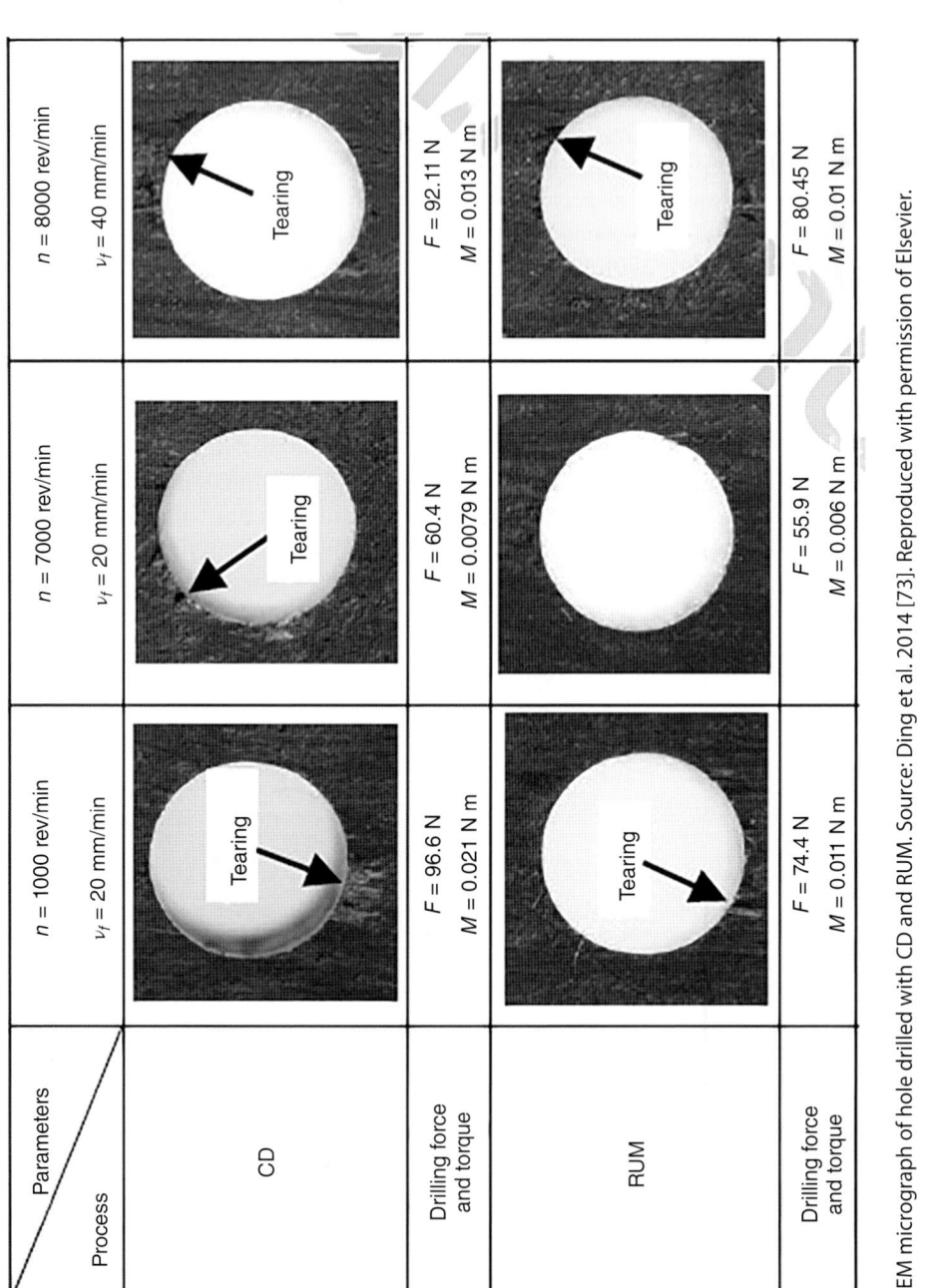

Figure 9.7 SEM micrograph of hole drilled with CD and RUM. Source: Ding et al. 2014 [73]. Reproduced with permission of Elsevier.

Phapale et al. [82] revealed that with the use of low water pressure, abrasive mass flow rate in AWJM exhibits minimum delamination.

In addition to delamination, various researchers categorized the kerf geometry and surface roughness in secondary processing of fiber reinforced PMCs by AWJM. Azmir and Ahsan [83] analyzed the effect of abrasives on kerf taper ratio in AWJM of glass epoxy composites. The researchers conducted experiments with aluminum oxide and garnet abrasive particles. It has been observed that the aluminum oxide abrasives exhibited low kerf taper ratio as compared to garnet abrasives. Shanmugam and Masood [84] studied kerf taper angle in machining of glass epoxy composite by AWJM. The author(s) observed that low standoff distance (SOD), minimum transverse speed, and high water pressure reached optimum values at minimum kerf taper angle. In another work, Azmir and Ahsan [85] studied the surface quality of glass epoxy composite by varying different process parameters. They revealed that a composite machined at low SOD yields better surface quality. This is because of high kinetic energy at low SOD, which produces a smooth machined surface. Reddy et al. [86] studied machining of CFRP composite by AWJM. They concluded that high water pressure in machining affects MRR and causes severe failure. Likewise, Kakinuma et al. [87] observed similar behavior in machining of CFRP composite by AWJM. Alberdi et al. [88] studied the machinability index of CFRP composite by varying the thickness of the work piece in AWJM. They concluded that machinability index of composites is significantly affected by the thickness of the material. Ibraheem et al. [89] recommended machining of GFRP composites at low SOD, i.e. better quality machined surface. Saleem et al. [90] compared AWJM of CFRP composite with conventional machining. SEM micrograph showed that a surface machined with conventional machining exhibits fiber pullout and matrix degradation (Figure 9.8a). On the other hand, the surface machined with AWJM has small streaks and craters as shown in Figure 9.8b.

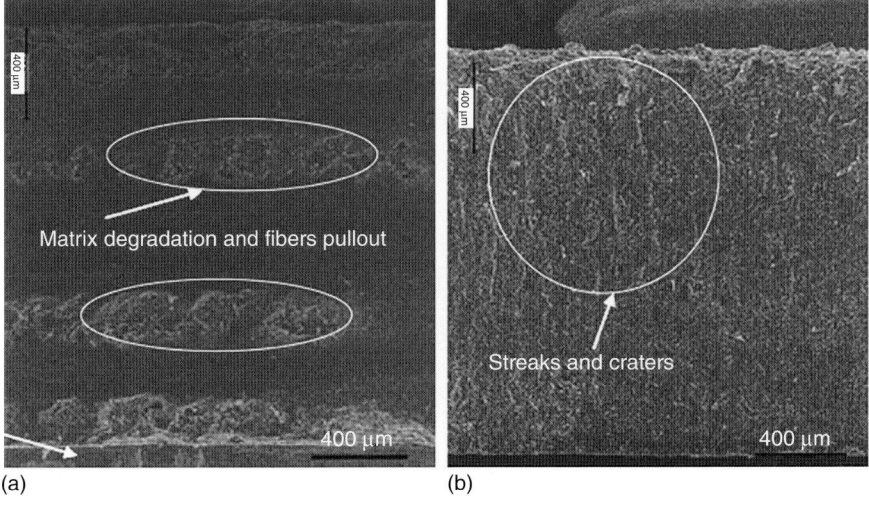

(a)   (b)

Figure 9.8 SEM micrograph of hole machined with (a) conventional machining and (b) AWJM. Source: Saleem et al. 2013 [90]. Reproduced with permission of Elsevier.

9.3.4 Electrical Discharge Machining of Reinforced Polymer Matrix Composites

EDM is widely used for machining of difficult-to-cut and electrically conductive materials to produce intricate and complex shapes. The material removal in EDM process is due to sparking action, which causes melting, vaporization, and thermal spalling of the work piece. The use of EDM process in secondary processing of fiber reinforced PMC, i.e. CFRP composite, is limited due to its semiconductive nature, low material removal, and thermal damage. The carbon in CFRP composite has high thermal conductivity, whereas the polymer phase of the composite exhibits low thermal conductive behavior [91]. Therefore, the secondary processing of CFRP by EDM becomes complex and distinct due to variation in physical properties. Apart from these constraints, various researchers have attempted machining of CFRP by the EDM process. Lau et al. [92] attempted machining of 2.5 mm thick CFRP by using copper and graphite electrodes. They revealed that machining of CFRP is feasible at low current density as compared to high current density. High current density caused deterioration of machined surface quality due to large size craters. Hocheng et al. [93] reported on the machining mechanism in secondary processing of CFRP by EDM process. SEM micrographs of the machined surface shows debris, resolidification, and formation of recast layer. The researchers recommended a low pulse current (less than 0.5 A as an optimum value) for minimum delamination and tear defects around the hole. Guu et al. [94] drilled holes on CFRP composite plate and observed the delamination, recast layer, and surface roughness characteristics of the machined surface. They concluded that output characteristics such as delamination, recast layer, and surface roughness become minimum at low pulse current due to low discharge in machining.

In another study, Habib et al. [95] reported the effect of carbon fiber direction on machining characteristics such as material removal rate (MRR), tool failure, and surface roughness. They revealed that machining mechanism in the direction parallel to carbon fiber orientation is due to melting and boiling. On the other hand, machining in the direction perpendicular to fibers causes high MRR due to fiber breaking and high impact of sparks. Ito et al. [96] observed Joule heating as the machining mechanism in secondary processing of CFRP by EDM process. This is because of short-circuiting of carbon fibers with electrode, which leads to high current flow and produces excessive heat in the machining zone. Therefore, the temperature of carbon fibers rises up to 3600 °C, which is significantly higher than the melting point of polymer matrix. Teicher et al. [97] drilled micro-holes and observed HAZ mechanism in CFRP composite. They concluded that HAZ occurs due to variation in physical properties of carbon fiber and polymer matrix, i.e. melting point. The excessive heat in machining causes deterioration of the machined surface and produces HAZ. Therefore, the researchers highly recommended low pulsed energy for secondary processing of CFRP composites by EDM process. Kumar et al. [98] proposed a novel technique for drilling of micro-holes on CFRP composite using EDM process. The researchers successfully attempted drilling of through holes by targeting the carbon fiber layer irrespective of the epoxy layers on the laminate surface. The tool electrode is accurately positioned on the carbon layer by incorporating a digital microscope. However, improper positioning of the tool electrode, i.e. partial engagement in carbon fiber and epoxy, caused poor quality of machined surface.

9.3.5 Electrochemical Discharge Machining of Reinforced Polymer Matrix Composites

In recent years, the ECDM process has become advantageous in the processing of conductive and nonconductive materials. ECDM process is capable of machining various difficult-to-cut materials irrespective of their properties [99–101]. Initially, the machining of glass epoxy and Kevlar epoxy composite was attempted by Jain et al. [102]. They concluded that machining characteristics such as MRR, overcut, and TWR are significantly affected by applied voltage. Additionally, MRR decreases with increase in fiber volume fraction of the composite. This is because of the higher discharge energy dissipated to the melt fibers fraction of the composite. In another study, Jain et al. [103] attempted secondary processing of glass epoxy and Kevlar epoxy composites by travel wire electrochemical spark machining (TW-ECSM) process. The researchers revealed that increase in applied voltage improves MRR due to high discharge energy in the machining zone. However, at high voltage the quality of machined surface becomes poor due to cracks and HAZ. Additionally, they concluded that TW-ECSM is a feasible method for secondary processing of fiber reinforced PMCs. Later on, Singh et al. [104] attempted machining carbon fiber epoxy composite by ECSM process. They revealed that the surface quality of a machined surface is significantly affected by tool feed rate in the ECSM process. Manna and Narang [105] studied micro-drilling of E-glass-fiber-epoxy composite by ECSM process. The surface quality of machined E-glass-fiber–epoxy composite becomes poor due to incomplete and poor cutting of fibers in ECSM process. Likewise, Malik and Manna [106] attempted slicing of E-glass-fiber-epoxy composite by TW-ECSM process. They concluded that slicing of E-glass-fiber-epoxy composite produced an uneven machined surface due to incomplete cutting of fibers. Antil et al. [107] drilled micro-holes in SiC reinforced PMC by ECDM process. The morphology of drilled micro-holes revealed the presence of cracks and HAZ due to high discharge energy and tearing of matrix.

The following findings have been drawn from comprehensive analysis of the secondary processing of fiber reinforced PMCs by nonconventional machining processes:

- Apart from low productivity, the Nd:YAG pulsed laser is highly susceptible for secondary processing of PMC due to low HAZ and surface damages.
- The surface quality and delamination factor in USM is well addressed by RUM, RUEM, UVC, and RMUD methods. The novel method, i.e. giving rotation to work piece in RMUD, emerged as a potential contender for secondary processing of fiber reinforced PMCs due to better surface quality.
- The thermally free cutting mechanism of AWJM has been successfully attempted for machining of PMCs. Additionally, the combination of optimum SOD, abrasive flow rate, and water pressure in AWJM produces better quality machined surface with minimum delamination and kerf rate.

- In EDM of CFRP, machining performance characteristics such as delamination, recast layer, and surface roughness become minimum at low pulse current.
- The targeting of carbon fiber layer irrespective of epoxy in micro-drilling of CFRP by EDM produces better surface quality holes.
- ECDM process emerged as a potential contender in secondary processing of nonconductive and partially conductive fiber reinforced PMCs. In the recent years, various researchers attempted machining of glass epoxy composite, Kevlar epoxy composite, E-glass-fiber-epoxy composite, CFRP, and SiC reinforced PMC by ECDM process.
- The irregularities in secondary processing of PMCs by ECDM such as HAZ and incomplete fiber cutting are significantly affected by tool feed rate and applied voltage.

9.4 Concluding Remarks

The superior mechanical properties of composites such as high strength to weight and stiffness to weight ratio make them optimal materials for various applications. The machining behavior of fiber reinforced PMCs by conventional and nonconventional machining processes is significantly different from that of metallic materials. The research work presented here gives a comprehensive state-of-the-art review and recent trends in secondary processing of fiber reinforced PMCs. The irregularities occurring in conventional drilling of PMCs were successfully suppressed by using VAD, HSD, and variation in drill bit geometry. Special drill geometries such as Jo drill, eight faceted drill, four faceted drill, and parabolic drill provide better quality machined surface in secondary processing of fiber reinforced PMCs. The fundamental challenge, i.e. drilling induced delamination, needs exploration for better understanding of its mechanism. Apart from machining surface damages, the selection of tool material is an important aspect in the secondary processing of fiber reinforced PMCs. The abrasive nature of carbon and glass fiber highly affects tool life in machining of PMCs. Moreover, the carbide tool, carbide insert, and PCD tool exhibit better performance in terms of tool life.

Nonconventional machining processes such as LBM, USM, AWJM, EDM, and ECDM were successfully attempted for secondary processing of composites. However, surface characteristics such as HAZ, surface damage, delamination, and recast layer were observed in secondary processing of composites by nonconventional machining processes. These irregularities were suppressed by different researchers using different techniques such as Nd:YAG pulsed laser, RMUD, and fiber layer targeting in machining. In addition to machining characteristics, the generation of fume particles in secondary processing of PMCs may cause serious health issues. Therefore, the exploration of sustainable and environment conscious machining in secondary processing of fiber reinforced PMCs is the need of the hour. Additionally, the machined surface irregularities caused by nonconventional machining process need to be explored by single and combined hybridization techniques such vibration assisted, magnetic assisted, grinding assisted, and additives mixed methods.

References

1 Soutis, C. (2005). Fibre reinforced composites in aircraft construction. *Progress in Aerospace Sciences* 41 (2): 143–151.

2 Antil, P., Singh, S., and Manna, A. (2018). Glass fibers/SiCp reinforced epoxy composites: effect of environmental conditions. *Journal of Composite Materials* 52 (9): 1253–1264.

3 Sinmazcelik, T., Avcu, E., Bora, M.Ö., and Çoban, O. (2011). A review: fibre metal laminates, background, bonding types and applied test methods. *Materials and Design* 32: 3671–3685.

4 Singh, S., Singh, I., and Dvivedi, A. (2013). Multi objective optimization in drilling of Al6063/10% SiC metal matrix composite based on grey relational analysis. *Proceedings of the Institution of Mechanical Engineers, Part B: Journal of Engineering Manufacture* 227 (12): 1767–1776.

5 Botelho, E.C., Silva, R.A., Pardini, L.C., and Rezende, M.C. (2006). A review on the development and properties of continuous fiber/epoxy/aluminum hybrid composites for aircraft structures. *Materials Research* 9: 247–256.

6 Singh, S., Singh, A., Singh, I., and Dvivedi, A. (2012). Optimization of the process parameters for drilling of metal matrix composites (MMC) using Taguchi analysis. *Advanced Materials Research* 410: 249–252.

7 Antil, P., Singh, S., and Manna, A. (2018). Analysis on effect of electroless coated SiCp on mechanical properties of polymer matrix composites. *Particulate Science and Technology.* 28: 1–8.

8 Singh, S. (2016). Study the drilling behaviour of aluminium 6061 metal matrix composites using Taguchi's methodology. *International Journal of Machining and Machinability of Materials* 18 (4): 327–340.

9 Ho-Cheng, H. and Dharan, C.K.H. (1990). Delamination during drilling in composite laminates. *Journal of Engineering for Industry* 112: 236–239.

10 Hocheng, H. and Tsao, C.C. (2003). Comprehensive analysis of delamination in drilling of composite materials with various drill bits. *Journal of Materials Processing Technology* 140: 335–339.

11 Bhattacharyya, D. and DPW, H. (1998). A study of drilling in Kevlar composites. *Composites Science and Technology* 58: 267–283.

12 Davim, J.P., Rubio, J.C., and Abrao, A.M. (2007). A novel approach based on digital image analysis to evaluate the delamination factor after drilling composite laminates. *Composites Science and Technology* 67: 1939–1945.

13 Davim, J.P. and Reis, P. (2003). Drilling carbon fiber reinforced plastics manufactured by autoclave – experimental and statistical study. *Materials and Design* 24: 315–324.

14 Hough, C.L., Lednicky, T.E., and Griswold, N. (1988). Establishing criteria for a computerized vision inspection of holes drilled in carbon fiber composites. *Journal of Testing and Evaluation* 16 (2): 139–145.

15 Davim, J.P. and Reis, P. (2003). Study of delamination in drilling carbon fiber reinforced plastics (CFRP) using design experiments. *Composite Structures* 59 (4): 481–487.

16 Durao, L.M.P., de Moura, M.F.S.F., and Marques, A.T. (2006). Numerical simulation of the drilling process on carbon/epoxy composite laminates. *Composites Part – A: Applied Science Manufacturing* 37: 1325–1333.

17 Khashaba, U.A. (2004). Delamination in drilling GFR–thermoset composites. *Composite Structures* 63: 313–327.

18 Liu, D., Tang, Y., and Cong, W.L. (2012). A review on mechanical drilling for composites laminates. *Composite Structures* 94: 1265–1279.

19 Singh, A.P., Sharma, M., and Singh, I. (2013). A review of modelling and control during drilling of fibre reinforced plastic composites. *Composites: Part B* 47: 118–125.

20 Park, K.Y., Choi, J.H., and Lee, D.G. (1995). Delamination-free and high efficiency drilling of carbon fiber reinforced plastics. *Journal of Composite Materials* 29: 1988–2002.

21 Tsao, C.C. and Hocheng, H. (2007). Parametric study on thrust force of core drill. *Journal of Materials Processing Technology* 192–193: 37–40.

22 Jain, S. and Yang, D.C.H. (1994). Delamination-free drilling of composite laminates. *Journal of Engineering for Industry* 116: 475–481.

23 Hocheng, H. and Tsao, C.C. (2005). The path towards delamination-free drilling of composite materials. *Journal of Materials Processing Technology* 167: 251–264.

24 Tsao, C.C. and Hocheng, H. (2004). Taguchi analysis of delamination associated with various drill bits in drilling of composite material. *International Journal of Machine Tools and Manufacture* 44 (10): 1085–1090.

25 Tsao, C.C. and Hocheng, H. (2005). Computerized tomography and C-Scan for measuring delamination in the drilling of composite materials using various drills. *International Journal of Machine Tools and Manufacture* 45: 1282–1287.

26 Wang, X., Wang, L.J., and Tao, J.P. (2004). Investigation on thrust in vibration drilling of drilling of fiber-reinforced plastics. *Journal of Materials Processing Technology* 148: 239–244.

27 Ramkumar, J., Aravindan, S., Malhotra, S.K., and Krishamurthy, R. (2004). An enhancement of machining performance of GFRP by oscillatory assisted drilling. *International Journal of Advanced Manufacture Technology* 23: 240–244.

28 Ramkumar, J., Malhotra, S.K., and Krishamurthy, R. (2004). Effect of work piece vibration on drilling of GFRP laminates. *Journal of Materials Processing Technology* 152: 329–332.

29 Arul, S., Vijayaraghavan, L., Malhotra, S.K., and Krishnamurthy, R. (2006). Influence of tool material on dynamics of drilling of GFRP composites. *The International Journal of Advanced Manufacturing Technology* 29: 655–662.

30 Rawat, S. and Attia, H. (2009). Characterization of the dry high speed drilling process of woven composites using machinability maps approach. *CIRP Annals – Manufacture Technology* 58: 105–108.

31 Gaitonde, V.N., Karnik, S.R., Campos Rubio, J. et al. (2008). Analysis of parametric influence on delamination in high-speed drilling of carbon fiber reinforced plastic composites. *Journal of Materials Processing Technology* 203: 431–438.

32 Karnik, S.R., Gaitonde, V.N., Campos Rubio, J. et al. (2008). Delamination analysis in high speed drilling of carbon fiber reinforced plastics (CFRP) using artificial neural network model. *Materials and Design* 29: 1768–1776.

33 Rubio, J.C., Abrao, A.M., Faria, P.E. et al. (2008). Effects of high speed in the drilling of glass fibre reinforced plastic: evaluation of delamination factor. *International Journal of Machine Tools and Manufacture* 48: 715–720.

34 Singh, S. (2016). Effect of modified drill point geometry on drilling quality characteristics of metal matrix composite (MMCs). *Journal of Mechanical Science and Technology* 30 (6): 2691–2698.

35 Piquet, R., Ferret, B., Lachaud, F., and Swider, P. (2000). Experimental analysis of drilling damage in thin carbon/epoxy plate using special drills. *Composites Part A: Applied Science and Manufacturing.* 31 (10): 1107–1115.

36 Mathew, J., Ramakrishnan, N., and Naik, N.K. (1999). Investigations into the effect of geometry of a trepanning tool on thrust and torque during drilling of GFRP composites. *Journal of Materials Processing Technology* 91 (1–3): 1.

37 El-Sonbaty, I., Khashaba, U.A., and Machaly, T. (2004). Factors affecting the machinability of GFR/epoxy composites. *Composite Structures* 63 (3–4): 329–338.

38 Debnath, K. and Singh, I. (2017). Low-frequency modulation-assisted drilling of carbon–epoxy composite laminates. *Journal of Manufacturing Processes* 25: 262–273.

39 Singh, I. and Bhatnagar, N. (2006). Drilling of uni-directional glass fiber reinforced plastic (UD-GFRP) composite laminates. *The International Journal of Advanced Manufacturing Technology* 27 (9–10): 870–876.

40 Rakesh, P.K., Singh, I., and Kumar, D. (2012). Drilling of composite laminates with solid and hollow drill point geometries. *Journal of Composite Materials* 46 (25): 3173–3180.

41 Hocheng, H., Puw, H.Y., and Huang, Y. (1993). Preliminary study on milling of unidirectional carbon fibre-reinforced plastics. *Composites Manufacturing* 4 (2): 103–108.

42 Rahman, M., Ramakrishna, S., and Thoo, H.C. (1999). Machinability study of carbon/PEEK composites. *Machining Science and Technology* 3 (1): 49–59.

43 Davim, J.P., Reis, P., and Antonio, C.C. (2004). A study on milling of glass fiber reinforced plastics manufactured by hand-lay up using statistical analysis (ANOVA). *Composite Structures* 64 (3–4): 493–500.

44 Davim, J.P. and Reis, P. (2005). Damage and dimensional precision on milling carbon fiber-reinforced plastics using design experiments. *Journal of Materials Processing Technology* 160 (2): 160–167.

45 Inoue, T., Hagino, M., Matsui, M., and Gu, L.W. (2009). Cutting characteristics of CFRP materials with end milling. In: *Key Engineering Materials*, vol. 407 (ed. F. Rui, Q. Lihong, C. Huawei, et al.), 710–713. Trans Tech Publications.

46 Azmi, A.I., Lin, R.J., and Bhattacharyya, D. (2013). Machinability study of glass fibre-reinforced polymer composites during end milling. *The International Journal of Advanced Manufacturing Technology* 64 (1–4): 247–261.

47 Kim, K.S., Kwak, Y.K., and Namgung, S. (1992). Machinability of carbon fiber–epoxy composite materials in turning. *Journal of Materials Processing Technology* 32 (3): 553–570.

48 Santhanakrishnan, G., Krishnamurthy, R., and Malhotra, S.K. (1992). Investigation into the machining of carbon-fibre-reinforced plastics with cemented carbides. *Journal of Materials Processing Technology* 30 (3): 263–275.

49 Davim, J.P. and Mata, F. (2005). Optimisation of surface roughness on turning fibre-reinforced plastics (FRPs) with diamond cutting tools. *The International Journal of Advanced Manufacturing Technology* 26 (4): 319–323.

50 Palanikumar, K., Karunamoorthy, L., and Karthikeyan, R. (2006). Assessment of factors influencing surface roughness on the machining of glass fiber-reinforced polymer composites. *Materials and Design* 27 (10): 862–871.

51 Palanikumar, K., Karunamoorthy, L., and Manoharan, N. (2006). Mathematical model to predict the surface roughness on the machining of glass fiber reinforced polymer composites. *Journal of Reinforced Plastics and Composites* 25 (4): 407–419.

52 Negarestani, R., Sundar, M., Sheikh, M.A. et al. (2010). Numerical simulation of laser machining of carbon-fibre-reinforced composites. *Proceedings of the Institution of Mechanical Engineers, Part B: Journal of Engineering Manufacture* 224 (7): 1017–1027.

53 Tagliaferri, V., Di Ilio, A., and Visconti, C. (1985). Laser cutting of fibre-reinforced polyesters. *Composites* 16 (4): 317–325.

54 Dell'Erba, M., Galantucci, L.M., and Miglietta, S. (1992). An experimental study on laser drilling and cutting of composite materials for the aerospace industry using excimer and CO_2 sources. *Composites Manufacturing* 3 (1): 14–19.

55 Lau, W.S., Lee, W.B., and Pang, S.Q. (1990). Pulsed Nd:YAG laser cutting of carbon fibre composite materials. *CIRP Annals – Manufacturing Technology* 39 (1): 179–182.

56 Dubey, A.K. and Yadava, V. (2008). Laser beam machining—a review. *International Journal of Machine Tools and Manufacture* 48 (6): 609–628.

57 Pan, C.T. and Hocheng, H. (1996). The anisotropic heat-affected zone in the laser grooving of fiber-reinforced composite material. *Journal of Materials Processing Technology* 62 (1–3): 54–60.

58 Uhlmann, E., Spur, G., Hocheng, H. et al. (1999). The extent of laser-induced thermal damage of UD and crossply composite laminates. *International Journal of Machine Tools and Manufacture* 39 (4): 639–650.

59 Fürst, A., Hipp, D., Klotzbach, A. et al. (2016). Increased cutting efficiency due to multi-wavelength remote-laser-ablation of fiber-reinforced polymers. *Advanced Engineering Materials* 18 (3): 403–408.

60 Fürst, A., Mahrle, A., Hipp, D. et al. (2017). Dual wavelength laser beam cutting of high-performance composite materials. *Advanced Engineering Materials* 19 (2): 1600356.

61 Mathew, J., Goswami, G.L., Ramakrishnan, N., and Naik, N.K. (1999). Parametric studies on pulsed Nd:YAG laser cutting of carbon fibre reinforced plastic composites. *Journal of Materials Processing Technology* 89: 198–203.

62 Leone C, Pagano A, Lopresto V, De Iorio I. Solid state Nd:YAG laser cutting of CFRP sheet: influence of process parameters on kerf geometry and HAZ. *Proceedings of 17th International Conference on Composite Materials – ICCM-17* (27–31 July 2009). Edinburgh International Convention Centre, Edinburgh, UK: IOM Communications.

63 Staehr, R., Bluemel, S., Jaeschke, P. et al. (2016). Laser cutting of composites—two approaches toward an industrial establishment. *Journal of Laser Applications* 28 (2): 022203.

64 Wolynski, A., Herrmann, T., Mucha, P. et al. (2011). Laser ablation of CFRP using picosecond laser pulses at different wavelengths from UV to IR. *Physics Procedia* 12: 292–301.

65 Leone, C., Genna, S., and Tagliaferri, V. (2014). Fibre laser cutting of CFRP thin sheets by multi-passes scan technique. *Optics and Lasers in Engineering* 53: 43–50.

66 Wu, C.W., Wu, X.Q., and Huang, C.G. (2015). Ablation behaviors of carbon reinforced polymer composites by laser of different operation modes. *Optics and Laser Technology* 73: 23–28.

67 Niino, H., Harada, Y., Anzai, K. et al. (2016). Laser cutting of carbon fiber reinforced plastics (CFRP and CFRTP) by IR fiber laser irradiation. *Journal of Laser Micro Nano Engineering* 11 (1): 104.

68 Nath, C., Lim, G.C., and Zheng, H.Y. (2012). Influence of the material removal mechanisms on hole integrity in ultrasonic machining of structural ceramics. *Ultrasonics* 52 (5): 605–613.

69 Hocheng, H. and Hsu, C.C. (1995). Preliminary study of ultrasonic drilling of fiber-reinforced plastics. *Journal of Materials Processing Technology* 48 (1–4): 255–266.

70 Cong, W.L., Pei, Z.J., Feng, Q. et al. (2012). Rotary ultrasonic machining of CFRP: a comparison with twist drilling. *Journal of Reinforced Plastics and Composites* 31 (5): 313–321.

71 Cong, W.L., Feng, Q., Pei, Z.J. et al. (2012). Rotary ultrasonic machining of carbon fiber-reinforced plastic composites: using cutting fluid vs. cold air as coolant. *Journal of Composite Materials* 46 (14): 1745–1753.

72 Cong, W.L., Pei, Z.J., Deines, T.W. et al. (2012). Rotary ultrasonic machining of CFRP composites: a study on power consumption. *Ultrasonics* 52 (8): 1030–1037.

73 Ding, K., Fu, Y., Su, H. et al. (2014). Experimental studies on drilling tool load and machining quality of C/SiC composites in rotary ultrasonic machining. *Journal of Materials Processing Technology* 214 (12): 2900–2907.

74 Liu, J., Zhang, D., Qin, L., and Yan, L. (2012). Feasibility study of the rotary ultrasonic elliptical machining of carbon fiber reinforced plastics (CFRP). *International Journal of Machine Tools and Manufacture* 53 (1): 141–150.

75 Kim, J.D. and Lee, E.S. (1996). A study of ultrasonic vibration cutting of carbon fibre reinforced plastics. *The International Journal of Advanced Manufacturing Technology* 12 (2): 78–86.

76 Debnath, K., Singh, I., and Dvivedi, A. (2015). Rotary mode ultrasonic drilling of glass fiber-reinforced epoxy laminates. *Journal of Composite Materials* 49 (8): 949–963.

77 Ramulu, M. and Arola, D. (1993). Water jet and abrasive water jet cutting of unidirectional graphite/epoxy composite. *Composites* 24 (4): 299–308.

78 Komanduri, R. (1997). Machining of fiber-reinforced composites. *Machining Science and Technology* 1 (1): 113–152.

79 Thongkaew, K., Wang, J., and Yeoh, G.H. (2016). An investigation of hole machining process on a carbon-fiber reinforced plastic sheet by abrasive waterjet. *Advanced Materials Research* 1136: 1113.

80 Mayuet, P.F., Girot, F., Lamíkiz, A. et al. (2015). SOM/SEM based characterization of internal delaminations of CFRP samples machined by AWJM. *Procedia Engineering* 132: 693–700.

81 Shanmugam, D.K., Nguyen, T., and Wang, J. (2008). A study of delamination on graphite/epoxy composites in abrasive waterjet machining. *Composites Part A: Applied Science and Manufacturing* 39 (6): 923–929.

82 Phapale, K., Singh, R., Patil, S., and Singh, R.K. (2016). Delamination characterization and comparative assessment of delamination control techniques in abrasive water jet drilling of CFRP. *Procedia Manufacturing* 5: 521–535.

83 Azmir, M.A. and Ahsan, A.K. (2009). A study of abrasive water jet machining process on glass/epoxy composite laminate. *Journal of Materials Processing Technology* 209 (20): 6168–6173.

84 Shanmugam, D.K. and Masood, S.H. (2009). An investigation on kerf characteristics in abrasive waterjet cutting of layered composites. *Journal of Materials Processing Technology* 209 (8): 3887–3893.

85 Azmir, M.A. and Ahsan, A.K. (2008). Investigation on glass/epoxy composite surfaces machined by abrasive water jet machining. *Journal of Materials Processing Technology* 198 (1–3): 122–128.

86 Reddy, S.M., Hussain, S., Srikanth, D.V., and Rao, M.S. (2015). Experimental analysis and optimization of process parameters in machining of RCFRP by AJM. *International Journal of Innovative Research in Science, Engineering, and Technology* 4: 7085–7092.

87 Kakinuma, Y., Ishida, T., Koike, R. et al. (2015). Ultrafast feed drilling of carbon fiber-reinforced thermoplastics. *Procedia CIRP* 35: 91–95.

88 Alberdi, A., Suárez, A., Artaza, T. et al. (2013). Composite cutting with abrasive water jet. *Procedia Engineering.* 63: 421–429.

89 Ibraheem, H.M., Iqbal, A., and Hashemipour, M. (2015). Numerical optimization of hole making in GFRP composite using abrasive water jet machining process. *Journal of the Chinese Institute of Engineers* 38 (1): 66–76.

90 Saleem, M., Toubal, L., Zitoune, R., and Bougherara, H. (2013). Investigating the effect of machining processes on the mechanical behavior of composite plates with circular holes. *Composites Part A: Applied Science and Manufacturing* 55: 169–177.

91 Sheikh-Ahmad, J.Y. (2016). Hole quality and damage in drilling carbon/epoxy composites by electrical discharge machining. *Materials and Manufacturing Processes* 31 (7): 941–950.

92 Lau, W.S., Wang, M., and Lee, W.B. (1990). Electrical discharge machining of carbon fibre composite materials. *International Journal of Machine Tools and Manufacture* 30 (2): 297–308

93 Hocheng, H., Guu, Y.H., and Tai, N.H. (1998). The feasibility analysis of electrical-discharge machining of carbon–carbon composites. *Material and Manufacturing Process* 13 (1): 117–132.

94 Guu, Y.H., Hocheng, H., Tai, N.H., and Liu, S.Y. (2001). Effect of electrical discharge machining on the characteristics of carbon fiber reinforced carbon composites. *Journal of Materials Science* 36 (8): 2037–2043.

95 Habib, S., Okada, A., and Ichii, S. (2013). Effect of cutting direction on machining of carbon fibre reinforced plastic by electrical discharge machining process. *International Journal of Machining and Machinability of Materials* 13 (4): 414–427.

96 Ito, A., Hayakawa, S., Itoigawa, F., and Nakamura, T. (2012). Effect of short-circuiting in electrical discharge machining of carbon fiber reinforced plastics. *Journal of Advanced Mechanical Design, Systems, and Manufacturing* 6 (6): 808–814.

97 Teicher, U., Müller, S., Münzner, J., and Nestler, A. (2013). Micro-EDM of carbon fibre-reinforced plastics. *Procedia CIRP* 6: 320–325.

98 Kumar, R., Agrawal, P.K., and Singh, I. (2018). Fabrication of micro holes in CFRP laminates using EDM. *Journal of Manufacturing Processes* 31: 859–866.

99 Singh, M. and Singh, S. (2018). Machining of difficult to cut materials by electrochemical discharge machining (ECDM) process: a state of art approach. In: *Advanced Manufacturing and Materials Science*, 139–149. Cham: Springer.

100 Antil, P., Singh, S., and Singh, P. (2018). Taguchi's methodology based electrochemical discharge machining of polymer matrix composites. *Procedia Manufacturing* 26: 469–473.

101 Antil, P., Singh, S., and Manna, A. (2018). SiC$_p$ glass fibers reinforced epoxy composites: wear and erosion behaviour. *Indian Journal of Engineering and Material Sciences* 25: 122–130.

102 Jain, V.K., Tandon, S., and Kumar, P. (1990). Experimental investigations into electrochemical spark machining of composites. *Journal of Engineering for Industry* 112 (2): 194–197.

103 Jain, V.K., Rao, P.S., Choudhary, S.K., and Rajurkar, K.P. (1991). Experimental investigations into traveling wire electrochemical spark machining (TW-ECSM) of composites. *Journal of Engineering for Industry* 113 (1): 75–84.

104 Singh, Y.P., Jain, V.K., Kumar, P., and Agrawal, D.C. (1996). Machining piezoelectric (PZT) ceramics using an electrochemical spark machining (ECSM) process. *Journal of Materials Processing Technology* 58 (1): 24–31.

105 Manna, A. and Narang, V. (2012). A study on micro machining of e-glass–fibre–epoxy composite by ECSM process. *The International Journal of Advanced Manufacturing Technology* 61 (9–12): 1191–1197.

106 Malik, A. and Manna, A. (2016). An experimental investigation on developed WECSM during micro slicing of e-glass fibre epoxy composite. *The International Journal of Advanced Manufacturing Technology* 85 (9–12): 2097–2106.

107 Antil, P., Singh, S., and Manna, A. (2018). Electrochemical discharge drilling of SiC reinforced polymer matrix composite using Taguchi's grey relational analysis. *Arabian Journal for Science and Engineering* 43 (3): 1257–1266.

10

Hybrid Glass Fiber Reinforced Polymer Matrix Composites: Mechanical Strength Characterization and Life Assessment

Parvesh Antil[1], Sarbjit Singh[2], and Manpreet Singh[2]

[1] Department of Basic Engineering, College of Agricultural Engineering & Technology, CCS HAU Hisar, Haryana, 125004, India
[2] Department of Mechanical Engineering, Punjab Engineering College, Chandigarh, 160012, India

10.1   Introduction

The development of PMCs as hybrid polymer matrix composites (HPMCs) started in 1990. These PMCs fundamentally comprise of two different materials. The main constituent is matrix in the form of polymer and second is reinforcement, which is added in different weight fractions in the form of particles, fillers, and fibers. Fiber reinforced polymer composites are classified as continuous and discontinuous depending upon the reinforcement. The properties of these composites can be varied as per industrial applications through combinations of matrix, reinforcement, and fabrication technique. Over the last decade, research fraternity is continuously working on the development of new polymer composite materials that can outshine conventional materials in terms of strength to weight ratio and tribological properties [1–3]. Chen et al. [4] fabricated potassium titanate whisker reinforced polyvinyl chloride (PVC) by hot-squeezing the antecedent fibers that contain the whiskers. The close control of whisker directions was acknowledged by placing the fibers in a foreordained manner. Ghoneim and Mahani [5] examined the mechanical and dielectric properties of polymethyl methacrylate-carbon black and polyester-carbon black composites by using high abrasion furnace carbon blacks. Fomitchov et al. [6] proposed a real-time cure monitoring laser ultrasonic system for graphite epoxy composite and used it over a resin transfer molding process. Ultrasonic generation at various parts of the composite was accomplished using an optical switch. The results proved that the curing process was capable of operating at high temperature as well as in high pressure environment. Soon-Gi (2002) [7] designed and fabricated electrically conductive glass fiber reinforced polymer (GFRP) by adding carbon powder or fiber to a vinyl ester resin matrix. The change in the electrical resistance of the composite was observed in order to investigate the auto diagnostic capacity and concluded that a composite reinforced with carbon powder has properties of damage diagnosis in terms of small strain stages. Huang et al. [8] evaluated the influence of variation in temperature and strain rate on the dynamic tensile properties of GFRP and found that the modified coated fiber bundle model can be extended to these properties and the bimodal Weibull statistical constitutive equation is effective in describing the stress–strain relations of GFRP under different strain rates and temperatures. Sekine and Beaumont [9] analyzed stress–corrosion cracking in unidirectional glass fiber reinforced polymer composite using the micromechanical theory. An equation was developed for macroscopic crack growth using the physical model developed by micromechanical theory. The observation revealed that below the threshold value of the stress integrity factor, matrix cracks do not appear. Shindo et al. [10] concluded that decrease in the rate of energy release is caused by damage initiation at the matrix at cryogenic temperature. Singh and Bhatnagar [11] investigated the effect of damage induced during drilling on the tensile properties of plastic composites reinforced with glass fibers. A mathematical model was proposed to analyze the relationship between the residual strength and process parameters of the drilling operation. Antil et al. [12] investigated the micro-drilling behavior of hybrid glass fiber reinforced PMC using an unconventional machining method. A state-of-the-art literature review on the machining possibilities of polymer matrix composites (PMC) using nonconventional machining was presented by Singh et al. [13]. Das et al. [14] studied the discrepancy in thermal diffusivity of plain woven fabric composite in a closed cycle cryo-refrigerator using applied modified Angstrom method. Sekine and Beaumont [15] investigated the fracture phenomenon using scanning electron microscopy (SEM) and concluded that the fracture surface of each fiber is categorized by a mirror region enclosed by a hackle region. The observed phenomenon states that the fracture is completed in two stages. The first stage includes slow fracture of some portion of the glass fiber followed by fast and unstable fracture across the fibers. Kim et al. [16] fabricated glass fiber reinforced composite using two types of compression molding, i.e. reform and sheet molding process. Using fiber content up to 40%, cutting force analysis and surface characterization were accomplished. Rakesh et al. [17] investigated the compressive strength of the developed glass fiber polymer composite with drilled holes. Drilling of the composites was performed using various process parameters such as the drill point geometry, feed rate, and cutting speed. Huh et al. [18] evaluated the fatigue behavior of glass fiber reinforced polymer composite by considering the orientation of reinforcing glass fibers. Gudonis et al. [19] examined the bond properties of GFRP reinforcement using two types of test methods: pullout and bending. The methods considered different confining conditions, stress–strain states of concrete, embedment lengths of bars, and casting directions. These factors may affect the bond–slip relationship obtained. Bey et al. [20] investigated the effect of fatigue loading on the mechanical behavior of sandwich composites. Two types of sandwich composites, i.e. $[0^0_4]$ and $[0^0/90^0_2/0^0]$, were tested using three points bending testing machine. Neuenschwander et al. [21] focused on nondestructive techniques to characterize PMCs using an air-coupled ultrasonic transducer. The author(s) used an acoustic lens having a lateral resolution up to 0.1 mm with elevation resolution of 2 μm to analyze the power of highly focused transducers for surface topography. Yong Tao et al. [22] worked on the failure behavior of the PMCs. The author(s) varied the strain rates to characterize the interfiber failure behavior of PMCs. Three strain rates, i.e. quasi static, 383, and 646 s^{-1}, were employed to characterize unidirectional glass fiber/epoxy composites. A new failure theory that included strain rate effects was used to evaluate the test results. The observations from the experimental data showed good agreement with the predicted failure and fracture angles. Ghasemi et al. [23] investigated the mechanical degradation of PMCs by analyzing the effect of thermal cycling. Two types of specimens, i.e. open hole and un-notched, were tested on the tensile testing machine and thermal cycling apparatus. The observations revealed that thermal cycling effect can be controlled by the hole diameters. Also, the diameter of the hole can play a significant role in the tensile strength of composites as larger holes reduce the tensile strength. Bigdeli and Fasano [24] enhanced the thermomechanical properties of PMCs by incorporating graphene nanoribbons as additional fillers and characterized the thermal conduction properties of composites. Using reverse nonequilibrium molecular dynamics simulation, a relation between thermal conductivity and thermal boundary resistance of graphene nanoribbons was reported. These relations could be helpful in predicting the thermal transport properties of PMCs that are used in various applications. Antil et al. [25] fabricated and analyzed the tribological behavior of hybrid glass fiber reinforced PMCs by using Taguchi's methodology.

10.2 Polymer Matrix Composites (PMCs)

The PMC comprises of two principle elements, i.e. matrix and reinforcement. The matrix helps in distribution of load to the reinforcement whereas reinforcement provides strength to the composite.

10.2.1 Matrix

The polymer is the important element for the production of PMC and the choice of matrix materials depends upon the mechanical and chemical properties of the polymer. Comparatively good mechanical strength, easy access, economic viability, and easy processability are some necessary requirements for a polymer to be considered as matrix. Araldite epoxy resins are widely used as polymer matrix for the fabrication of PMCs. In the present work, Araldite AW106 resin and hardener HV053U mixed in a ratio of 5 : 4 were used as polymer matrix. The advantages of AW106 resin/HV053U hardener such as ease of application, good resistance to static and dynamic loads, high shear and peel strength, and electrical insulating nature influence its choice as matrix [26]. The typical properties of AW106 resin/HV053U hardener are shown in Table 10.1.

10.2.2 Reinforcement

Silicon carbide (SiC) particles have attained an influencing position in terms of the additional reinforcement provided in composite fabrication. It is because the

Table 10.1 Properties of AW106 resin/ HV053U hardener [26].

Property	Test value
Ultimate tensile strength (MPa)	33 MPa
Elongation (%)	9
Thermal conductivity (W/m K)	0.22

addition of SiC to the matrix significantly improves the strength, modulus, and wear resistance of the fabricated PMCs. The SiC particles are easily available and have good wettability with the matrix. The SiC particles used for the present work are black/gray SiC (chemically grade 99% pure) along with 0.1–0.2% of free carbon (C) and ferrous oxide (Fe_2O_3). In the present research work three types of SiC reinforcement, i.e. 37 μm (400 mesh), 44 μm (320 mesh), and 63 μm (220 mesh), have been used with three levels of weight fraction, i.e. 5%, 10%, and 15%.

10.2.3 Fabrication of HGFRPC

The fabrication of either metal matrix composite (MMC) or PMC finishes with the accumulation of two basic elements, i.e. matrix and reinforcement, depending upon the type of composite used. In the current work, hybrid and normal PMCs were prepared by using hand lay-up methodology. The normal polymer matrix contains glass fiber as reinforcement and Araldite epoxy polymer as matrix whereas hybrid polymer matrix includes additional reinforcement in the form of abrasive SiC particles having variable weight and size. Glass fiber as reinforcement was used in two forms specified as E glass CSM (chopped strand mat) and S glass fabric mat with grades of 450 GSM and 400 GSM, respectively. These glass fibers were used in five alternate layers in which E glass fiber was layered three times and S glass fibers two times. Chemically tested and graded SiC was used in three variable average sized particles of 63 μm (220 mesh), 44 μm (320 mesh), and 37 μm (440 mesh) to act as secondary reinforcement in the fabrication of hybrid PMC. The matrix for normal composite was prepared with an epoxy–hardener ratio of 10 : 8. A mold of mild steel was prepared to hold the layered components during curing. The fabrication of normal PMC started with placing of plastic sheets at the surface of the mold to avoid any adhesion of matrix with the mold. A layer of the prepared matrix was spread over a plastic sheet as per the desired dimensions of fiber mat and then the first sheet of fiber mat was placed. After placing the fiber mat, a layer of matrix was uniformly spread over the mat and alternate layer of S glass was placed. The process was repeated until it reached the desired number of layers of reinforcement. On an average, 60 g of matrix was layered on each sheet. After proper placements of fiber mats, again a plastic sheet was placed above the last fiber mat. The layered matrix and reinforcement were then cured at room temperature for the next eight hours and after passage of cure time, the fabricated composite was taken for postprocessing. For the fabrication of hybrid composite, the whole process

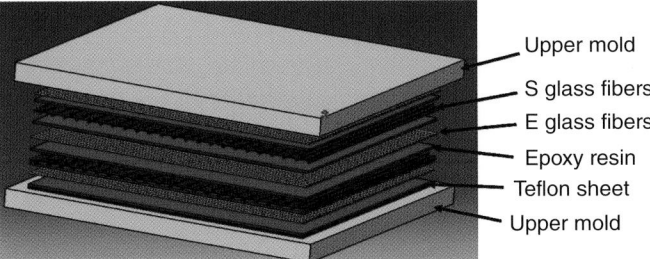

Upper mold

S glass fibers

E glass fibers

Epoxy resin

Teflon sheet

Upper mold

Figure 10.1 Schematic representation of the fabrication process [1].

was the same except for the matrix. A new matrix was prepared by accumulating the desired size and weight fraction of SiC particles. The SiC particles were mixed in the matrix by manual stirring. One key observation from preparing the matrix was that the matrix preparation should end within 15 minutes because exothermic reaction increases the temperature of the matrix and converts the liquid phase to a semisolid dense phase. This change in phase makes it hard for the matrix to bond properly with the reinforcement. The SiC particles were used in three weight fractions of 5%, 10%, and 15%.

The mechanical strength of the PMCs is highly influenced by the fiber volume fraction. In the present work, the fiber volume fraction is obtained as follows [27]:

$$V_f = \left[\frac{\rho_m \times W_f}{(\rho_m \times W_f) + (\rho_f \times W_m)} \right] \tag{10.1}$$

Here ρ_m, ρ_f are the densities of the matrix and fibers; W_f, W_m are the weights of fibers and the matrix whereas the fiber volume fraction is denoted as V_f. By using the values of these factors, the value of V_f was observed as 12.73%. The modeled representation of the fabrication process is shown in Figure 10.1.

10.2.4 Morphology of Normal and Hybrid PMC

The exploration of state of fibers, SiC particles, and epoxy polymer is accomplished by SEM as presented in Figure 10.2. The test specimens were cut out from the fiber plate transverse to the fiber direction. The variation in matrix reinforcement bond behavior is primarily observed from morphological inspection. The key difference between normal and hybrid PMCs is observed as the gap existing between the matrix and reinforcement. The incorporation of SiC reduces the gap, which is relatively higher in normal PMC. The incorporation of SiC particles reduces the interparticle distance, which improves the bond strength of the composite. The better bond strength in hybrid PMC improves the resistance against fracture as well. Figure 10.2a,b shows the SEM of normal and hybrid composites at lower magnification of 100× in which the bonding status of matrix and reinforcement is visible. A higher magnification of 1500× as shown in Figure 10.2c,d shows the actual gaps present between matrix and reinforcement. Morphological analysis revealed that glass fibers, SiC particles, and epoxy polymer are efficiently bonded to each other and also that the epoxy polymer is uniformly layered over

(a)　　　　　　　　　　　　　　　　(b)

(c)　　　　　　　　　　　　　　　　(d)

Figure 10.2 (a) SEM of normal composite at 100× ; (b) SEM of hybrid composite at 100×; (c) SEM of normal composite at ×1500; (d) SEM of hybrid composite at 1500× [1].

the fibers. The SiC particles precisely reduced the gaps present in the normal composite and led to decrease in interparticle distance. The normal composite at higher magnification confirms the improper bonding between the matrix and reinforcement.

10.2.5 Mechanical Strength Analysis

The performance of any composite depends upon its ability to resist deformation or fracture under various loading conditions. The ability to resist load is generally termed as mechanical strength. In the present chapter, the mechanical strength characterization of fabricated composites was analyzed by a universal testing machine (UTE 40) on the basis of the performance of composite in terms of compressive, flexural, impact, and tensile strength. The tensile properties of fabricated PMC were analyzed on the universal testing machine as per ASTM D3039 standard. During tensile test, the composite specimen was tested five times for each weight fraction having a gauge length of 5 cm. The crosshead pin and load cell were adjusted at 0.5 mm/min and 2 kN respectively. The compressive testing of the composite was also analyzed on the same UTE 40 by introducing a fixture in order to avoid bending during testing. Three point bending test based on ASTM D790 standard was adopted to analyze the flexural strength of the composite. The specimens with dimensions 80 mm × 13 mm × 6 mm were cut from

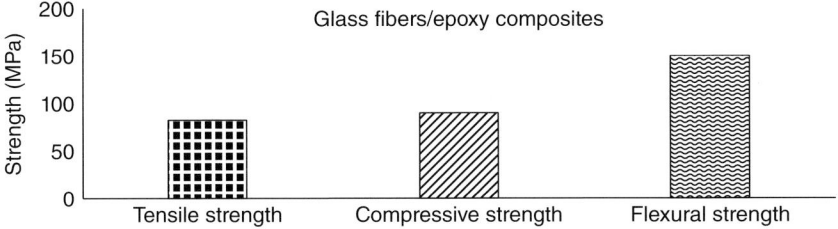

Figure 10.3 Composite without SiC as reinforcement.

the composite panel and tested on UTE 40. At least five samples were tested for each composite. ITD 50 impact analysis machine with ASTM D256 was used to analyze the impact properties of the fabricated composites. The mechanical strength of composites was tested in two categories, i.e. composite having no SiC reinforcement and composite having SiC reinforcement domains. Based on the observed results for composite having no SiC reinforcement, the outcomes were plotted in the form of a bar graph. The average value for tensile strength is 83 MPa, flexural strength is 149 MPa, compressive strength is 91 MPa, and impact strength is 3.55 J/cm (Figure 10.3). The effect of incorporating SiC particle as secondary reinforcement is plotted and shown in Figure 10.4a–d. The graph signifies the influence of variation in size and weight fraction adopted during the fabrication of the composite. The observed test results for tensile test show the improvement in strength of the composite by almost 25% with increase in weight fraction whereas the strength tends to decrease up to 18% with increase in particle size of SiC. A nearly related form is perceived in other test results where strength improves with increased weight fraction and reduces with increase in particle size. Altogether, the composites reinforced with smaller particle size of 37 μm come out to be the best in terms of resistance against failure. The key aspects that influence the strength of the composite are found to be the interfacial area and the interparticle distance. These aspects are mostly influenced by the profile, size, and volume fraction of the reinforcing materials [28].

Composites having comparatively small size of SiC of 37 μm have a greater interfacial area as compared to other composites having SiC particles of comparatively bigger size of 44 μm and 63 μm each. The interfacial area varies with the variations in particle size and increase with small size and vice versa. It is because a smaller particle size enables higher amount of reinforced SiC to be accommodated in the available area. As a result, the higher amount of reinforced particles decreases the interparticle distance and results in improved resistance against loading. Earlier research available regarding influence of particle size revealed that smaller size reinforcement improved the strength and microstructure because of reduced interparticle distance [29]. The composites with higher particle size fail to resist constant stress concentration because these being lesser in number occupy extra interfacial area [30]. It is also evident that higher particle size has possibilities to induce imperfections in the form of cavities and crack because of the high arithmetic prospect of having a size bigger than the precarious size [31]. As a result, strength of the composite decreases with large size reinforcement because of decreased resistance against failure. The research

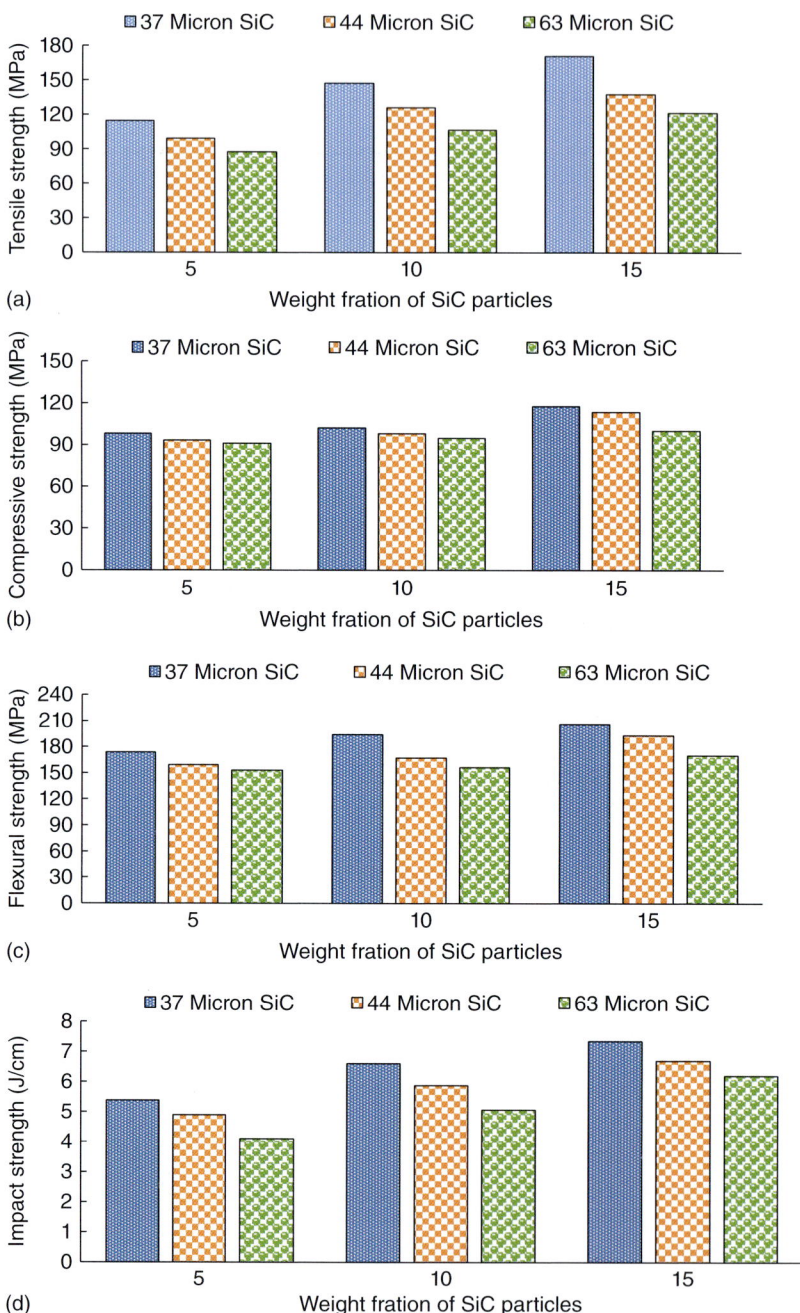

Figure 10.4 (a) Tensile strength vs. weight fraction of SiC, (b) compressive strength vs. weight fraction of SiC, (c) flexural strength vs. weight fraction of SiC, and (d) impact strength vs. weight fraction of SiC.

observations regarding weight fraction revealed that resistance against loading improves with increased weight fraction as increased reinforcement reduces interparticle distance. However, in the case of low weight fraction, the matrix debonds at an early stage because the reinforced particle fails to cover up the gap induced due to increased interparticle distance [32]. Besides, the strength of the composite depends upon the capability for stress transfer within the composite. Stress transfer improves with improved matrix reinforcement interaction but higher particle size reduces interfacial action and initiates cracking [33]. The effective stress transfer mechanism then improves bond strength because small particle size has more total surface area, which results in better resistance capability [34]. Another reason for reduced strength might be the micro cracks that are formed because of the internal stress developed during curing stage [35]. These stresses are typically originated by the chemical contraction during curing period and result in cracking of the matrix [36].

10.2.6 Energy Dispersive Spectroscopy

Energy dispersive spectroscopy (EDS) was used to analyze and confirm the presence of SiC particles in hybrid PMC. Randomly, one test specimen was cut from the fabricated hybrid PMC plate to perform EDS analysis using field emission scanning electron microscopy (FESEM). The observed percentage content of the elements that constitute the hybrid PMC is presented in Table 10.2. The FESEM image of the cut-out test specimen, element confirmation by EDS, and mapping analysis are shown in Figure 10.5. Mapping exploration demonstrates the three-dimensional dispersal of components of PMC existing as glass fiber, epoxy polymer, and SiC particles. The variable shades plots illustrate the existence of different constituents present in a similar zone. During investigation, the prime objective is to assess the existence of SiC as shown in Figure 10.5, which is also confirmed by mapping analysis. Even though partial inconsistency exists in the reinforcement, hybrid PMC shows a nearer and robust relationship with the

Table 10.2 Elemental contribution characterized by EDS.

Element	Atomic %	Weight %
C	97.01	92.13
Na	0.20	0.36
Mg	0.08	0.16
Al	0.24	0.51
Si	1.21	2.71
Cl	0.56	1.57
K	0.04	0.12
Ca	0.35	1.10
Fe	0.06	0.26

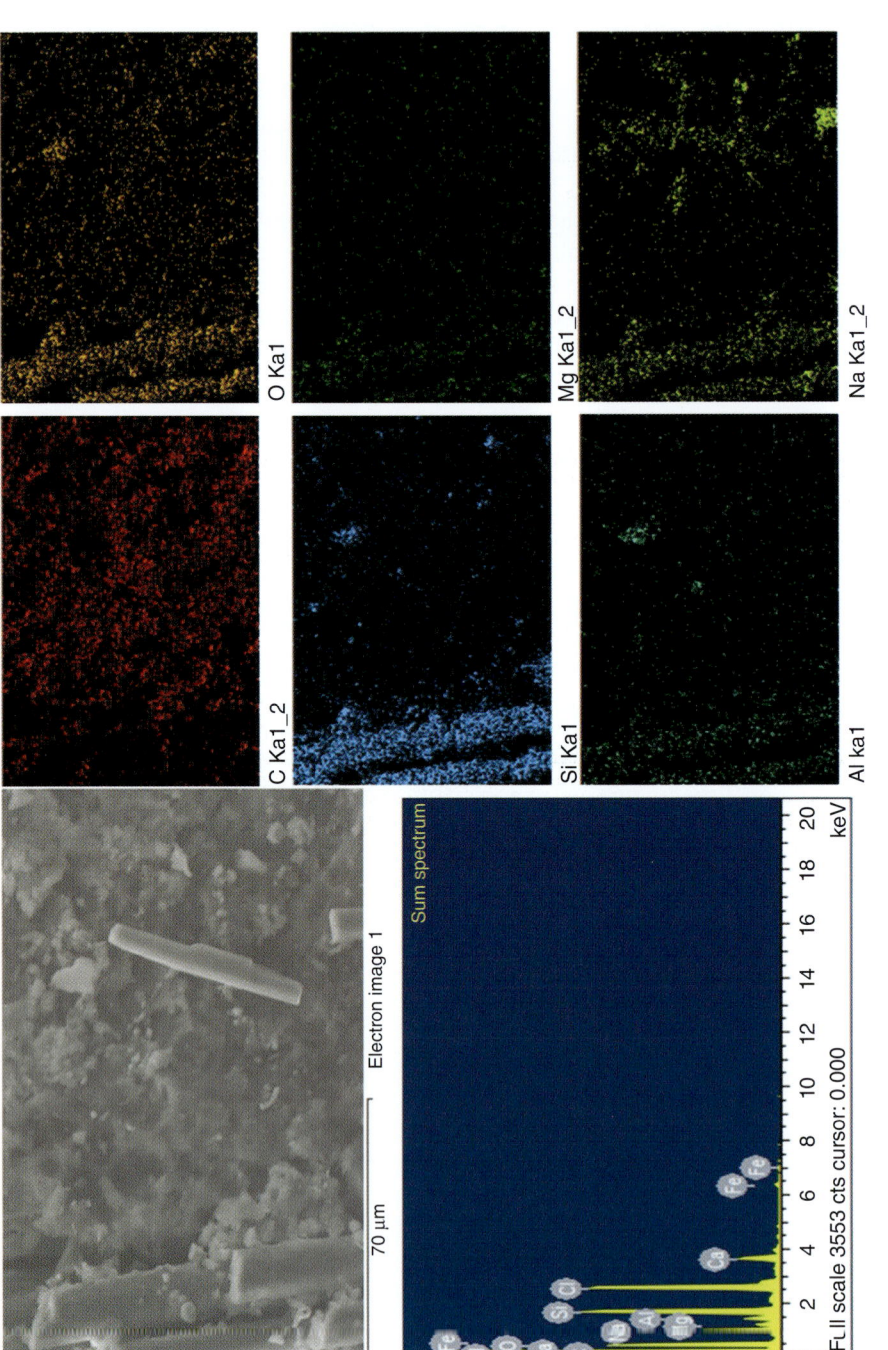

Figure 10.5 EDS mapping analysis of composite reinforced with SiC particles.

epoxy polymer as compared to a normal composite. Table 10.2 shows the atomic and weight percentage of constituents of hybrid PMC. The presented data show that carbon is a leading constituent of hybrid PMC with the highest atomic and weight percentage.

10.3 Environmental Degradation of PMCs

The usage of epoxy polymer-based composite materials is growing rapidly in aviation and chemical industries. However, the materials used in these industries are subjected to various chemical and other degrading abrasive elements. These composites when used in construction projects tolerate various environmental variations including variable temperature, rain, and sometimes acid rain in certain areas of Europe, east United States, and southeast Canada. These composites when used in shipping industries are subjected to corrosive and reactive fluids, which degrade the strength of the composite. Considering the applications of these composites in harsh and unadorned environment, some tests similar to those conditions have been performed to analyze the degrading behavior of these materials. In order to match those environmental conditions, chemicals such as sulfuric acid, sodium chloride, and distilled water were used to make solutions such as acid rain, seawater, and normal rain, respectively. The solution having similarities with acid rain refers to the blend of sulfuric acid and normal water possessing advanced level of hydrogen ions. This solution is prepared when water molecules react with sulfur dioxide and nitrogen oxide and result in the formation of acid. Here, sulfur dioxide reacts with hydrogen radicals to get oxidized [37]. This oxidized sulfur dioxide yet again reacts with oxygen to generate sulfur trioxide as shown in Eq. (10.2):

$$HOSO_2 + O_2 \rightarrow HO_2 + SO_3 \qquad (10.2)$$

Sulfuric acid is produced by the reaction of sulfur trioxide with water molecules present in the atmosphere as presented in Eq. (10.3):

$$SO_3 + H_2O \rightarrow H_2SO_4 \qquad (10.3)$$

Sulfuric acid (H_2SO_4) is the main element in acid rain, which possesses a pH value of 4.0. In the present case, the acid solution was prepared by keeping the pH value of solution at 3.0 to analyze the composite behavior at a much adversative level. The H_2SO_4 chemical with concentration of 0.01% by weight was used to prepare an acid rain-like solution. As per earlier observation regarding application in marine and shipping industries, the composite material was also tested in a solution containing a blend of sodium chloride (NaCl) and water, keeping the pH similar to that of seawater. For that requirement, pH of 8.1 was achieved by mixing NaCl having a concentration of 35 g/l with normal water. For the normal rain purpose, the composites were dipped in a container having distilled water. After preparing the solutions, testing was conducted in glass containers at room temperature to evade any reaction.

The composite materials were tested before immersing into solutions to analyze the mechanical strength. Based on the mechanical strength evaluation, the composite material having 15% weight fraction of SiC 400 mesh was selected for analysis. The composites dipped in acid, seawater, and rainwater solution were analyzed regularly in a span of three, six, and nine months. Three types of strength, i.e. tensile strength, flexural strength, and compressive strength, were examined on a universal testing machine (UTE 40) as per ASTM standards. The variation in observed strength at variable conditions is shown in Figure 10.6. The experimentally observed data for tested strength of composites immersed in acid, seawater, and rainwater for variable intervals of time are shown in Figure 10.6a–c. The graph values divulged that the composite immersed in acid solution showed extreme degrading of 10.80% in tensile strength for 270 days followed by seawater solution (6.74%) and rainwater (4.90%). The flexural strength shows depreciation of 9.01% for the specimen dipped in acid solution but the composite dipped in seawater and rainwater shows comparatively less depreciation by 5.40% and 4.21% respectively. An almost similar trend was reported by compressive strength analysis where strength is decreased by 7.22%, 5.5%, and 4.1% for acid, seawater,

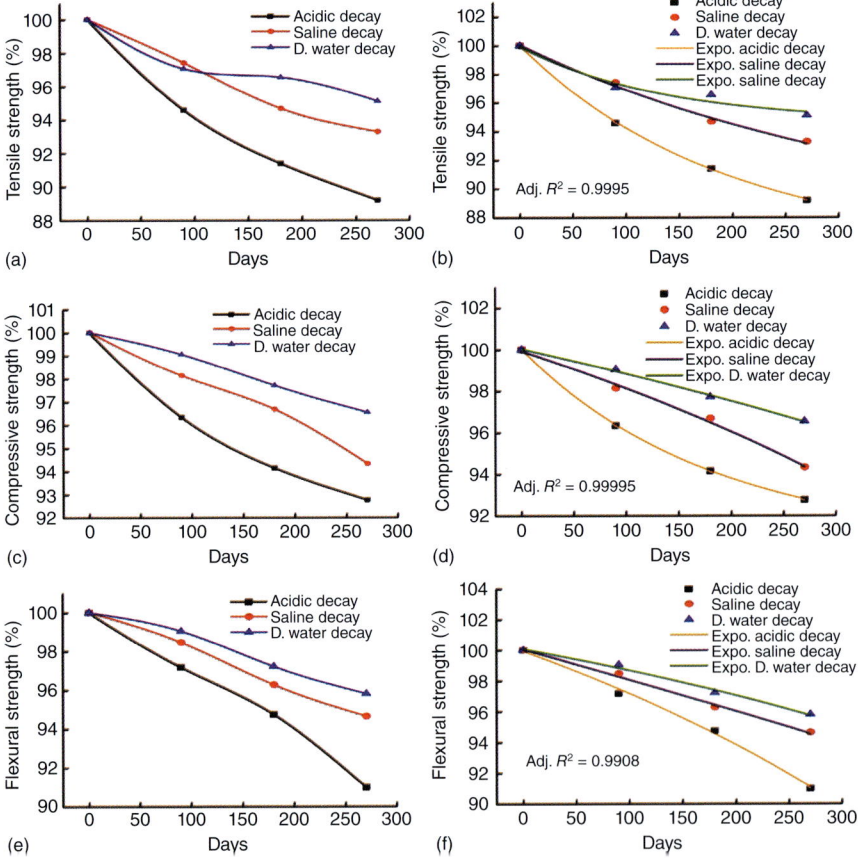

Figure 10.6 Experimental decrease in (a) tensile strength, (b) compressive strength, and (c) flexural strength; exponential decrease in (d) tensile strength, (e) compressive strength, and (f) flexural strength [1].

and rainwater solutions. Figure 10.6d–f reveals an exponential declining curve for individual strength at several intervals of time, which was plotted using Eq. (10.4):

$$Y = A_1 \times \exp(-x/t_1) + y_0 \qquad (10.4)$$

The composites immersed in acid solution and seawater acidic and saline solution suffered higher degradation as compared to rainwater because degradation of matrix was initiated when chloride ions seeped from these solutions to chemically react with metal ions from glass fibers. [38]. Another reason for the deprivation of composite surface could be the voids and grooves that were engorged by the moisture content.

10.4 Life Assessment of PMCs

The prominence of components made up of these composite materials depends upon their durable performance under loading. These materials are gaining high eminence in aviation industries. Previous research investigation revealed that the factor of safety level in aviation components is mentioned as 2.0 [39]. By considering the same, life assessment of composite material is calculated as the time when the material will indicate decrease in overall strength by half of its actual strength. The assessment of deterioration enactment of composite material was done by exponential functions. These functions are extensively used to analyze several circumstances such as degrading behavior, quantity evaluation, and investment calculation. The functions are accurate in evaluating the forecast behavior based on the interpretations delivered.

This chapter includes an exponential equation as shown in Eq. (10.5) to assess the life of composites immersed in acid solution, which is rated as a severe environment for composites. The investigational values of deterioration in tensile strength of the material in a time period of 0–9 months are shown in columns 1 and 2 of Table 10.3.

$$Y = AB^x \qquad (10.5)$$

Here, Y and x are denoted as predicted value and explanatory variable respectively.

Table 10.3 Life assessment of composites.

X	Y_e	log Y_e	XY_e	X^2	Predicted value (y_p)
0	100	2	0	0	99.08
90	94.6	1.9758	177.822	8100	94.72
180	91.4	1.9609	352.962	32400	91.39
270	89.2	1.9503	526.962	72900	86.56
$\sum X = 540$	$\sum Y_e = 375.2$	$\sum \log Y_e = 7.88$	$\sum XY_e = 1057.36$	$\sum X^2 = 113400$	

Taking log to Eq. (10.1)

$$\log Y = \log A + x \log B$$

By substituting the values, $y = \log Y$, $a = \log A$, and $b = \log B$, the equation develops as

$$y = a + bx$$

The unknown values of a, b can be obtained from Table 10.3 using subsequent equations:

$$\Sigma y_e = na + b \Sigma x \tag{10.6}$$

$$\Sigma xy_e = a \Sigma x + b \Sigma x^2 \tag{10.7}$$

Using Eqs. (10.6) and (10.7),

$$y_p = 99.083 \times (0.9995)^x \tag{10.8}$$

If y_p is 50%, then $= \frac{\log(0.5046)}{\log(0.9995)}$; $x = 1368$.

Correspondingly, using Eq. (10.5) the strength of composite at variable period can be obtained. The life assessment analysis revealed that a component made up of this composite material will decay up to 50% of its strength in 1368 days when used under acid rain.

Equally, by computing values of constants for seawater and rainwater, the lifespan of composite was found to be 2600 and 3751.12 days respectively. The observation depicted in Figure 10.7 shows that extreme and least deterioration in tensile strength is witnessed in that composite which was tested under acid rain and normal rain respectively. Table 10.4 reveals the relative investigation of life assessment. In acid rain environment, the material degraded to half of its strength in around 1368 days; however, in rainwater an identical phase arrived after 3751.12 days.

10.4.1 Morphological Inspection

The initiation of surface cracks was accomplished with waning hardness of the uppermost layer composite surface immersed in experimental fluids. The uppermost layer here denotes the epoxy matrix, which emanates from direct interaction with the experimental solution. The resistance against cracks induced due to this phenomenon was improved by amalgamation of SiC particles with epoxy polymer. This amalgamation increases the hardness of the composite by transmitting load from the moderately soft matrix to reinforced SiC particles. The resistance of the composite surface is significantly influenced by the size of reinforced particles. The higher particle size decreases the interfacial area, which reduces the amount of secondary reinforcement in the matrix. The reduced amount of SiC fails to perform as load carrier and as a result increases the stress in matrix. But smaller reinforcement removes these obligations and improves resistance of the composite because of increased interfacial area. The solution having pH value of 3.0, which is less than acid rain, significantly acts as bond breaker between the

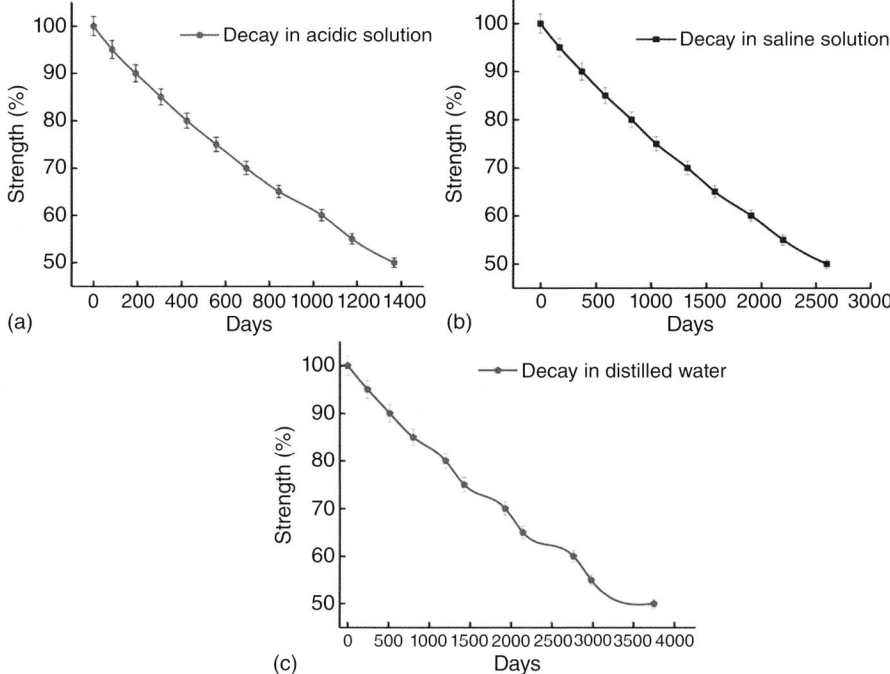

Figure 10.7 Periodic decay in tensile strength under (a) acidic solution, (b) saline solution, and (c) distilled water [1].

Table 10.4 Relative exploration of life predictions.

Solution	Exponential equation	Predicted days (till half strength)
Acidic	$Y = 99.083 \times (0.9995)^x$	1368
Saline	$Y = 99.5359 \times (0.99973)^x$	2600
Distilled water	$Y = 99.8159 \times (0.9998)^x$	3751.12

reinforcement and the matrix. When the composite specimen is immersed in acid solution, voids and grooves are induced over the surface and these defects are higher in number as compared with the surface of a specimen dipped in other solutions having high pH. The degradation rate of the surface is considerably increased because the induced voids diminish the interfacial strength. But in case of a composite having relatively smaller reinforcement, the resistance was much improved. It is due to improved interfacial area, which holds enormous SiC particles as compared to bigger reinforcement. The increased number of SiC particles improves resistance against matrix degradation followed by void formation for a comparatively long duration. Surface analysis of composites tested under conditions of acid rain, seawater, and normal rain is shown in Figures 10.8–10.10. The morphology revealed that the composite immersed in acid rain solution showed

Figure 10.8 Composite dipped in acidic solution.

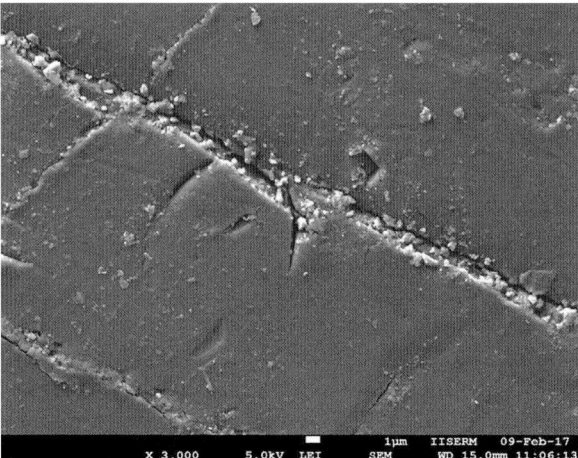

Figure 10.9 Composite dipped in saline solution.

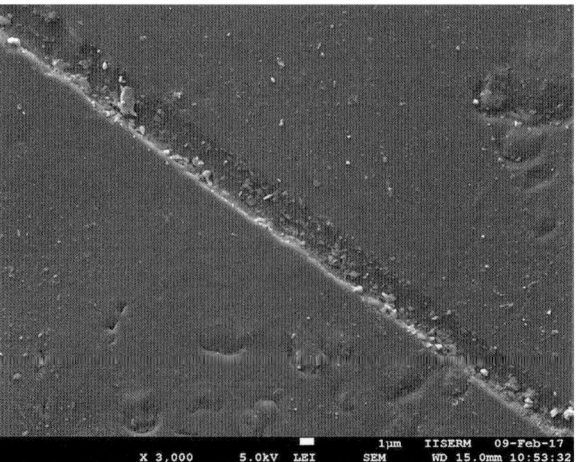

Figure 10.10 Composite dipped in distilled water.

higher degradation as compared to others. It is because the solution with lesser pH persuades comparatively high voids over the composite surface. These voids result in debonding of matrix and reinforcement because of decrease in interfacial strength between glass fibers, SiC, and epoxy polymer. These cavernous and widespread grooves initiate cracks over the surface and finally induce peeling off and pulling off of fibers. Figure 10.8 shows the composite surface degraded in acid solution. The morphology clearly indicates deep degradation of matrix in a large area. The grooves in larger area decrease the bond strength, which leads to untimely damage under variable loading. The SEM of composites immersed in seawater is shown in Figure 10.9. Morphological analysis revealed that linear matrix tearing is present over the composite surface, although overall matrix seems to have more bonding with reinforcement as compared with acid solution. The matrix tearing prompted on the surface under saline solution is contracted as equated with acid solution but greater in number and deeper than rainwater as shown in Figure 10.10. The reduced blows in the composite immersed in rainwater help in producing healthier bond strength and resistance against failure. The nonexistence of cavities on the external layer of the composite improves interfacial bond strength which induces durable behavior of composites.

10.5 Conclusions

The important conclusions from the present chapter are summarized and presented as follows:

- Composites with or without SiC as secondary reinforcement were fabricated using hand lay-up methodology and tested for mechanical strength using UTE 40 machine.
- The mechanical strength evaluation study confirmed that secondary reinforcement of SiC particles significantly improves the mechanical properties of composites. The observed results for the influence of size of reinforcement revealed that strength of composites reduces with increased particle size. In the present work, the composites having SiC particle of 37 μm average size produced enhanced resistance against failure.
- The results obtained for weight fraction and particle size of the secondary reinforcement revealed that increase in weight fraction shows improved mechanical properties whereas mechanical properties diminish at comparatively higher particle size.
- The strength evaluation of composites was also analyzed under adversarial climate changes. Based on the test results, extreme deprivation in tensile strength was observed in a solution having a pH value of 3.0, which is more acidic than normal acid rain followed by seawater and rainwater. The strength of the composite was reduced to half in 1368 days, while composites sustained against seawater and rainwater for 2600 and 3751.12 days respectively before degrading to half the strength.
- The bond strength between fiber, SiC, and epoxy was analyzed by scanning electron microscopy. SEM images validated lesser interparticle distance and improved bond behavior.

- EDS demonstrated the influence of numerous components in fabricating composites, and mapping analysis provides evidence about the uniformity of accumulation in reinforcements.

References

1 Antil, P., Singh, S., and Manna, A. (2017). Glass fibers/SiCp reinforced epoxy composites: effect of environmental conditions. *Journal of Composite Materials* 52 (9): 1253–1264.

2 Antil, P., Singh, S., and Manna, A. (2018). Analysis on effect of electroless coated SiC_p on mechanical properties of polymer matrix composites. *Particulate Science and Technology* https://doi.org/10.1080/02726351.2018.1444691.

3 Antil, P., Singh, S., and Manna, A. (2018). Effect of reinforced SiC particulates of different grit size on mechanical and tribological properties of hybrid PMCs. *Material Today Proceedings* 5 (2): 8073–8079.

4 Chen, L.F., Hong, Y.P., Zhang, Y., and Qiu, J.L. (2000). Fabrication of polymer matrix composites reinforced with controllably oriented whiskers. *Journal of Materials Science* 35: 5309–5312.

5 Ghoneim, M. and Mahani, R.M. (2001). Electrical and mechanical properties of some composites with polymer matrix. *International Journal of Polymeric Materials and Polymeric Biomaterials* 50 (2): 141–161.

6 Fomitchov, P.A., Kim, Y.K., Kromine, A.K., and Krishnaswamy, S. (2002). Laser ultrasonic array system for real-time cure monitoring of polymer-matrix composites. *Journal of Composite Materials* 36 (15): 1889–1901.

7 Soon-Gi, S. (2002). Self-diagnosis of GFRP composites containing carbon powder and fiber as electrically conductive phases. *Metals and Materials International* 7 (6): 605–611.

8 Huang, Z., Nie, X., and Xia, Y. (2004). Effect of strain rate and temperature on the dynamic tensile properties of GFRP. *Journal of Materials Science* 39 (10): 3479–3482.

9 Sekine, H. and Beaumont, P.W.R. (2006). Micro-mechanical theory of macroscopic stress-corrosion cracking in unidirectional GFRP. *Journal of Material Science* 41 (14): 4604–4610.

10 Shindo, Y., Narita, F., and Sato, T. (2006). Analysis of mode II interlaminar fracture and damage behavior in end notched flexure testing of GFRP woven laminates at cryogenic temperatures. *Acta Mechanica* 187 (1): 231–240.

11 Singh, I. and Bhatnagar, N. (2006). Drilling-induced damage in uni-directional glass fiber reinforced plastic (UD-GFRP) composite laminates. *The International Journal of Advance Manufacturing Technology* 27 (9): 877–882.

12 Antil, P., Singh, S., and Manna, A. (2017). Electrochemical discharge drilling of SIC reinforced polymer matrix composite using Taguchi's grey relational analysis. *Arabian Journal for Science and Engineering* 43 (3): 1257–1266.

13 Singh, M. and Singh, S. (2018). Machining of difficult to cut materials by electrochemical discharge machining (ECDM) process: a state of art approach.

Advanced Manufacturing and Materials Science https://doi.org/10.1007/978-3-319-76276-0_14.

14 Das, K., Kamaruzzaman, S.M., Middya, T.R., and Datta, S. (2009). Study of variation of thermal diffusivity of advanced composite materials of E-glass fiber reinforced plastic (GFRP) in temperature range 5–300 K. *Bulletin Material Science* 32 (1): 89–92.

15 Sekine, H. and Beaumont, P.W.R. (2010). Micromechanically derived lowest threshold stress intensity factor for macroscopic stress-corrosion cracking in unidirectional GFRP composites. *Journal of Material Science* 45: 5988–5992.

16 Kim, D., Kim, Y.H., Gururaja, S., and Ramulu, M. (2010). Processing and fiber content effects on the machinability of compression molded random direction short GFRP composites. *International Journal of Automotive Technology* 11 (6): 849–855.

17 Rakesh, P.K., Inderdeep, S., and Kumar, D. (2011). Compressive behavior of glass fiber reinforced plastic laminates with drilled hole. *Advance Materials Research* 410 (1): 349–352.

18 Huh, Y.-H., Lee, J.-H., Kim, D.-J., and Lee, Y.-S. (2012). Effect of stress ratio on fatigue life of GFRP composites for WT blade. *Journal of Mechanical Science and Technology* 26 (7): 2117–2120.

19 Gudonis, E., Kacianauskas, R., Gribniak, V. et al. (2014). Mechanical properties of the bond between GFRP reinforcing bars and concrete. *Mechanics of Composite Materials* 50 (4): 457–466.

20 Bey, K., Tadjine, K., Khelif, R. et al. (2015). Mechanical behavior of sandwich composites under three-point bending fatigue. *Mechanics of Composite Materials* 50 (6): 747–756.

21 Neuenschwander, J., Furrer, R., and Roemmeler, A. (2016). Application of air-coupled ultrasonics for the characterization of polymer and polymer-matrix composite samples. *Polymer Testing* 56: 379–386.

22 Tao, Y., Chen, H., Yao, K. et al. (2017). Experimental and theoretical studies on inter-fiber failure of unidirectional polymer-matrix composites under different strain rates. *International Journal of Solids and Structures* 113: 37–46.

23 Ghasemi, A.R. and Moradi, M. (2017). Effect of thermal cycling and open-hole size on mechanical properties of polymer matrix composites. *Polymer Testing* 59: 20–28.

24 Bigdeli, M.B. and Fasano, M. (2017). Thermal transmittance in graphene based networks for polymer matrix composites. *International Journal of Thermal Sciences* 117: 98–105.

25 Antil, P., Singh, S., and Manna, A. (2018). SiC$_p$/glass fibers reinforced epoxy composites: wear and erosion behavior. *Indian Journal of Engineering and Materials Sciences* 25 (2): 122–130.

26 Araldite (AW 106/HV 953U) (2011). Adhesive and tooling. Publication No. A 230 e GB. http://www.farnell.com/datasheets/24297.pdf (accessed 10 March 2015).

27 El, Messiry, M. (2013). Theoretical analysis of natural fiber volume fraction of reinforced composites. *Alexandria Engineering Journal* 52 (3): 301–306.

28 Deepa, K.S., Sebastian, M.T., and James, J. (2007). Effect of inter particle distance and interfacial area on the properties of insulator conductor composites. *Applied Physics Letters* 91: 202904. https://doi.org/10.1063/1.2807271.

29 Slipenyuk, A., Kuprin, V., Milman, Y. et al. (2006). Properties of P/M processed particle reinforced metal matrix composites specified by reinforcement concentration and matrix-to-reinforcement particle size ratio. *Acta. Material Papers* 54 (1): 157–166.

30 Finot, M., Shen, Y.L., Needleman, A., and Suresh, S. (1994). Micromechanical modeling of reinforcement fracture in particle-reinforced metal matrix composites. *Metallurgical and Materials Transactions A* 25 (11): 2403–2420.

31 Song, M. and Xiao, D.H. (2008). Modeling the fracture toughness and tensile ductility of SiCp/Al metal matrix composites. *Material Science and Engineering: A* 474 (1–2): 371–375.

32 Bartczak, Z., Argon, A.S., Cohen, R.E., and Weinberg, M. (1999). Toughness mechanism in semi-crystalline polymer blends: II. High-density polyethylene toughened with calcium carbonate filler particles. *Polymer* 40 (9): 2347–2365.

33 Zhang, S., Cao, X.Y., Ma, Y.M. et al. (2011). The effects of particle size and content on the thermal conductivity and mechanical properties of Al_2O_3/high density polyethylene (HDPE) composites. *EXPRESS Polymer Letters* 5 (7): 581–590.

34 Fu, S.-Y., Feng, X.-Q., Lauke, B., and Mai, Y.-W. (2008). Effects of particle size, particle/matrix interface adhesion and particle loading on mechanical properties of particulate–polymer composites. *Composites: Part B* 39 (6): 933–961.

35 Ersoy, N., Garstka, T., Potter, K. et al. (2010). Development of the properties of a carbon fibre reinforced thermosetting composite through cure. *Composites: Part A* 41 (3): 401–409.

36 Sorrentino, L., Bellini, C., Capriglione, D., and Ferrigno, L. (2015). Local monitoring of polymerization trend by an inter digital dielectric sensor. *The International Journal of Advance Manufacturing Technology* 79 (5–8): 1007–1016.

37 John, S., Pandis, H., and Spyros, N. (1998). *Atmospheric Chemistry and Physics – From Air Pollution to Climate Change*. Wiley. ISBN: 978-0-471-17816-3.

38 Wang, J., Ganga Rao, H., Liang, R. et al. (2015). Durability of glass fiber-reinforced polymer composites under the combined effects of moisture and sustained loads. *Journal of Reinforced Plastics and Composites* 34 (21): 1739–1754.

39 Zrua, L. (1993). Reliability based safety factor for aircraft composite structures. *Computers and Structures* 48 (4): 745–748.

11

Fire Performance of Natural Fiber Reinforced Polymeric Composites

Divya Zindani[1], Kaushik Kumar[2], and João Paulo Davim[3]

[1] National Institute of Technology Silchar, Department of Mechanical Engineering, Silchar, 788010, India
[2] Birla Institute of Technology, Mesra, Department of Mechanical Engineering, Ranchi, 835215, India
[3] University of Aveiro, Department of Mechanical Engineering, Campus Santiago, Aveiro, 3810-193, Portugal

11.1 Introduction

Synthetic fibers-based polymeric composites possess excellent mechanical characteristics and therefore have been used intensively and widely for various engineering applications [1]. However, poor recyclability of synthetic fiber reinforced polymeric composites results in adverse environmental impacts. Owing to the environmental implications, synthetic fibers have been replaced with natural fibers as reinforcements. Natural fibers have umpteen environmental benefits such as biodegradability, low pollutant emission, and recyclability [2]. Further, such fibers also possess excellent mechanical characteristics such as higher strength and high specific stiffness [3]. Mechanical performance of natural fibers has been enhanced by taking care of their hydrophobic nature through various chemical treatments [4, 5]. Moisture absorption capability of fibers has also been reduced by modification of polarity of natural fibers through alkali treatment or acetylation [6].

With numerous advantages, natural fibers tend to suffer from a major drawback of low fire resistance. Higher flammability of the natural fiber reinforced composites is mainly due to the presence of cellulose in natural fibers and hydrocarbon in the polymers. Polymers impact the environment adversely on decomposing [7]. The decomposition process occurs at temperatures ranging 300–500 °C resulting in the generation of combustible gas, char, and smoke. Releasing smoke impairs the visibility in the surrounding environment and thereby poses serious risks to humans [8]. Furthermore, there is degradation in structural integrity and hence the load bearing ability of the natural fiber reinforced polymeric composites on heating due to the softening and creep behavior of fiber reinforcement and polymeric matrix material [9]. With stricter norms and regulations in the infrastructure domain, it becomes necessary to inspect the susceptibility of the composites to fire. Other application areas such as aviation industry require different tests pertaining to flammability such as smoke density and vertical burn test on the fiber reinforced polymeric composite structures; cone calorimeter

Reinforced Polymer Composites: Processing, Characterization and Post Life Cycle Assessment,
First Edition. Edited by Pramendra K. Bajpai and Inderdeep Singh.
© 2020 Wiley-VCH Verlag GmbH & Co. KGaA. Published 2020 by Wiley-VCH Verlag GmbH & Co. KGaA.

parameters procedure has been followed for adjudging the suitability of composite structures for building applications. Because of the stricter norms in the various application areas it is necessary to investigate the flame retardant treatments in order to enhance the burning resistance of natural fiber reinforced polymeric composite materials. Studies have claimed that the burning resistance of such composites can be enhanced with the addition of flame retardants or fibers with flame retarded characteristics, but that comes by compromising with the mechanical properties [10, 11]. Therefore, it still remains a challenge to create a balance between the mechanical and flammability performance of natural fiber reinforced composite materials.

One of the major reasons behind lesser attention toward flammability aspects of natural reinforced fiber composites is that the different experiments required for measurement of fire performance are cost intensive and consume a lot time. Furthermore, some of the tests require special arrangements and advanced facilities to be implemented in the household and to get the relevant information [12]. It is also difficult for the experimental space to truly depict the complex real life scenario and subsequently develop a model to estimate the thermal structural correlation, and this is another topic of research interest that deserves attention.

The present chapter aims to discuss the fire performance of natural fiber reinforced polymeric composite materials. After the introduction, the chapter briefs the readers with the flammability characteristics of natural fibers and thermal properties of polymers. Next is an outline on the fire retardant phenomenon with the different methods of flame retardants. In the subsequent section the fire performance of natural fiber reinforced polymeric composite materials for transportation interiors and building materials is discussed. The chapter finally ends with the concluding remarks.

In order to comprehend the present research milieu in the domain of flammability of composites, databases of peer reviewed journals such as Web of Science and Scopus were searched. Keywords used during the search were "Flammability" and "Composites." The search yielded a total of 2477 and 2549 articles respectively. Analysis has been done in the categories of subject, number of articles with publication years, regional distribution around the globe, and the organizations. The results have been depicted in Figures 11.1–11.4 respectively. The figures clearly indicate the growing awareness in the subject and the importance gained by the same in recent years.

11.2 Flammability Aspects and Thermal Properties of Natural Fibers and Natural Fiber Reinforced Polymeric Composites

Pectin, hemicellulose, cellulose, wax, lignin, inorganic ash, and some small amounts of secondary metabolites are the constituting components of a cellulose-based plant fiber [13]. Crystallinity of cellulose is on the higher side

Figure 11.1 Distribution of articles on the basis of subject area. Source: http://wcs.webofknowledge.com.

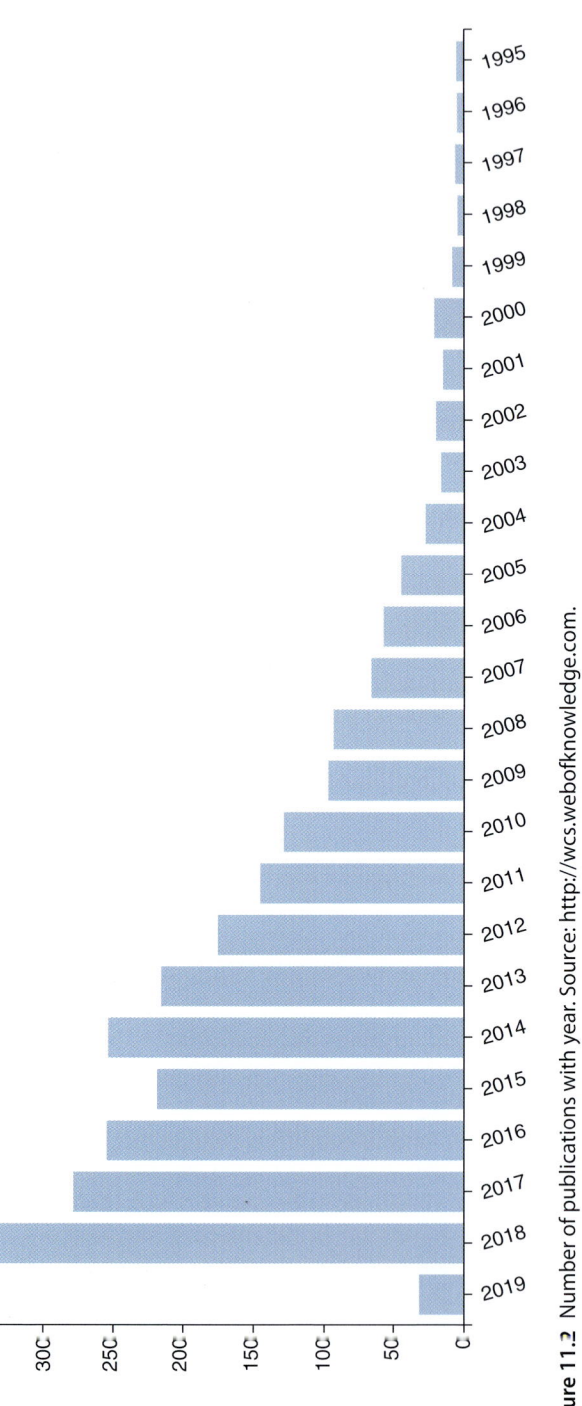

Figure 11.2 Number of publications with year. Source: http://wcs.webofknowledge.com.

Figure 11.3 Regional distribution of the articles published around the globe. Source: http://wcs.webofknowledge.com.

Figure 11.4 Distribution of articles on the basis of universities. Source: http://wcs.webofknowledge.com.

including that of glucose. Micro-cellulose fibrils are cemented together by the hemicellulose. Cellulose linked together as units forms the main structural component of the fiber cell [11]. The following are the steps of thermal decomposition of cellulose content in a fiber cell: desorption of water absorbed, formation of dehydrocellulose as a result of evolution of water, generation of char and other volatiles with the decomposition of dehydrocellulose, levoglucosan generation, and finally its degradation resulting in char, tar, gases, and nonflammable volatiles [14, 15]. Hemicellulose, on the other hand, decomposes at temperatures ranging 200–260 °C and yields incombustible gases. Lignin decomposes in a wide range of temperature ranging 160–400 °C. Lignin is responsible for promoting the degradation of cellulose content in the fiber cell, thereby resulting in dehydration and ultimately the formation of char. Strong bonds such as that between ether and carbon break at higher temperatures whereas lower temperatures initiate the breakage of weak bonds [16]. Heat release rates have been found to be lower for lignin rich fibers from bast such as flax and hemp in comparison to the plant fibers from leaf such as abaca and cabuya [17]. The flammability characteristics of plant fibers have been found to be affected greatly by the presence of chemicals on the surface of fibers. Formation of levoglucosan as a result of thermal decomposition results in increased flammability. Higher crystallinity of the fibers is one of the major reasons that can be attributed to the higher levels of levoglucosan. Flammability is however diminished with the highly oriented fibrils as these tend to reduce the process of pyrolysis [8].

Animal-based fibers are mainly constituted by the amino acid groups and therefore play a dominant role in the determination of burning behavior and thermal decomposition. Wool, one of the main types of animal fibers, is constituted with eighteen α-amino acids. Aspartic acids, glutamic acid, lysine, and cysteine are the building blocks for α-amino acids [18]. A pair of cysteine residues results in the formation of a disulfide bond, which plays a predominant role in making wool stable against environmental degradation [19]. Moisture evaporation from wool takes place at around 100 °C and at around 210 °C, initiating an endothermic reaction. The endothermic reaction results in the formation of gaseous products such as hydrogen sulfide. Formation of char then takes place as a result of breaking of disulfide bonds.

Effects of thermal behavior of natural fibers on the flammability and thermal decomposition of the natural fiber reinforced polymeric composites have been investigated by various researchers. Investigations have revealed that with the inclusion of natural fibers in the polymeric matrices, total heat release rates, heat release rates, horizontal and vertical charring rates, and production of smoke got substantially reduced [20–23]. Furthermore, the natural fiber reinforced composites failed to achieve the desired vertical burn grade; for instance, polymeric composites with hemp fabric revealed a trivial effect of hemp fabric on flammability [24]. Studies have revealed the inadequacy of natural fiber reinforced composites to meet the strict regulations for various engineering applications. In a quest to achieve the desired performance levels concerning the fire properties, umpteen investigations are under progress pertaining to fire retardation.

11.3 Fire Retardants

A material possessing fire retardant characteristics will have very less tendency to ignite, and the rate of burning would also be slow [7]. Fire retardation is very much a crucial aspect for natural fiber-based polymeric composites as they burn easily, once their constituting elements reach the ignition temperature and therefore fail to meet the flammability requirements. Slowing down the chain branching reaction during the combustion process can aid in achieving the required levels of flame retardation in natural fiber reinforced polymeric composite materials. Reduction in flammability of either the fiber reinforcement or polymeric matrix or the entire composite will aid in enhancing the fire retardation of the polymeric composite material. The following section illuminates the readers with the various methods to achieve flame retardation that have aided in minimizing the flammability of natural fiber polymeric composite materials.

11.4 Flame Retardants

Additive type and reactive type are the two major categories of flame retardants that can be included in a composite material. Additive flame retardants are introduced into the composites physically during their fabrication such as the addition of some mineral fillers and other organic compounds. Additive type of flame retardants have been incorporated both for thermoplastics and thermosets. On the other hand, reactive flame retardants chemically modify the chemical structure of the polymeric composite materials and hence their flammability characteristics. Flame retardants can either be included in the backbone of the polymer or in the functional group of the polymeric matrix [25]. Some of the major flame retardants that have been effective in the reduction of the flammability of natural fiber polymeric composite materials have been discussed in the following subsection.

11.4.1 Mineral Flame Retardants

Hydroxides of aluminum and magnesium are some of the metal hydroxides that aid in enhancing the flame retardant characteristics of polymeric composite materials. These metal hydroxides release significant amount of water, which aids in suppression of the combustion process [10]. The presence of metal hydroxides cools down the polymeric matrix as they absorb heat when they undergo endothermic decomposition. The process of endothermic decomposition takes place in the temperature range of 300–320 °C [26, 27]. Increase in decomposition temperature and limiting oxygen limits were reported by El-sabbagh et al. [26] on the addition of 20–30 wt% $Mg(OH)_2$. Limiting oxygen limit was reported to be 28% with the polymeric composite having 0.5 part flax and 0.3 part metal hydroxides by weight. Increased tensile stiffness was also reported with the addition of flame retardant. However, due to the weak adhesion between the polymeric matrix and the flame retardant, the tensile strength was found to be reduced to around 31%. Another major disadvantage associated with the flame retardant is the need for significant levels of metal hydroxide addition to achieve the desired flame resistance.

11.4.2 Bulging of Flame Retardants

In order to explore the replacement of mineral-based flame retardants, researchers have extensively searched for alternatives. With the continual research, flame retardants with bulging properties or "intumescent" flame retardants were revealed. Halogen-free retardants have been explored that are effective in forming char and are at the same time eco-friendly [28]. Swollen multicellular char is formed by the intumescent flame retardants that are capable of protecting the surrounding polymer from the flame and hence the radiant heat flux [29]. Therefore, the self-sustained combustion of polymeric material is suppressed by the intumescent char as it reduces the heat between the underlying material and the flame [30, 31]. Furthermore, halogen-free flame retardants present low toxicity on combustion [32].

Nitrogen and ammonia are produced by the intumescent flame retardants based on ammonium polyphosphate (APP). The presence of ammonia and nitrogen dilutes the fuel gases. There has been a good deal of research in exploring the utilization of intumescent flame retardant for minimizing the flammability of natural fiber-based polymeric composite materials [10, 21, 23, 33–35].

Fire performance of kenaf-based polypropylene composites was compared for different kinds of intumescent flame retardants based on APP. Improved dispersion of intumescent particles and kenaf fibers was revealed through morphological analysis. The twin screw extruder used aided in the excellent blending and hence homogeneous distribution of intumescent flame retardant particles. Homogeneous distribution resulted in decrease in heat release rates and sustained combustion. The char formed after the cone calorimeter test could provide a better insulation for the underlying elements of the composite material. Furthermore, the fabricated composite had higher flexural and tensile stiffness. However, vertical burn test revealed that the kenaf-polypropylene composite with the intumescent flame retardant particles burned continuously with dripping. The tensile strength of the fabricated composite was also found to be lower.

Wool has higher temperature of combustion and limiting oxygen limit in comparison to other natural fibers. Investigations on the flammability properties as well as the fire reaction properties of wool were carried out by Kim et al. [36]. Significant reduction in heat release rates were reported for the combined wool and intumescent retardant particles in comparison to that of neat polypropylene. The reduced content of intumescent retardant particles from 20 to 15 wt% could achieve the desired V-0 rating and was also able to prohibit sustained burning of the composite material [34]. The presence of wool was also responsible for the self-extinguishing behavior under vertical flame application. The conclusion could be drawn as the other natural fibers could not give similar performance levels. The cysteine amino group present in wool is the source of cysteic acid, which combines with phosphoric acid, and the combination then acts as a catalyst for phosphorylation. Char formation as well as dehydration is promoted by the acidic catalysts. Therefore, the acidic catalysts aid in formation of rigid char around the wool-polypropylene composite material and therefore prevent the flame from moving through the composite material. However, a weaker adhesion between APP particles and wool was reported by various authors. A weaker adhesion signifies reduced tensile strength of the composite specimen. Hence, it becomes quintessential to balance the mechanical properties with the fire retardancy characteristics, so that the application domain could be expanded.

A double step compounding process for polypropylene, intumescent fire retardant particles, and antioxidant was followed to result in flame retardant polypropylene [37]. The interference between talc and intumescent flame retardant particles is avoided by preparation of two types of combinations, one being the combination of talc and polypropylene and the other being polypropylene, antioxidant, and intumescent flame retardant particles. Significant reductions in flammability were reported with the usage of flame retardant polypropylene for the fabrication of flax and wool composites [38]. The composites fabricated using flax and flame retardant polypropylene were reported to have higher flexural and tensile strength in comparison to the composites made using flax and polypropylene. The enhanced mechanical properties could be attributed to better interfacial adhesion between the matrix and the fibers and also to the proper dispersion of the intumescent flame retardant particles and talc. These particles, which otherwise would have diminished the interfacial bonding, are surrounded by the polypropylene matrix. Hence, the composites fabricated using double mixing step process have a better interfacial adhesive strength between the polymeric matrix and fibers in comparison to that fabricated using a single-step compounding process.

Progressive research work has been carried out to develop new flame retardants; however, one of the major challenges lies in investigating their detrimental effect on the interfacial adhesion. Further, understanding of the reaction between fire retardant particles and fibers still requires further investigations. Investigations are also required to be carried out for better understanding of the effect of flame retardants on the fire properties as well as mechanical properties of the fibers and polymers.

11.5 Fire Performance for Usability as Materials in Transportation

Natural fiber reinforced composites have found a wide range of applications in automotive and aerospace industries. Their usability has been explored mainly for enhancing the interiors. The potential of natural fiber reinforced polymeric materials for such applications has also been evidenced through the various flame retardant methods. However, transportation industries have very stringent rules and regulations and candidate materials must pass the various tests in order to be used befittingly for such applications. As for instance, smoke density measurements, vertical burning, and heat release are the tests prescribed by the Code of Federal Regulation (CFR) for aircraft interiors.

The flammability of fabricated flax-epoxy composites was compared with that of glass fabric-polymeric composite [39] and was found to be higher. A more rigorous burning of flax-epoxy composites in comparison to the glass counterpart was reported from the 60 seconds vertical burn tests. The performance of flax-epoxy composite after the inclusion of intumescent flame retardant particles revealed a fire performance comparable to that of the glass-epoxy composites. It took 14 and 15 seconds respectively for the flax-intumescent fire retardant composite and glass-intumescent fire retardant composite to extinguish. Both the composites were able to pass therefore the CFR vertical burn test. The peak heat release rates of flax-intumescent fire retardant composite specimen was revealed to be 269.4 kW/m^2, which was similar to that of the glass-intumescent fire retardant composite (269.2 kW/m^2). The significant reduction in flammability was attributed to the more intumescent char of flax-intumescent fire retardant composite. The presence of lignin and therefore the reaction between intumescent fire retardant and flax promoted the effectiveness of char formation.

Studies have revealed a lacuna of flammability tests as per the Federal Aviation Regulations (FAR) regulation on natural fiber reinforced composite materials. Furthermore, investigations on mechanical properties as well as the fire performance should also be done for the newly developed bio-based reinforcements as well as the related polymeric matrix.

Natural fiber reinforced composite materials have long been used in various automotive parts owing to their high specific mechanical properties and long-term sustainability. These composite have been used increasingly for door panels, bolsters, instrument panels, pillars, and seat back panels [40, 41]. Acceptable flammability test results have suitably been prescribed in Federal Motor Vehicle Safety Standard (FMVSS). The standards have specified the burn resistance of materials to be used for automotive interiors [42]. As for instance, the stipulated fire performance level, i.e. horizontal burning rate should not be higher than 102 mm/min. The flammability of latex jute felt composite has been evaluated by Fatima and Mohanty [43]. The horizontal burning rate was revealed to be lower than 9.77 mm/min with the inclusion of 1 wt% sodium phosphate. The fabricated composite specimen was able to meet the regulatory requirement of 102 mm/min.

11.6 Fire Performance for Usability as Building Materials

Honeycomb structures have excellent stiffness, thermal and acoustic insulation, and strength to weight ratios and therefore have been used widely for various practical applications [44]. The suitability of natural fibers to construct sandwich panels has been explored widely by the researchers. These natural fiber reinforced composite materials have proved their potentiality to be used as facades, roofs, floors, and decks, and in balcony construction [45]. However, in order to justify the requirements of structural integrity as well as the functional properties at higher temperatures it becomes quintessential to employ various flame retardant methods. As in the case of transportation, the infrastructure domain also has its standards, i.e. National Building Codes. These codes enlist the required tests and their limits for the materials to provide the minimum fire safety levels. Evaluation of time to ignition and heat release rates was conducted on the composite materials for determination of the group numbers [46]. Flax-based polypropylene composite was able to clear pass flammability test (UL-94, V-1) and hence its suitability for honeycomb structure was justified. The manufactured core showed an ignition time of about 15 seconds, total heat release rate of 69 MJ/m^2 and heat flux of 450 kW/m^2. The group number was reported to be three signifying the suitability of the material to be used for wall and ceilings. Investigations on different honeycomb structures revealed that there was no significant effect on the fire properties of sandwich panels.

Another critical issue for application acceptability of natural fiber polymeric composite is that of post-fire mechanical behavior. The load bearing capacity of natural fiber polymeric composite subjected to fire is related to the combustion, thermal degradation, and heat transfer mechanism. Tensile properties for different plant fiber-based composites, i.e. flax, hemp, and jute, under the influence of heat were evaluated by Bhat et al. [47]. It was revealed that the plant fiber-based polymeric composite reached the softening stage very early and failed at a heat flux of around 35 kW/m^2. Severe weakness of natural fiber reinforced polymeric composites was also reported before significant thermal decomposition could occur. Evaporation of water, delamination, and softening are some of the major root causes behind lower tensile strengths. Impact behavior of composites under the influence of heat was also investigated by researchers. A comparative analysis of the same was carried out by Rajaei et al. [39] for flax and glass polymeric composite materials. The composite specimens were heated at different temperatures of 100, 200, and 300 °C. From the drop test findings it was reported that the energy absorption of glass reinforced polymeric composite increased whereas that of flax composite decreased. The increase was attributed to the increased damping effect due to partial delamination and the decrease was due to the degradation of flax fiber at 300 °C. Therefore, it was concluded that the flax-based epoxy composites were able to show potential to be used for modest impact conditions.

11.7 Summary

One of the major disadvantages associated with natural fiber-based polymeric composite materials is that of high flammability, which is mainly due to the presence of hydrocarbon polymers and fibers that are combustible. The present chapter has presented the flammability characteristics of natural fiber reinforced polymeric materials and therefore has made an effort to understand their burning behavior. The chapter has also highlighted the various flame retardant methods and stressed on the importance of flame retardant particles being effective in reducing the heat release. However, the adverse impact of such particles on the mechanical properties of the fabricated composites presents a major challenge. The use of polymeric matrix with high fire resistance and the treatment of bio-based fibers can aid in minimizing the loss of mechanical properties. However, it was revealed that it is still difficult to meet the prescribed flame requirements for various engineering applications. This issue opens up another frontier for research toward material selection and selection of proper flame retardant methods and manufacturing process that can ultimately result in an optimal composite with high mechanical performance and high fire resistance.

References

1 Feih, S., Boiocchi, E., Mathys, G. et al. (2011). Mechanical properties of thermally-treated and recycled glass fibres. *Composites Part B: Engineering* 42 (3): 350–358.

2 Kim, N.K., Lin, R.J.T., and Bhattacharyya, D. (2014). Extruded short wool fibre composites: mechanical and fire retardant properties. *Composites Part B: Engineering* 67: 472–480.

3 Mohanty, A.K., Misra, M.A., and Hinrichsen, G. (2000). Biofibres, biodegradable polymers and biocomposites: an overview. *Macromolecular Materials and Engineering* 276 (1): 1–24.

4 Kalia, S., Kaith, B.S., and Kaur, I. (2009). Pretreatments of natural fibers and their application as reinforcing material in polymer composites—a review. *Polymer Engineering and Science* 49 (7): 1253–1272.

5 Keener, T.J., Stuart, R.K., and Brown, T.K. (2004). Maleated coupling agents for natural fibre composites. *Composites Part A: Applied Science and Manufacturing* 35 (3): 357–362.

6 Wongsriraksa, P., Togashi, K., Nakai, A., and Hamada, H. (2013). Continuous natural fiber reinforced thermoplastic composites by fiber surface modification. *Advances in Mechanical Engineering* 5: 685104.

7 Price, D., Anthony, G., and Carty, P. (2001). Polymer combustion, condensed phase pyrolysis and smoke formation. In: *Fire Retardant Materials* (ed. A.R. Horrocks and D. Price), 1–30. Cambridge: Elsevier.

8 Bhattacharyya, D., Subasinghe, A., and Kim, N.K. (2015). Natural fibers: their composites and flammability characterizations. In: *Multifunctionality of Polymer Composites*, Chapter 4 (ed. K. Friedrich and U. Breuer), 102–143. Oxford, UK: Elsevier.

9 Mouritz, A.P. and Gibson, A.G. (2007). *Fire Properties of Polymer Composite Materials*, vol. 143. Springer Science & Business Media.

10 Arao, Y., Nakamura, S., Tomita, Y. et al. (2014). Improvement on fire retardancy of wood flour/polypropylene composites using various fire retardants. *Polymer Degradation and Stability* 100: 79–85.

11 Azwa, Z.N., Yousif, B.F., Manalo, A.C., and Karunasena, W. (2013). A review on the degradability of polymeric composites based on natural fibres. *Materials and Design* 47: 424–442.

12 Nguyen, Q.T., Tran, P., Ngo, T.D. et al. (2014). Experimental and computational investigations on fire resistance of GFRP composite for building façade. *Composites Part B: Engineering* 62: 218–229.

13 Dittenber, D.B. and GangaRao, H.V. (2012). Critical review of recent publications on use of natural composites in infrastructure. *Composites Part A: Applied Science and Manufacturing* 43 (8): 1419–1429.

14 Das, O. and Sarmah, A.K. (2015). Mechanism of waste biomass pyrolysis: effect of physical and chemical pre-treatments. *Science of the Total Environment* 537: 323–334.

15 Horrocks, A.R. (1983). An introduction to the burning behaviour of cellulosic fibres. *Journal of the Society of Dyers and Colourists* 99 (7–8): 191–197.

16 Ferdous, D., Dalai, A.K., Bej, S.K., and Thring, R.W. (2002). Pyrolysis of lignins: experimental and kinetics studies. *Energy and Fuels* 16 (6): 1405–1412.

17 Kozłowski, R. and Władyka-Przybylak, M. (2008). Flammability and fire resistance of composites reinforced by natural fibers. *Polymers for Advanced Technologies* 19 (6): 446–453.

18 Christoe, J.R., Denning, R.J., Evans, D.J. et al. (2000). Wool. In: *Kirk-Othmer Encyclopedia of Chemical Technology* (ed. O. Kirk), 1–41. Hoboken, NJ: Wiley.

19 Benisek, L. (1975). Flame retardance of protein fibers. In: *Flame-Retardant Polymeric Materials* (ed. M. Lewin, S.M. Atlas and E.M. Pearce), 137–191. Boston, MA: Springer.

20 Bertini, F., Canetti, M., Patrucco, A., and Zoccola, M. (2013). Wool keratin-polypropylene composites: properties and thermal degradation. *Polymer Degradation and Stability* 98 (5): 980–987.

21 Jeencham, R., Suppakarn, N., and Jarukumjorn, K. (2014). Effect of flame retardants on flame retardant, mechanical, and thermal properties of sisal fiber/polypropylene composites. *Composites Part B: Engineering* 56: 249–253.

22 Kim, N.K., Dutta, S., and Bhattacharyya, D. (2018). A review of flammability of natural fibre reinforced polymeric composites. *Composites Science and Technology* 162: 64–78.

23 Subasinghe, A. and Bhattacharyya, D. (2014). Performance of different intumescent ammonium polyphosphate flame retardants in PP/kenaf fibre composites. *Composites Part A: Applied Science and Manufacturing* 65: 91–99.

24 Szolnoki, B., Bocz, K., Soti, P.L. et al. (2015). Development of natural fibre reinforced flame retarded epoxy resin composites. *Polymer Degradation and Stability* 119: 68–76.

25 Lu, S.Y. and Hamerton, I. (2002). Recent developments in the chemistry of halogen-free flame retardant polymers. *Progress in Polymer Science* 27 (8): 1661–1712.

26 El-sabbagh, A., Steuernagel, L., and Ziegmann, G. (2013). Low combustible polypropylene/flax/magnesium hydroxide composites: mechanical, flame retardation characterization and recycling effect. *Journal of Reinforced Plastics and Composites* 32 (14): 1030–1043.

27 Lee, C.H., Salit, M.S., and Hassan, M.R. (2014). A review of the flammability factors of kenaf and allied fibre reinforced polymer composites. *Advances in Materials Science and Engineering* 2014: 1–8.

28 Bai, G., Guo, C., and Li, L. (2014). Synergistic effect of intumescent flame retardant and expandable graphite on mechanical and flame-retardant properties of wood flour-polypropylene composites. *Construction and Building Materials* 50: 148–153.

29 Camino, G., Costa, L., and Martinasso, G. (1989). Intumescent fire-retardant systems. *Polymer Degradation and Stability* 23 (4): 359–376.

30 Bourbigot, S. and Duquesne, S. (2009). Intumescence-based fire retardants. In: *Fire Retardancy of Polymeric Materials* (ed. C.A. Wilkie and A.B. Morgan), 129–162. Boca Raton, FL: CRC Press.

31 Rajaei, M., Wang, D.Y., and Bhattacharyya, D. (2017). Combined effects of ammonium polyphosphate and talc on the fire and mechanical properties of epoxy/glass fabric composites. *Composites Part B: Engineering* 113: 381–390.

32 Camino, G. and Lomakin, S. (2001). Intumescent materials. In: *Fire Retardant Materials* (ed. A.R. Horrocks and D. Price), 318–336. Cambridge, UK: Woodhead Publishing.

33 Das, O., Kim, N.K., Sarmah, A.K., and Bhattacharyya, D. (2017). Development of waste based biochar/wool hybrid biocomposites: flammability characteristics and mechanical properties. *Journal of Cleaner Production* 144: 79–89.

34 Kim, N.K. and Bhattacharyya, D. (2016). Development of fire resistant wool polymer composites: mechanical performance and fire simulation with design perspectives. *Materials and Design* 106: 391–403.

35 Subasinghe, A.D.L., Das, R., and Bhattacharyya, D. (2016). Parametric analysis of flammability performance of polypropylene/kenaf composites. *Journal of Materials Science* 51 (4): 2101–2111.

36 Kim, N.K., Lin, R.J.T., and Bhattacharyya, D. (2015). Effects of wool fibres, ammonium polyphosphate and polymer viscosity on the flammability and mechanical performance of PP/wool composites. *Polymer Degradation and Stability* 119: 167–177.

37 Bhattacharyya, D. and De, S.K.G.K. (2012). *Fire Retardant Polypropylene*. Google Patents.

38 Kim, N.K., Lin, R.J.T., and Bhattacharyya, D. (2017). Flammability and mechanical behaviour of polypropylene composites filled with cellulose and protein based fibres: a comparative study. *Composites Part A: Applied Science and Manufacturing* 100: 215–226.

39 Rajaei, M., Kim, N.K., and Bhattacharyya, D. (2018). Effects of heat-induced damage on impact performance of epoxy laminates with glass and flax fibres. *Composite Structures* 185: 515–523.

40 Ahmad, F., Choi, H.S., and Park, M.K. (2015). A review: natural fiber composites selection in view of mechanical, light weight, and economic properties. *Macromolecular Materials and Engineering* 300 (1): 10–24.

41 Dunne, R., Desai, D., Sadiku, R., and Jayaramudu, J. (2016). A review of natural fibres, their sustainability and automotive applications. *Journal of Reinforced Plastics and Composites* 35 (13): 1041–1050.

42 Chai, M. (2014). Flammability performance of bio-derived composite materials for aircraft interiors. Doctoral dissertation. ResearchSpace@ Auckland.

43 Fatima, S. and Mohanty, A.R. (2011). Acoustical and fire-retardant properties of jute composite materials. *Applied Acoustics* 72 (2–3): 108–114.

44 Banerjee, S. and Bhattacharyya, D. (2011). Optimal design of sandwich panels made of wood veneer hollow cores. *Composites Science and Technology* 71 (4): 425–432.

45 Ticoalu, A., Aravinthan, T., and Cardona, F. (2010). A review of current development in natural fiber composites for structural and infrastructure applications. In: *Proceedings of the Southern Region Engineering Conference (SREC 2010)* (ed. A. Ticoalu, T. Aravinthan and F. Cardona), 113–117. Engineers Australia.

46 Banu, D., Feldman, D., Haghighat, F. et al. (1998). Energy-storing wallboard: flammability tests. *Journal of Materials in Civil Engineering* 10 (2): 98–105.

47 Bhat, T., Kandare, E., Gibson, A.G. et al. (2017). Tensile properties of plant fibre-polymer composites in fire. *Fire and Materials* 41 (8): 1040–1050.

12

Post Life Cycle Processing of Reinforced Thermoplastic Polymer Composites

N.H. Salwa[1], S.M. Sapuan[2], M.T. Mastura[3], and M.Y.M. Zuhri[2]

[1] Institute of Tropical Forest and Forest Products (INTROP), Universiti Putra Malaysia (UPM), Department of Biocomposites Technology and Design, Serdang, Selangor, 43400, Malaysia
[2] Universiti Putra Malaysia (UPM), Department of Mechanical and Manufacturing Engineering, Faculty of Engineering, Serdang, Selangor, 43400, Malaysia
[3] Universiti Teknikal Malaysia Melaka (UTem), Department of Manufacturing Engineering Technology, Faculty of Mechanical and Manufacturing Engineering Technology, Hang Tuah Jaya, Durian Tunggal, Melaka, 76100, Malaysia

12.1 Introduction

Expanding development on environmental awareness toward achieving product sustainability has encouraged remarkable efforts in exploiting more environmentally friendly materials in product design. Crude oil is admitted globally as a major unsustainable resource where its consumption has caused increased concentrations of carbon dioxide (CO_2) and methane (CH_4) in the atmosphere, which has largely exceeded the natural levels. In the present day, most plastics consumed are manufactured from crude oil and other fossil fuels, such as natural gas and coal. Furthermore, their end-life prediction and, hence, their perseverance in the environment, go so much beyond the predictable one. CO_2 from fossil fuel combustion should also be taken into consideration apart from resource depletion issues. Certainly, this represents a major contributor to global warming (GW) and if this is not addressed, it could have social, economic, and environmental consequences in the future [1]. For an engineer to design a product, a list of properties including stiffness, strength, density, and working temperature must be considered to ensure that the material selected best suits the intended purpose and the respective production technology. This list of requirements needs to embrace potential environmental issues too, which involve low energy consumption, low carbon emissions, and recyclability. A product design focusing on environmental concerns would use materials with lower environmental impacts and choose cleaner production processes. Moreover, hazardous and toxic materials are avoided in the design while at the same time maximizing the efficiency of the energy used for production. The strategy also concerns how the product is used as well as the product's waste management and recycling. All possible considerations on the effects during its production process, through its use, and

Reinforced Polymer Composites: Processing, Characterization and Post Life Cycle Assessment,
First Edition. Edited by Pramendra K. Bajpai and Inderdeep Singh.
© 2020 Wiley-VCH Verlag GmbH & Co. KGaA. Published 2020 by Wiley-VCH Verlag GmbH & Co. KGaA.

finally its end of life should be taken on board in designing a product. Life cycle assessment (LCA) methodology is a useful tool in designing a product to address the environmental concerns. It is a very thorough technique in determining the total environmental impacts, bringing the environmental impacts into one consistent framework.

12.2 Polymer Composites

Composites have been developed since six decades ago and were previously commercially used in large quantities only in the defense industry. Later, the applications were broadened to other products such as bodies of sports cars, boat exterior, and sporting goods. These reinforced polymer composites are favored over conventional materials such as glass and metal, mostly for their light weight and suitable mechanical properties. Polymer composites with the matrix made from polymer materials are incorporated with suitable reinforcement fiber or fillers, i.e. glass, carbon, Kevlar, boron, or natural fiber, to further enhance the composites' performance. The reinforcing fibers play an important role in composites to prevent cracks or deflects at the matrix–fiber interface. A specific composite material can be customized and designed for a specific load and this attracts various applications in the industries nowadays. Composites have been predominantly utilized in aerospace, automotive, marine, building and construction, furniture, packaging, telecommunications, and railway industries. The main advantages of composite materials are light weight, high strength, high stiffness, aesthetics, corrosion resistance, and ability to cope with extreme stresses over long periods [2]. Besides, composites are also preferred over conventional materials for their resistance to chemicals and heat and also for their electrical insulation properties. Composites termed as fiber reinforced plastics are a blend of reinforcement fiber and polymer matrix materials. Generally, composites are compounds composed of two or more constituent materials. These materials when combined are stronger than the individual materials by themselves. More environmentally friendly composite materials are getting higher attention in the recent years and research reveals that bio-composites are materials with great potential to be the solutions in addressing the needs of sustainability in product design [3–5]. A report stated that bio-composites have great advantages because of their high performance in mechanical properties, many processing advantages, low cost and light weight, availability, and renewable, cheap, environmentally friendly features, recyclability, and degradability [3].

"Bio-composites" are composites where either the matrix or fiber, or both matrix and fiber, are derived from biological resources. Bio-composites can be classified as in Figure 12.1.

The first two categories are not entirely environmentally friendly since one of the constituents is still made from unsustainable resources. The third category where both constituents are derived from renewable resources is termed as "green bio-composites," and they are believed to be the best alternative to address environmental concerns [3]. The properties of green bio-composites can

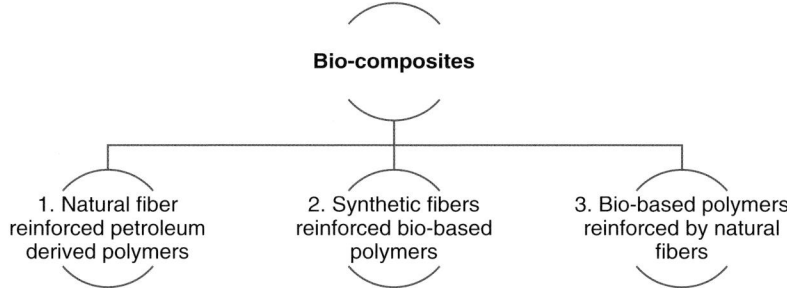

Figure 12.1 Bio-composites classification. Source: Adapted from Al-oqla et al. 2017 [3] and Mitra 2014 [6].

be customized according to the specific type of applications through appropriate selection of fibers, biopolymer matrix, additives, and manufacturing methods. The important considerations in designing green bio-composites are fiber selection (type, harvest time, extraction method, aspect ratio, treatment, and fiber content), biopolymer matrix selection, interfacial strength, fiber dispersion, fiber orientation, composite manufacturing process, and porosity [7]. There are four most important criteria emphasized in developing green bio-composites with stable mechanical properties: (i) homogeneous dispersion of the fibers; (ii) excellent interaction between the matrix and the fibers; (iii) low porosity of the matrix; (iv) optimized percentage of fibers to reinforce the material [7].

12.2.1 Thermoplastic Polymer

Numerous types of petroleum-based thermoplastics polymer materials can be utilized for a variety of applications according to the material characteristics required. Among them are polypropylene (PP) for mechanical properties and water vapor barrier, polyethylene (PE) for sealing and water vapor barrier, ethylene vinyl alcohol (EVOH) for barrier properties to gas and water vapor, and polyethylene terephthalate (PET) and polyamide (nylon) for aroma/oxygen barrier with stiffness [8]. Generally, thermoplastic polymer materials can be heated and formed repetitively and used in the applications of injection moldings. On the other hand, thermosetting materials are formed through an initial heating and curing and cannot be melted and reformed. During curing, they form three-dimensional molecular chains, called cross-linking. These materials include polyesters, melamine, phenolics, cyanate esters, bis(maleimido)diphenylmethane (BMI), polyimide, polyurethane, polyester, vinylesters, and epoxides. Thermoplastic polymers are not cross-linked, and the properties of the monomer units and the degree of entanglement of the polymer chains contribute to their strength and stiffness. Thermoplastics have many advantages including unlimited shelf life, short process cycle times, ability to be reformed, increased moisture resistance, increased toughness, and impact resistance. However, the main drawback of thermoplastics materials is the cost, which is higher than that of the conventional thermosetting materials [9]. Thermoplastic biopolymer made from starch is named thermoplastics starch

Table 12.1 Characteristics of thermoplastics starch.

Source of starch	Density	Water content (%)	Amylose content (%)	Crystallinity (%)
Wheat	1.44	13–19	26–27	36
Maize/Corn	1.5	12–13	26–28	39
Potato	1.54–1.55	18–19	20–25	25
Sago	—	10–20	24–27	—
Sugar palm	1.54	15	37.60	—

Source: Sanyang et al. 2018 [11] and Gurunathan et al. 2015 [12]. Reproduced with permission of Springer Nature.

(TPS), which is the resulting product of plasticized starch. It possesses similarity in characteristics with petroleum-based thermoplastic materials that allow the application of similar processing such as extrusion, compression molding, and injection molding [10, 11]. Noteworthy characteristics of TPS are gathered in Table 12.1.

12.2.2 Reinforcing Fibers in Composites

Polymer used as matrix in composites, also termed as resin, plays a significant role in allowing loads to be distributed/transferred, to assist in governing the chemical properties of the composites, to carry interlaminar shear, and to hold the reinforcement together as binder, which permits the applied force to be transferred to the reinforcement. Polymer matrix also helps protect or prevent the reinforcing fiber from mechanical damage due to crack propagation within the material during fabrication and in the finished product or component [2, 13]. Synthetic fibers such as carbon or glass fiber reinforce composite materials, provide stiffness and enough strength, and govern the unique properties of the final composite materials [14–17]. Regarding environmental concerns, the main advantages of using natural fibers over conventional synthetic fibers are biodegradability, renewability, recyclability, and nontoxic nature [10, 18, 47]. Other qualities such as easy availability, low cost, and remarkable properties such as low density, light weight, sufficient specific strength, and stiffness are very appealing to provide excellent reinforcement alternatives to the synthetic fibers [8, 18, 19]. In addition, exploitation of natural fibers as reinforcement in composites could also reduce tool wear in machining and render them non-abrasive and easy to manufacture [7, 18].

Mechanical properties of a fiber to be used as reinforcement in polymer composites can be contributed by many factors including fiber–matrix adhesion, volume fraction of the fibers, fiber aspect ratio (l/d), and fiber orientation [8]. Furthermore, critical requirements in selecting the fibers in bio-composite fabrication are higher degree of polymerization, cellulose content, and lower microfibrillar angle where these attributes could produce composite materials with higher tensile modulus and strength [20].

Thermoplastic polymers reinforced with natural fibers have been widely studied for industrial applications. New environmental legislation in the European

Table 12.2 Natural fiber and thermoplastic matrix used in applications.

Thermoplastic matrix	Reinforcing fiber/filler	References
Blended cassava starch and chitosan	Kraft fibers	[25]
Sugar palm starch	Sugar palm fiber	[26]
Cationic starch (and polyvinyl alcohol)	Bagasse pulp beaten fiber slurry (and magnesium stearate)	[27]
Sugar palm starch	Agar	[28]
Sugar palm starch/Agar	Seaweed/Sugar palm fiber	[10, 29]
Wood powder	poplar fibers	[30]
Cassava starch (CS)	cassava bagasse (CB) and sugar palm fiber (SPF)	[31]
Poly(3-hydroxybutyrate-*co*-valerate) (PHBV)	Wheat straw, brewing spent grains, and olive mills	[32]
Cassava starch	Rich-fiber lentil flour	[33]
Polycaprolactone (CAPA 6800) matrix with graphite	Pine cone fiber	[34]
Poly(lactic acid) (PLA)	Bamboo cellulose nanowhiskers (BCNW)	[35]

Union (EU) and other countries and also the effort of industry players to play a role in achieving the goal of sustainability have led to greater use of eco-friendly materials. Particularly in the automotive industry, these natural fibers reinforced thermoplastic polymer materials are used to make a variety of interior parts such as seat bottoms, heat restraints, back cushions, and underfloor body panels [21]. There are also studies on the potential of these bio-composites in application for components with higher performance requirement such as suspension component, which interlinks two wheels and functions as a sway-restriction bar or automotive anti-roll bar (ARB), car bumper beam, and parking brake lever [22–24]. Recent studies on promising applications of natural fiber reinforced thermoplastics composites in short-life packaging and consumer products are shown in Table 12.2 [10, 26, 28, 29, 32, 36].

12.2.3 Green Bio-composites

Composites are composed of two or more constituent materials as the matrices and fibers. These materials when combined are stronger than individual materials by themselves. More environmentally friendly composite materials are getting greater attention in the recent years and research has revealed that bio-composites are materials with great potential to be the solutions in addressing needs of sustainability in product design [4]. Studies have proved that bio-composites have great advantages because of high performance in mechanical properties, many processing advantages, low cost and light weight, availability, and renewable, cheap, environmentally friendly features, recyclability, and degradability [3, 5].

Bio-based composites from natural fibers and bio-resins are getting increasing attention in the recent years. According to La Rosa and Cicala [21], the bio-content in bio-resins is in the range of 20–50%; however, these are not biodegradable. Examples of thermoplastic non-biodegradable bio-resins are bio-polyolefin, bio-nylons, cellulose, and lignin-based polymers. A new class of totally biodegradable "green" composites could be obtained by combining natural fiber and biodegradable bio-based polymer (La Rosa and Cicala [21]). The benefits of using green composites produced with natural fibers and bio-derived matrices, compared to traditional composites, their outstanding properties, and their market opportunities are increasing for many industrial sectors [21]. A study reveals that the consumers seemed more positive with a fully bio-based product than partial ones and for that reason, the future bio-based plastics should be made of 100% bio-based materials, i.e. natural fiber reinforced biopolymer composites (and/or bio-based additives) [37]. Green bio-composites in the market have an enormous opportunity with the growing emphasize on the environmental issues and growing demand for green material, with an advantage of the evolving process technologies. They possess significant strengths: (i) environmental impacts reduction; (ii) reduced cost; (iii) composites weight reduction; (iv) development of the rural communities; (v) biodegradability; and (vi) recyclability [21].

The life cycle of a product begins during the conceptual design stage. The ability to reduce the environmental impact of each stage decreases throughout the product life cycle as the product progresses through the life cycle. It is estimated that 70–80% of the total cost of a product is obligated during the design stage where product designers focus on material selection to reduce environmental impacts. Concurrently, the manufacturing engineer's attention would be on energy reduction. These two goals may not be compatible with each other without the ability to comprehend the impact of the material on the manufacturing process. This situation would be even more demanding if the designers are not aware of how their selection contributes to the manufacturing processes [38]. The same goes to the production of green bio-composite material whereby all possible considerations are addressed on the effects during its production process, its use phase, and finally its end of life in designing the environmentally friendly material.

12.3 Life Cycle Assessment (LCA)

12.3.1 Definition

ISO 14040:2006 defines LCA as the "compilation and evaluation of the inputs, outputs and of the potential environmental impacts due to a product-system throughout its life-cycle" [39]. The LCA method can be applied to estimate the impact of a final product by accounting for the impacts across its full life cycle from resource extraction to end of life [38]. LCA can be utilized as a supporting tool for design to learn and assess the technical solutions to be employed in the production process to minimize the impacts originating not only from the production itself but also from the phases of use and end of life. LCA could as well

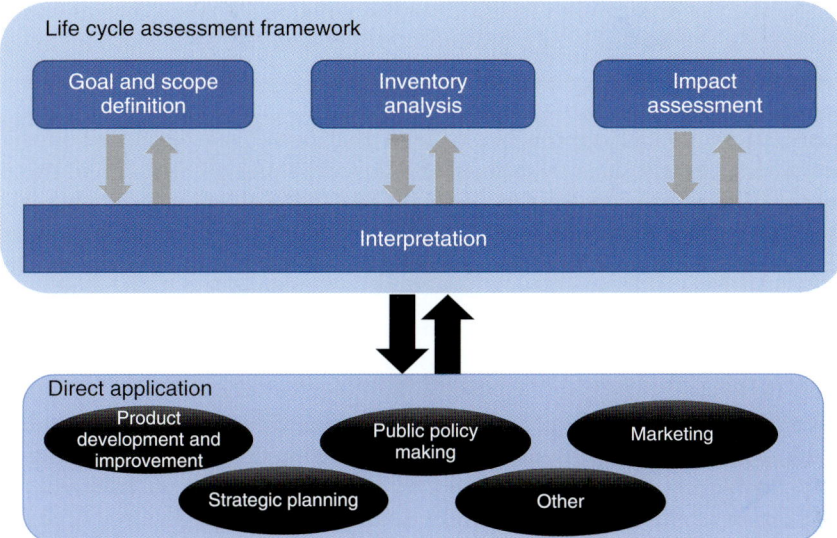

Figure 12.2 LCA procedure according to ISO 14040.

be the comparative assessment tool, on which much research has already been done [39]. LCA was previously used as a reflective tool but has been expanded to become an action-oriented decision-making tool, in order to aid designers and producers to reduce environmental impacts. Brundage et al. [38] clarified that currently LCA is used for product/process selection, design, and optimization, and also possibly to attach it with simulation techniques and design tools to help organizations with utmost consciousness of the environmental consequences of their decision and actions both on and off site. The traditional LCA technique assumes the traditional life cycle perspective, from material extraction to disposal. However, when considering all information used in each stage, this perspective might overlook some important factors such as environmental impact of manufacturing that could improve design choices [38]. Figure 12.2 depicts the LCA framework according to ISO 14040.

LCA is a framework that can be used to assess the environmental impacts of a product throughout its life, starting from the extraction of raw materials from the earth and ending at the waste products being returned to the earth. An LCA involves collecting information on the inputs and outputs, such as emissions, waste, and resources, of a process (life cycle inventory [LCI]) and translating these to environmental consequences (using impact assessment methodologies) such as contribution to climate change, smog creation, eutrophication, acidification, and human and ecosystem toxicity [4].

LCA principally includes six important phases in a product life cycle: (i) materials extraction; (ii) manufacturing and waste production; (iii) packaging; (iv) transportation; (v) product use; and (vi) product disposal. The Engineer Manufacture of Product activity describes that the process of making the product encompasses the acquisition of stock materials, equipment, and tooling [38]. Detailed fundamental information is needed regarding the manufacturing processes, the materials, and the energy used to calculate the amount of emissions and waste created during the life cycle of a product [21]. In general, there are four main steps in performing the LCA technique, namely, (i) goal and purpose; (ii) LCI; (iii) life cycle impact assessment (LCIA); and (iv) interpretation of results. Furthermore, there are two approaches to LCA that have been developed in the past few years, namely, the attributional and the consequential. The aim of these approaches is to provide answers to questions relating to different system modeling [38]. The attributional LCA (A-LCA) provides information about the impacts of the processes employed to produce, consume, and dispose of a product. On the other hand, contributional life cycle assessment (CLCA) provides information about the consequences of changes in the level of output, consumption, and disposal of a product, including effects both inside and outside the life cycle of the product [1].

Most of the LCA studies done for biopolymers to date employ the attributional approach and this approach involves accounting the impacts of a product system without considering any external influences. It attempts to provide information on the portion of global burdens that can be directly associated with a product (throughout its life cycle). Ideally, this contains processes that are directly linked by flows to the unit processes that supply the functional unit (FU). Inventory data from the material suppliers or the average data obtained, for example, the average electricity mix of a region, is used in the LCA study. Some LCA studies on bio-based products also use the consequential approach (CLCA) in order to avoid the problem of allocation in multifunctional bio-based systems and also includes the activities that have indirect implications on the product system [40]. CLCA is also known as the change-oriented approach, which tries to provide information on the environmental impacts that occur, directly or indirectly, as a consequence of a decision. This is usually represented by changes in demand for a product. The product system analyzed by this approach consists of processes that are affected by cause-and-effect chain with the origin from a particular decision. CLCA uses data from the suppliers, only if it is not constrained or uses marginal data. CLCAs analyze the product system but have in mind the market mechanisms, state of advancement of technologies, and the co-products indirectly affecting the utilization of the considered product in the case of a multifunctional product system. This approach would very much influence the policy and decision making. An example of CLCA approach is the study by Alvarenga et al. [41] which assessed the effects on indirect land use change (ILUC) caused by expansion of sugarcane plantation. They emphasized that the effects of ILUC should be considered when assessing new bio-based products, as some projected environmental benefits could have been null and void if there is low control on deforestation caused by ILUC. There are a few CLCA studies done so far for bio-based plastics – each of the attributional life cycle assessment (ALCA) and CLCA studies has its own goal and system boundaries. The choice of LCA approach, system boundary, methodology, type of LCA (screening, comprehensive), and so on lies solely on the producer and he/she needs to justify the decisions before taking these results into the eco-design framework [40]. The LCA method has expanded into a wide range of applications both at the downstream and upstream activities such as eco-labeling, energy systems, product designs, transportation, and food production, which further highlight its functionality and effectiveness [4].

The following are some of the software suitable for LCA of plastics [21]:

1. SimaPro, Pre-Product Ecology Consultants, Amersfoort, NL, www.pre.nl/ simapro/.
2. Boustead, Boustead Consulting Ltd, West Grinstead, UK, www.boustead-consulting.co.uk.
3. GaBi, Institute for Polymer Testing and Polymer Science (IKP), Polymer Institute University of Stuttgart www.lbp-gabi.de.
4. Umberto, Institute for Environmental Informatics, Hamburg, www.umberto .de/english/ LCA [21].

12.3.1.1 Goal and Scope

LCA begins with a description of the reasons for implementing the study, i.e. the goal of the study and the scope of work to be defined with respect to the functional unit, system boundaries, assumptions, and limitations of the study. A functional unit is defined by the functional requirements of a product system for a certain period. Whereas system boundaries usually describe whether an LCA study represents a complete analysis (cradle-to-grave) or a partial analysis (cradle-to-gate or gate-to-grave). Mansor et al. [4] and Parameswaranpillai and Vijayan [42] explained that the main elements in the first phase of LCA operation comprise of (i) definition of the goal of the project, i.e. the reason for the study and the intended application; (ii) definition of scope of the study or system boundary; and (iii) determination of the functional unit. Both cradle-to-grave and cradle-to-gate approaches could be applied in this early phase of assessing the environmental impact. Cradle-to-grave includes all life cycle phases together, starting with the production phase that comprises of material and component production, use phase, and EOL processing of the product. On the other hand, cradle-to-gate is a partial product life cycle involving resource extraction and production phase to the factory gate before it is sent to the manufacturer for application or to the consumers [4]. In the two latest revised versions of LCA standards, which are ISO 14040 (principles and framework) and ISO 14044 (requirements and guidelines), the errors and inconsistencies are removed as also the readability of the past standards enhanced. The standards also emphasize that cradle-to-grave approach is not the only approach to perform LCA; other approaches such as cradle-to-gate studies, gate-to-gate studies, and specific part of the phase in the life cycle may also be operated in LCA studies with proper rationalization [4]. Figure 12.3 describes the product life cycle with the cradle-to-grave approach.

A complete and detailed LCA system boundary draws a perimeter around the activities that are included within its assessment. This usually starts with raw material extraction and considers each step of material transformation, their intermediate states, and finally the synthesis of the final product. The scope or system boundaries represent the perimeter of exchanges between a process chain and the environment. Processing stages start and end in this boundary and within this defined periphery, material flow and transformation take place to produce the final product of interest. During each transformation stage, other chemicals and energy are required as input [43]. Research revealed that manufacturing activities cause distressing degradation of the natural resources

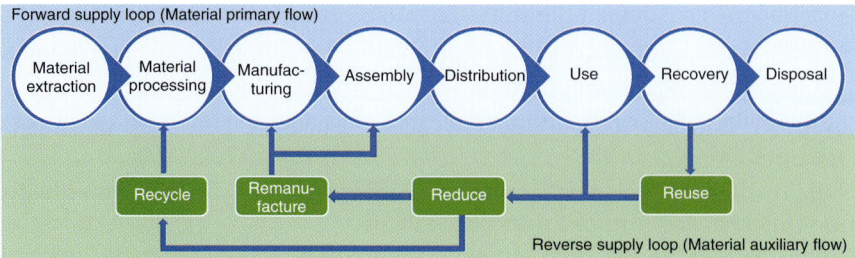

Figure 12.3 Material flow of forward and reverse supply loop in a product life cycle. Source: Adapted from Brundage 2018 [38].

as well as generate dangerous effects to the general society. Consequently, industries are demanded to evaluate their performances toward realizing the three sustainability dimensions, i.e. economic development, social development, and environmental protection [44]. Manufacturing has produced a huge impact on the environment, such as high energy consumption, waste generation, and greenhouse gas emissions. In the design phase of a product, a product designer may consider an alternative attribute with less energy intensive process but does not compromise on the specification's requirement.

12.3.1.2 Life Cycle Inventory (LCI)

LCI is the phase to quantify the usage of resources and materials and the consumption of fuels and energy, as well as the transport associated with a product in its life cycle of the analyzed functional unit [43, 45]. LCI is crucial in LCA. In this phase, data is collected to calculate material use, energy input, and pollutant emissions during the entire life cycle of a product or process. These data can be obtained from companies engaged in product fabrication and processing activities and also from published databases [42]. Compiled data on emissions and resources consumed are recorded that can be attributed to a specific product. Several databases have been developed that allow users to insert new information or data specific to their products and processes. The most comprehensive international LCI database is Ecoinvent v3 (www.ecoinvent.org/database/) [21]. From the list of LCI compiled, contributions to different environmental impact categories, such as global warming potential (GWP), acidification, eutrophication, or human toxicity, are generated. Any other processes or activities outside the LCA boundary will be ignored and thus environmental damages caused by them are not counted. A set of LCI consists of inputs and outputs that represent the flow of material and energy used within the technical structure of the LCA as well as emissions that are generated from the processes. The consequences of the inputs and emissions which are gathered in LCI need to be translated into environmental impacts. The next phase, i.e. the impact assessment phase, would provide this translation.

Apart from knowledge or information access to process (and chemistry) details, data selection and compilation can demand an extensive amount of time and effort. The reliabilities of any LCA investigation and its outcome are entirely dependent on the quality and adequacy of data used in constructing LCA models. Therefore, precaution and extra steps in data selection are advisable. It is important to note that data selectivity, quality, and adequacy are the first challenge in applying LCA [43]. Khoo et al. [43] proposed eight criteria to serve as guidance in data selection for the development of LCI. This will hopefully help maintain the confidence of LCA applications and its associated results.

Criterion 1 – *Information source*: This includes inventory gathering and compilation. Primary or firsthand industrial data would serve as the most reliable source and quality of information to support any LCA objectives and its outcomes.

Criterion 2 – *System boundary*: System boundary defined, including the scope of data gathered, for any LCA investigations will have an effect on its overall results. Inventory data is only considered for processes that are within the technical periphery of the LCA study or system boundary.

Criterion 3 – *Process data*: In the case where industrial or process-based data is unavailable from LCA datasets, technical information is needed to generate the input–output details required for a good, high quality LCA investigation. Technological variations can alter the input values used in LCA.

Criterion 4 – *Data completeness/comparability*: In the preparation of LCI for each different synthesis route, the same types of pollutants should be presented for comparison.

Criterion 5 – *Representativeness*: For the case where an exclusively designed, or proprietary, process is employed to produce a product/chemical, the "import" of LCI from other types of methods or processes for making the same chemical should not be applied. The outcome of making use of unrelated, irrelevant data for any LCA might lead to erroneous conclusions.

Criterion 6 – *Scale*: Data generated from experimental records should not be directly applied in the LCA of industrial scale-up considerations. The reason lies in un-optimized processes carried out at laboratory scales (discovery stage) where manufacturing factors such as environmental regulations, energy use, waste management, safety, and scalability are not taken into consideration.

Criterion 7 – *Location*: Location-dependent LCI are highly recommended for such cases where biomass resources are utilized in the LCA system, and these resources are produced from a different geographical area or region.

Criterion 8 – *Transparency, Traceability, Reproducibility*: This involves the need for proper inventory documentation. "Transparency" of data requires assumptions or calculations used in obtaining any values to be properly stated and documented. Recorded details of how data is obtained, methods of calculations, and any set of assumptions are all documented for enough data transparency. All information should be "traceable" to its source; and "reproducibility" of data is highly dependent on appropriate documentation procedures.

12.3.1.3 Life Cycle Impact Assessment (LCIA)

The third phase of the LCA framework is the LCIA, which focuses on evaluating and understanding the environmental impacts established by the LCI analysis. ISO 14040 outlines the impact assessment phase with the following components: (i) Classification: assigning inventory results to impact

Table 12.3 LCA of common impact and damage categories.

Damage category	Impact category	Description of impact
Human health	Carcinogen/non-carcinogen	Human toxicity (contributes to toxic conditions to human)
	Respiratory organics/inorganics	Human toxicity (contributes to toxic conditions to human)
	Ozone layer depletion	Contributes to depletion of stratospheric ozone
Eco-system quality	Aquatic eco-toxicity	Contributes to toxic conditions to plants and animals
	Terrestrial eco-toxicity/acidification/nitrification	Contributes to toxic conditions to plants and animals
	Aquatic acidification	Contributes to acid deposition in water
	Aquatic eutrophication	Provision of nutrients contributes to biological oxygen demand
Climate change	Global warming	Contributes to atmospheric absorption of infrared radiation
Resources use	Abiotic/nonrenewable energy	Contributes to depletion of nonrenewable resources
	Biotic/mineral extraction	Contributes to depletion of renewable resources

Source: Adapted from Mansor et al. 2015 [4].

categories; (ii) characterization: modeling inventory data within impact categories; the characterization process involves defining characterization factors to convert each pollutant emission into equivalent potentials represented by a reference substance (e.g. CO_2 equivalent); (iii) weighting: aggregating inventory data in very specific cases to combine the impact categories into a single score [42]. Impact categories represent the negative effects to the environment due to substances emitted and resources used that caused the damage. The impact categories are grouped into major ones and termed as "Damage categories" and they represent the environmental sectors suffering the damage. Table 12.3 recapitulates the commonly used impact categories in LCA and their damage categories accordingly.

An extensive variety of LCIA methodologies have been developed, based on midpoint and/or endpoint indicators such as CML 2002, Eco-Indicator 99, EDIP (1997e2003), EPS2000, Impact 2002þ, LIME, LUCAS, ReCiPe, Swiss Ecoscarcity or Ecological scarcity, and TRACI MEEuP methodology [21]. Output emissions are organized into classes according to the effect they have on the environment, and each emission might belong to more than one classification. In the case of chemicals emitted that contribute to the greenhouse effect or to ozone layer depletion, they are divided between those two classes, whereas nitrogen oxides emissions may simultaneously belong to several classes, such as aquatic toxicity, acid rain, and eutrophication. Assigning and converting LCI results into numerical indicator results is called characterization [21].

Table 12.4 ISO/TR 14047 (2003) environmental impact classification factors.

Environmental impact classification factor	Description
Acidification potential (AP)	Is a consequence of acids being emitted to the atmosphere and subsequently deposited in surface soils and water. AP classification factors are mainly based on the contributions of SO_2, NOx, HCl, NH_3, and HF and expressed as SO_2 equivalent
Aquatic toxicity potential	Calculated based on the maximum tolerable concentrations of different toxic substances in water by aquatic organisms
Human toxicity potential	Calculated in kilogram and takes into account the release of materials toxic to humans in air (A), water (W), and soil (S) based on human toxicological factors
Eutrophication potential (EP)	Pollution state of aquatic ecosystems where the overfertilization of water and soil has turned into an increased growth of biomass. EP is calculated in kilogram based on a weighted sum of the emission of nitrogen and phosphorus derivatives. The classification factors for EP are expressed as phosphate equivalents
Global warming potential (GWP)	Calculated for each different greenhouse gas, including CO_2, N_2O, CH_4, and volatile organic compounds (VOCs). GWP is expressed as CO_2 equivalent
Nonrenewable/abiotic resource depletion potential	Calculated for fossil fuels, metals, and minerals by dividing the quantity of resource used by the estimated total world reserves of that resource
Ozone depletion potential	The potential of depletion of the ozone layer due to the emissions of chlorofluorocarbon compounds and other halogenated hydrocarbons
Photochemical oxidants creation potential	Photochemical smog is caused by the degradation of VOCs and nitrogen and is expressed in kilogram of ethylene

Source: La Rosa and Cicala 2015 [21]. Reproduced with permission of Elsevier.

ISO/TR 14047 (2003) defines the environmental impact classification factors and each is briefly explained in Table 12.4.

Other factors such as land use and loss of biodiversity are considered in natural fiber production [21].

12.3.1.4 Life Cycle Results Interpretation

Interpretation of the results is the fourth phase of LCA. In this phase, the findings from inventory analysis and impact assessment are combined to reach conclusions and make recommendations. A sensitivity analysis can be carried out to understand how model parameters influence the LCA results or how certain critical parameters help reduce environmental impacts [42]. The results can be presented in impact categories or grouped together into damage categories. Normalization can be performed by dividing impact category scores by the average person's annual contribution, allowing for the combination of categories, whose scores now have no units. Since some environmental impacts can be considered more important than others, a weight can also be assigned to the normalized impact score [4].

12.4 LCA Studies on Bio-composites

The life cycle of a product made from reinforced thermoplastic starch polymer composites may begin with the plant's cultivation for the sources of fiber and thermoplastic starch biopolymer. The production of natural fiber and starch biopolymer are two different product cycle systems; later the intermediate products are transported to the gate of composites manufacture and afterwards for product application. Figure 12.4 illustrates the overall life cycle phases for natural fiber reinforced biopolymer composites.

12.4.1 LCA Bio-composites

To obtain products with environmentally sustainable properties, application of life cycle thinking (LCT) to their design is essential. LCA of fiber reinforced composite materials would assist in improving the characterization of composite products, in terms of environmental performance, and compare them with other classes of materials. Generally, LCAs of bio-based polymer composites

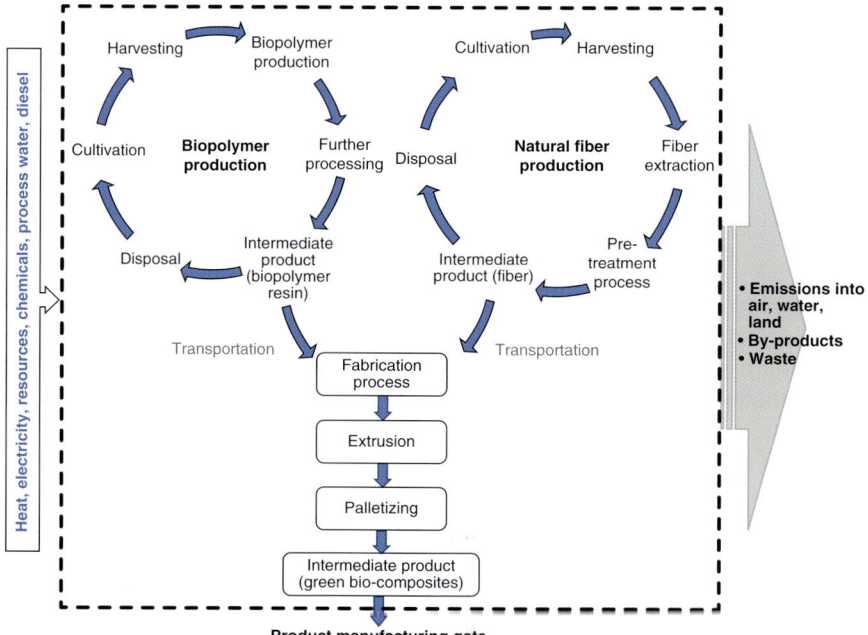

Figure 12.4 System boundary of cradle-to-gate model for manufacturing natural fiber reinforced bio-composites [4, 43, 46].

have shown favorable results in terms of environmental impacts and energy use compared to petroleum-based products [21]. An environmental assessment study of bio-composites with PP or polylactic acid (PLA) matrices reinforced with natural fibers (cotton fibers, cellulose, jute fiber, and kenaf) and glass fiber using LCA methodology was done by Czaplicka-Kolarz et al. [47]. Four methods of environmental impact assessment were employed, namely the Intergovernmental Panel on Climate Change (IPCC), cumulative energy demand (CED), Ecological Footprint, and Ecoindicator99H/A, to determine the indicators of greenhouse gas emission, renewable and nonrenewable energy demand, land use as well as the influence of composites on human health, ecosystem quality, and fossil resource consumption, respectively. The study revealed that reinforcing the polymer matrix with cotton fibers was the least advantageous to the environment, while the lowest environmental impact was from the composites of PLA and PP with cellulose fiber [47]. Rodriguez et al. [48] recently performed LCA study on bio-composite materials made from banana fiber extracted from Colombian region plantation and compared it with polyester resin. They also investigated the life cycle cost (LCC) on the cradle-to-gate scope bio-composites and found that it has lower cost and environmental impact than polyester, considering the scope and boundaries of the study and the production of the same selected component. Another recent study of composites using LCA was of two types of bamboo fibers, i.e. bamboo original fibers (BOF) and bamboo viscose fibers (BVF), and flax fibers as reference were used to reinforce PP. The composites were fabricated into laminates using manual stacking and hot pressing. Environmental performance assessment conducted showed that the composites reinforced with bamboo fibers were more eco-friendly as compared to the composites with flax fibers. The results of LCA analysis indicate that laminate preparing procedures caused the most significant environmental burden in the life cycles of these composites. The long-distance transportation significantly increased the environmental impacts of flax fibers while fiber dispersion process was the primary contributor of environmental impacts for bamboo fiber reinforced composites. They also compared three EOL options of the bio-composites, i.e. landfill, mechanical recycling, and incineration, and in terms of material and energy recovery, recycling, and incineration were discovered to be superior to landfill [49].

12.4.1.1 LCA Natural Fiber
Natural fibers possess good reinforcing capability when properly compounded with polymers. The main advantage of using natural fiber composites is the reduction of the environmental impact because of the use of nonrenewable energy/material sources, lower pollutant emissions, lower greenhouse gas emissions, enhanced energy recovery, and EOL biodegradability of components. However, some fibers may require equivalent or higher energy consumption due to strong reliance on agrochemicals [21]. Gu et al. [49] proposed a natural fiber product system that includes four phases: (i) cultivating and harvesting corps, (ii) extracting fibers, (iii) fabricating laminates, and (iv) disposing wastes. The system boundaries for bamboo and flax fibers studied in Gu et al. [49] are shown in Figure 12.5.

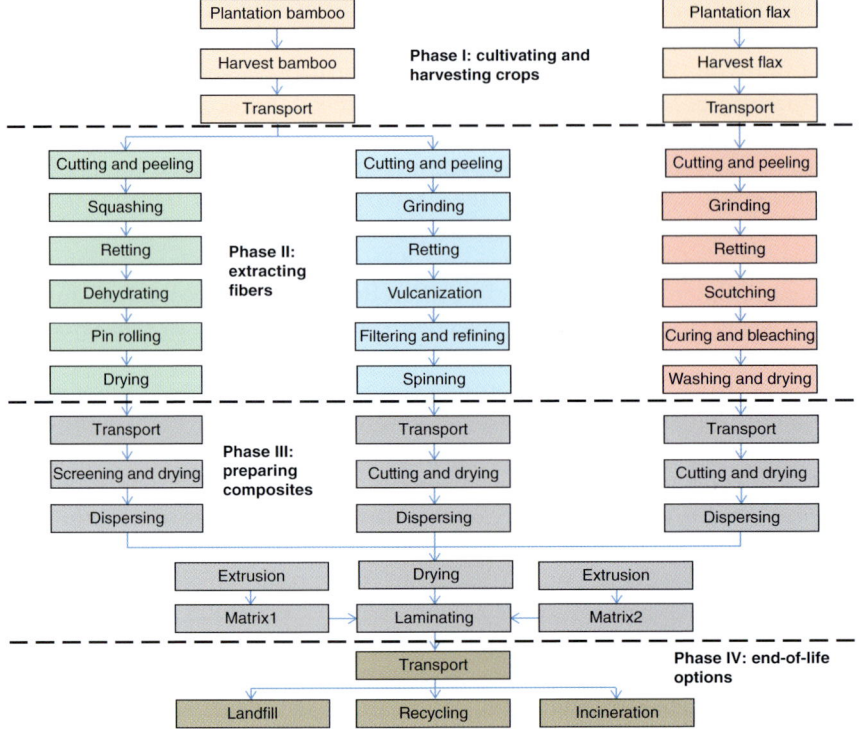

Figure 12.5 System boundaries of bamboo and flax fiber. Source: Gu et al. 2018 [49]. Reproduced with permission of Elsevier.

12.4.2 LCA Biopolymer

LCAs of biopolymers derived from agricultural products would include the farming phase that comprises of the fuel required for farming activities such as plowing, agrochemical application, and harvesting. For other supplies required in the farming activities such as fertilizer, herbicides, and pesticides also their manufacturing process and transportation should be considered. Land use and water consumption may also be important factors, as are nitrogen-based emissions from fertilizer use. Other processes to include in the system for typical biopolymers include milling and production [50]. Groot and Borén [51] studied cradle-to-gate analysis LCA of PLA biopolymers from sugarcane with functional unit of 1 tonne of material at the factory gate in Thailand. The materials assessed are L-lactide, D-lactide, polymer poly-L-lactic acid (PLLA), nucleated polylactic acid (nPLA) (PLLA with low levels of polymer poly-D-lactic acid [PDLA]), and stereocomplex polylactic acid (scPLA) (PLLA with increased levels of PDLA). The study took into account the sugarcane agricultural system, industrial activities related to auxiliary chemicals, distribution of raw materials, processing of sugarcane into sugar, and the final production of lactic acid, lactides, and PLA biopolymers. These PLA biopolymers from sugarcane were discovered to have lower emissions of greenhouse gases, and less use of material resources and non-renewable energy, compared to fossil-based polymers. GWP in L-lactide production is 300–600 kg CO_2 equiv/tonne and for PLLA 500–800 kg CO_2 equiv/tonne. Being based on an agricultural system, the bio-based PLA gives rise to higher contributions to acidification, photochemical ozone creation, eutrophication, and farm land use compared to the fossil polymers [51]. Figure 12.6 shows the production system from cultivation to PLA biopolymer.

Ingrao et al. [1] in their study aimed at performing A-LCA to identify environmental hotspots in the life cycle of expanded PLA trays for fresh food packaging, thereby representing a valid tool to identify more sustainable alternatives such as the utilization of second-generation PLA granules. The study highlighted that the total damage is equal to 1.85 mpt and is mainly due to production (for almost 49.7%) and transport (for 25.43%) of the PLA granules; the electricity consumption for their processing (for 12.2%); and delivery (for 5.94%) of the produced trays. PLA granule production is the most impacting phase and other relevant contributions to the environmental impact associated with the investigated system are from the transportation of the granules to the tray manufacturing plant and the electricity consumption for the processing of the granules into trays.

This is because the delivery stage involves long distances and different transport means, including, in the examined case, a freight ship, thereby producing significant impacts in terms of exploitation on nonrenewable energy resource and GHG emission. Previously, the same group of researchers studied the application of LCA with the aim of assessing both environmental impacts and improvement potentials in the life cycle of foamy PS trays for fresh meat packaging [39].

12.5 LCA Limitations

Polymer composites would be a feasible option with the occurrence of decomposition in landfills. Value assessments through incineration and composting need to be made, for example the trade-off between accumulating carbon in a landfill vs. carbon and methane emissions to the atmosphere respectively [52]. LCA could be a great tool in assessing the values judgment and estimating the impact to the environment produced from the product system. In ALCA of natural fiber reinforced bio-composites, a few limitations are realized and the following are among the major ones according to Venkatachalam et al. [40]:

 i. Inconsistencies in the LCA results due to the assumptions about system boundaries and allocation methods.
 ii. Lack of information on the data sources used in the LCA of biopolymers.
 iii. Different boundary settings (between natural and technical systems), different feedstocks, and different production technologies between studies.
 iv. The incorporation of land use change (LUC) between studies is not consistent.
 v. Discrepancies in energy consumption due to geographical differences and corresponding energy mixes.

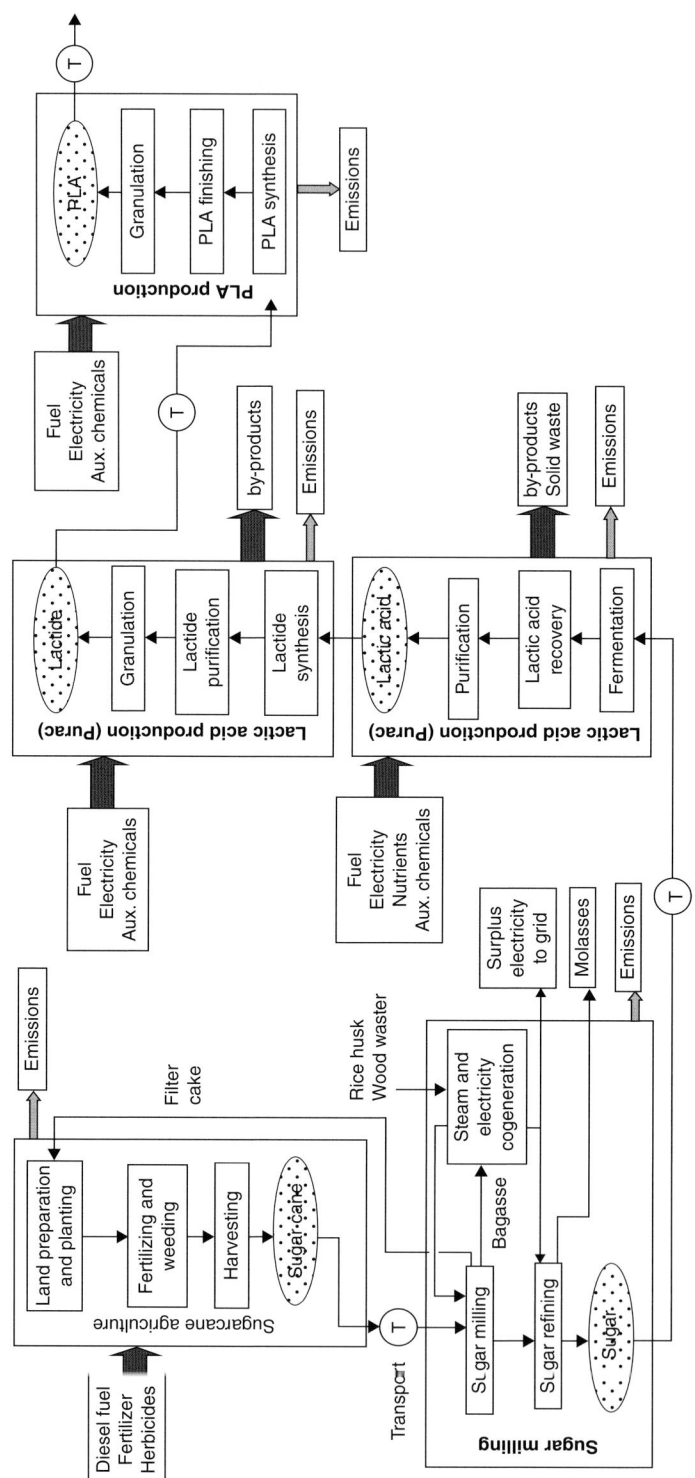

Figure 12.6 Schematic production flow chain of PLA biopolymer from cultivation of crop. Source: Groot et al. 2010 [51]. Reproduced with permission of Springer Nature.

vi. Studies that looked only at selective impact categories drew conclusions based on material preferences (GWP and non-renewable energy use (NREU) of biopolymers better than conventional polymers).

vii. Including EoL and use phase provides a comprehensive estimate of the total impacts but there are large uncertainties associated with consumer behavior in use phase and inventory data in EoL processes.

viii. Potential benefits of biopolymers with regards to different impact categories cannot be realized until material and energy consumption of farming and production processes are reduced.

ix. Given the fact that biopolymers are relatively new to the global plastics market, any comparison with the conventional polymers must take the status of technologies and market presence into account.

x. Environmental impacts quoted for conventional fossil-based polymers are not consistent between studies, which can have an adverse effect on the results.

Some important recommendations to improve the product development and result in a sustainable production of bio-based plastics through LCA are also proposed by the same researchers. Most importantly, the competing bio-based products at the national and international level have to be carefully identified to obtain average (primary from manufacturers, secondary from literature) and marginal data as much as possible. A thorough cost–benefit and market analysis needs to be conducted too for the product along with the economic and technical feasibility of implementing eco-design strategies as well as sourcing of the raw materials; cultivation of biomass and the complete value chain are to be thoroughly investigated (including direct and indirect elementary flows). Last but not least, scenario and sensitivity analysis has to be done for both the ALCA and CLCA approaches to address the uncertainties in the methodology, inventory data, and process parameters [40].

12.6 Conclusions

The main benefit of using LCA in composite production is that it provides a comprehensive way of determining the total environmental impact for designing new materials and processes. LCA methodology is a very useful tool to drive the choice of materials and processes toward a more sustainable production system. LCA is a useful tool for choosing clean production processes, avoiding hazardous and toxic materials, maximizing the efficiency of the energy used for production and for the product in use, and designing for waste management processes, avoiding hazardous and toxic materials. Nevertheless, there are still several critical aspects that need to be looked at – the main limitation of LCA methodology is data uncertainty. Existing data bases for fiber reinforced composites, green or traditional based, are uncertain and need to be continually updated and corrected. Data uncertainty also derives from the fact that studies are carried out on regional sites and the results are to some extent subject to country-specific circumstances.

Acknowledgment

The authors wish to express the highest appreciation to the Public Service Department (JPA), Malaysia, for the study sponsorship to the main author and financial support from the Ministry of Education, Malaysia, through Universiti Putra Malaysia Grant Scheme HiCOE (6369107).

References

1 Ingrao, C., Gigli, M., and Siracusa, V. (2017). An attributional life cycle assessment application experience to highlight environmental hotspots in the production of foamy polylactic acid trays for fresh-food packaging usage. *Journal of Cleaner Production* 150: 93–103.

2 Sapuan, S.M. (2017). Materials selection for composites: concurrent engineering perspective. In: *Composite Materials: Concurrent Engineering Approach*, vol. 219. Elsevier Inc.

3 Al-Oqla, F.M., Almagableh, A., and Omari, M.A. (2017). Design and fabrication of green biocomposites. In: *Green Biocomposites*, Green Energy and Technology (ed. M. Jawaid, M. Salit and O. Alothman), 45–67. Cham: Springer. ISBN: 978-3-319-49381-7.

4 Mansor, M.R., Salit, M.S., Zainudin, E.S. et al. (2015). Life cycle assessment of natural fiber polymer composites. In: *Agricultural Biomass Based Potential Materials* (ed. K. Hakeem, M. Jawaid and O.Y. Alothman), 121–141. Cham: Springer.

5 Mukherjee, T. and Kao, N. (2011). PLA based biopolymer reinforced with natural fibre: a review. *Journal of Polymers and the Environment* 19 (3): 714–725.

6 Mitra, B.C. (2014). Environment friendly composite materials: biocomposites and green composites. *Defence Science Journal* 64 (3): 244–261.

7 Ramesh, M., Palanikumar, K., and Hemachandra Reddy, K. (2017). Plant fibre based bio-composites: sustainable and renewable green materials. *Renewable and Sustainable Energy Reviews* 79: 558–584.

8 Su, Y., Yang, B., Liu, J. et al. (2018). Prospects for replacement of some plastics in packaging with lignocellulose materials: a brief review. *BioResources* 13 (2): 4550–4576.

9 Salit, M.S. (2014). Tropical natural fibres and their properties. In: *Tropical Natural Fibre Composites*, Engineering Materials, 15–38. Singapore: Springer.

10 Jumaidin, R., Sapuan, S.M., Jawaid, M. et al. (2017). Effect of seaweed on mechanical, thermal, and biodegradation properties of thermoplastic sugar palm starch/agar composites. *International Journal of Biological Macromolecules* 99: 265–273.

11 Sanyang, M.L., Ilyas, R.A., Sapuan, S.M., and Jumaidin, R. (2018). Sugar palm starch-based composites for packaging applications. In: *Bionanocomposites for Packaging Applications* (ed. M. Jawaid and S. Swain), 125–147. Cham: Springer International Publishing.

12 Gurunathan, T., Mohanty, S., and Nayak, S.K. (2015). A review of the recent developments in biocomposites based on natural fibres and their application perspectives. *Composites Part A: Applied Science and Manufacturing* 77: 1–25.

13 Huzaifah, M.R.M., Sapuan, S.M., Leman, Z., and Ishak, M.R. (2017). Comparative study on chemical composition, physical, tensile, and thermal properties of sugar palm fiber (*Arenga pinnata*) obtained from different geographical locations. *BioResources* 12 (4): 9366–9382.

14 Al-Oqla, F.M., Salit, M.S., Ishak, M.R., and Aziz, N.A. (2015). Selecting natural fibers for bio-based materials with conflicting criteria. *American Journal of Applied Sciences* 12 (1): 64–71.

15 Johansson, C., Bras, J., Mondragon, I. et al. (2012). Renewable fibers and bio-based materials for packaging applications – a review of recent developments. *BioResources* 7 (2): 1–47.

16 Sanyang, M.L., Sapuan, S.M., Jawaid, M. et al. (2016). Effect of sugar palm-derived cellulose reinforcement on the mechanical and water barrier properties of sugar palm starch biocomposite films. *BioResources* 11 (2): 4134–4145.

17 Sapuan, S.M., Kho, J.Y., Zainudin, E.S. et al. (2011). Materials selection for natural fiber reinforced polymer composites using analytical hierarchy process. *Indian Journal of Engineering and Materials Science* 18 (4): 255–267.

18 Singh, A.A., Afrin, S., and Karim, Z. (2017). Green composites: versatile material for future. In: *Green Biocomposites*, Green Energy and Technology (ed. M. Jawaid, M. Salit and O. Alothman), 29–44. Cham: Springer.

19 Alkbir, M.F.M., Sapuan, S.M., Nuraini, A.A., and Ishak, M.R. (2016). Fibre properties and crashworthiness parameters of natural fibre-reinforced composite structure: a literature review. *Composite Structures* 148: 59–73.

20 Saba, N., Jawaid, M., Sultan, M.T.H., and Alothman, O.Y. (2017). Green biocomposites for structural applications. In: *Green Biocomposites*, Green Energy and Technology (ed. M. Jawaid, M.S. Salit and O.Y. Alothman), 1–27, Cham. Springer.

21 La Rosa, A.D. and Cicala, G. (2015). LCA of fibre-reinforced composites A2. In: *Handbook of Life Cycle Assessment (LCA) of Textiles and Clothing*, Chapter 14, Woodhead Publishing Series in Textiles (ed. S.S. Muthu), 301–323. Woodhead Publishing.

22 Davoodi, M.M., Sapuan, S.M., Ahmad, D. et al. (2011). Concept selection of car bumper beam with developed hybrid bio-composite material. *Materials and Design* 32 (10): 4857–4865.

23 Mansor, M.R., Sapuan, S.M., Zainudin, E.S. et al. (2014). Conceptual design of kenaf fiber polymer composite automotive parking brake lever using integrated TRIZ-morphological chart-analytic hierarchy process method. *Materials and Design* 54: 473–482.

24 Mastura, M.T., Sapuan, S.M., Mansor, M.R., and Nuraini, A.A. (2018). Materials selection of thermoplastic matrices for 'green' natural fibre composites for automotive anti-roll bar with particular emphasis on the environment. *International Journal of Precision Engineering and Manufacturing – Green Technology* 5 (1): 111–119.

25 Nattapon, K., Orapin, K., and Natta, L. (2012). Biodegradable foam tray from cassava starch blended with natural fiber and chitosan. *Industrial Crops and Products* 37 (1): 542–546.

26 Sahari, J., Sapuan, S.M., Zainudin, E.S., and Maleque, M.A. (2013). Mechanical and thermal properties of environmentally friendly composites derived from sugar palm tree. *Materials and Design* 49: 285–289.

27 Huang, C., Zhu, Q., Li, C. et al. (2014). Effects of micronized fibers on the cushion properties of foam buffer package materials, Fibrillated cellulose cushion. *BioResources* 9 (4): 5940–5950.

28 Jumaidin, R., Sapuan, S.M., Jawaid, M. et al. (2016). Characteristics of thermoplastic sugar palm starch/agar blend: thermal, tensile, and physical properties. *International Journal of Biological Macromolecules* 89: 575–581.

29 Jumaidin, R., Sapuan, S.M., Jawaid, M. et al. (2017). Thermal, mechanical, and physical properties of seaweed/sugar palm fibre reinforced thermoplastic sugar palm starch/agar hybrid composites. *International Journal of Biological Macromolecules* 97: 606–615.

30 Luo, Y.Y., Xiao, S.L., and Li, S.L. (2017). Effect of initial water content on foaming quality and mechanical properties of plant fiber porous cushioning materials. *BioResources* 12 (2): 4259–4269.

31 Edhirej, A., Sapuan, S.M., Jawaid, M., and Zaharia, N.I. (2017). Cassava/sugar palm fiber reinforced cassava starch hybrid composites: physical, thermal and structural properties. *International Journal of Biological Macromolecules* 101: 75–83.

32 Berthet, M.A., Angellier-Coussy, H., Machado, D. et al. (2015). Exploring the potentialities of using lignocellulosic fibres derived from three food by-products as constituents of biocomposites for food packaging. *Industrial Crops and Products* 69: 110–122.

33 Ochoa-Yepes, O., Medina-Jaramillo, C., Guz, L., and Famá, L. (2018). Biodegradable and edible starch composites with fiber-rich lentil flour to use as food packaging. *Starch – Stärke* 70: 7–8.

34 Jha, K., Chamoli, S., Tyagi, Y.K., and Maurya, H.O. (2018). Characterization of biodegradable composites and application of preference selection index for deciding optimum phase combination. *Materials Today: Proceedings* 5: 3353–3360.

35 Qian, S., Zhang, H., Yao, W., and Sheng, K. (2018). Effects of bamboo cellulose nanowhisker content on the morphology, crystallization, mechanical, and thermal properties of PLA matrix biocomposites. *Composites Part B* 133: 203–209.

36 Ilyas, R.A., Sapuan, S.M., Ishak, M.R., and Zainudin, E.S. (2017). Effect of delignification on the physical, thermal, chemical, and structural properties of sugar palm fibre. *BioResources* 12 (4): 8734–8754.

37 Sijtsema, S.J., Onwezen, M.C., Reinders, M.H.J. et al. (2016). Consumer perception of bio-based products – an exploratory study in 5 European countries. *NJAS - Wageningen Journal of Life Sciences* 77: 61–69.

38 Brundage, M.P., Bernstein, W.Z., Hoffenson, S. et al. (2018). Analyzing environmental sustainability methods for use earlier in the product lifecycle. *Journal of Cleaner Production* 187: 877–892.

39 Ingrao, C., Giudice, A.L., Bacenetti, J. et al. (2015). Foamy polystyrene trays for fresh-meat packaging: life-cycle inventory data collection and environmental impact assessment. *Food Research International* 76: 418–426.

40 Venkatachalam, V., Spierling, S.V., Horn, R., and Endres, H.J. (2018). LCA and eco-design: consequential and attributional approaches for bio-based plastics. *Procedia CIRP* 69: 579–584.

41 Alvarenga, R.A.F., Jo, D., De Meester, S. et al. (2013). Life cycle assessment of bioethanol-based PVC: Part 2: Consequential approach. *Biofuels, Bioproducts and Biorefining* 7: 396–405.

42 Parameswaranpillai, J. and Vijayan, D. (2014). *Life Cycle Assessment (LCA) of Epoxy-Based Materials*. Wiley-VCH.

43 Khoo, H.H., Isoni, V., and Sharratt, P.N. (2018). LCI data selection criteria for a multidisciplinary research team: LCA applied to solvents and chemicals. *Sustainable Production and Consumption* 16: 68–87.

44 Gbededo, M.A., Liyanage, K., and Garza-Reyes, J.A. (2018). Towards a life cycle sustainability analysis: a systematic review of approaches to sustainable manufacturing. *Journal of Cleaner Production* 184: 1002–1015.

45 Ingrao, C., Tricase, C., Cholewa-Wójcik, A. et al. (2015). Polylactic acid trays for fresh-food packaging: a carbon footprint assessment. *Science of the Total Environment* 537: 385–398.

46 Montalbo-Lomboy, M., Schrader, J.A., and Grewell, D. (2016). Cradle-to-gate life cycle assessment of bioplastic horticulture containers and comparison to standard petroleum-plastic containers. In: J.A. Schrader, H.A. Kratsch, and W.R. Graves, (Eds). Bioplastic Container Cropping Systems: Green Technology for the Green Industry. *Sustainable Hort. Res. Consortium*, Ames, IA, USA.

47 Czaplicka-Kolarz, K., Burchart-Korol, D., and Korol, J. (2013). Environmental assessment of biocomposites based on LCA. *Polimery* 58 (6): 476–481.

48 Rodriguez, L.J., Orrego, C.E., Ribeiro, I., and Pecas, P. (2018). Life-cycle assessment and life-cycle cost study of banana (*Musa sapientum*) fiber biocomposite materials. *Procedia CIRP* 69: 585–590.

49 Gu, F., Zheng, Y., Zhang, W. et al. (2018). Can bamboo fibres be an alternative to flax fibres as materials for plastic reinforcement? A comparative life cycle study on polypropylene/flax/bamboo laminates. *Industrial Crops and Products* 121: 372–387.

50 Yates, M.R. and Barlow, C.Y. (2013). Life cycle assessments of biodegradable, commercial biopolymers—a critical review. *Resources, Conservation and Recycling* 78: 54–66.

51 Groot, W.J. and Borén, T. (2010). Life cycle assessment of the manufacture of lactide and PLA biopolymers from sugarcane in Thailand. *International Journal of Life Cycle Assessment* 15 (9): 970–984.

52 Zgola, M.L., Gregory, J.R., Olivetti, E.A., and Kirchain, R.E. (2010). Life cycle analysis of plastics for packaging: PVC and PET. *Proceedings of the 2010 IEEE International Symposium on Sustainable Systems and Technology, ISSST 2010*.

13

Reprocessing and Disposal Mechanisms for Fiber Reinforced Polymer Composites

Vijay Chaudhary[1], Khushi Ram[2], and Furkan Ahmad[2]

[1]Amity University, Amity School of Engineering, Mechanical Engineering Department, Sector 125, Noida, 201313, Uttar Pradesh, India
[2]Netaji Subhas University of Technology (Formerly NSIT, University of Delhi), MPAE Division, Sector-3, Dwarka, 110078, New Delhi, India

13.1 Introduction

Fiber reinforced polymer composites are being widely used for manufacturing products in every sector such as automobiles, medicine, construction, aerospace, and shipping. Favorable properties of fiber reinforced polymer composites such as high density, high strength with light weight, good resistance to wear stands these composites in the application of structural and nonstructural components. A serious concern of pollution due to synthetic plastic wastages increases the attention of material researchers toward the natural fiber-based polymer composites. The biodegradability of natural fiber-based polymer composites enables them to replace synthetic fiber-based polymer composites [1–3].

Apart from application of polymer composites, manufacturing of polymer composite includes the selection of processing method, selection of reinforced fiber, and selection of polymer matrix. Processing methods can be categorized as hand lay-up, compression molding, injection molding, pultrusion method, resin transfer molding, tube rolling, etc. [2]. Reinforced fibers can be categorized on the basis of their synthetic or biodegradable nature. Synthetic fibers include glass, carbon, aramid, Kevlar, etc. Natural or biodegradable fibers are extracted from different natural resources such as plants (seed fibers, leaf fibers, bast or skin fibers, fruit fibers, and stalk fibers), animals (hair, silk, and feather), and minerals (asbestos). Some common natural fibers are sisal, banana, and agave, which are plant fibers, while cotton and kapok are examples of the plant seed, and hemp, flax, jute, kenaf, ramie, vine, and rattan are extracted from bast surrounding the stem of the plant. Coir fibers are extracted from the coconut fruit while bamboo and grass fibers are collected from the stalk of the plants [4]. To provide the aesthetic appearance of a composite structure and to surround the fiber, polymer matrix (thermoset or thermoplastic) plays a vital role and always protects the composite structure against environmental damages due

Reinforced Polymer Composites: Processing, Characterization and Post Life Cycle Assessment,
First Edition. Edited by Pramendra K. Bajpai and Inderdeep Singh.
© 2020 Wiley-VCH Verlag GmbH & Co. KGaA. Published 2020 by Wiley-VCH Verlag GmbH & Co. KGaA.

to temperature, moisture, harmful gases, and chemical attacks. Protection of matrix always enhances the life cycle of the developed fiber reinforced polymer composites [5].

After the successful manufacturing and consumption of fiber reinforced polymer composites, the problem of waste management arises due to their long useful life. To optimize the life of fiber-based polymer composites, it is necessary to know about the reprocessing and disposal mechanism of fiber reinforced polymer composites. Recycling of fiber reinforced polymer composites creates a strong alternative to producing the new product with the help of recycled polymer composite parts. Recycling of fiber reinforced polymer composite helps in plastic waste management. Various recycling techniques have been developed by the researchers for successful recycling of fiber reinforced polymer composite components. Reprocessing or recycling of fiber reinforced polymer composites degrades the mechanical properties of the developed recycled product. Degradation of properties of fiber reinforced polymer is due to various governing factors such as shearing forces at high temperatures in mechanical recycling that develop the internal residual stresses in polymer composite and high temperatures that promote the rupture of the polymer chains. Still, material scientists and researchers are continuing their research on many improvements in the recycling methods of fiber reinforced polymer composites. Various authors have studied the recycling of fiber reinforced polymer composites and based the results on their research findings. Some authors evaluated the mechanical properties after the recycling of composites [6–9].

Lea et al. [10] developed a sugarcane fiber reinforced polypropylene composite. The authors performed the mechanical recycling of composite material using grinding process and then reprocessed the composite material using injection molding process. The authors concluded that the recycled composite showed 20% decrement in tensile strength as compared to the original composite material. Sun et al. [11] performed the recycling of carbon fiber from carbon fiber reinforced polymer composites using electrochemical technique. Experimental design of the electrochemical method includes the process parameters of different concentrations of NaCl (3%, 10%, and 20%) and applied current (4, 10, 20, and 25 mA). The authors evaluated the effect of process parameters on the recycling efficiency in terms of the mechanical properties of carbon fiber. The authors concluded that, after recycling of carbon fibers, fibers lose 20% tensile strength as compared to the virgin carbon fiber. They also confirmed that recycling of fibers using electrochemical technique is an effective, easy, and economical process. Ashton et al. [12] carried out the recycling of multilayered polymeric material using micronization technique. Experimentation of research includes the micronization of composite material and after that manufacturing of recycled material through extrusion and injection molding process. The authors also performed the tensile test and dynamic mechanical thermal analysis (DMTA) on recycled composite material. They concluded that the recycled composite showed low level of interfacial adhesion at the fiber and matrix interface, which showed a loss of ductility, but tensile strength showed an enhancement of 18.43% as compared to the virgin composite; DMTA showed a decrement in storage

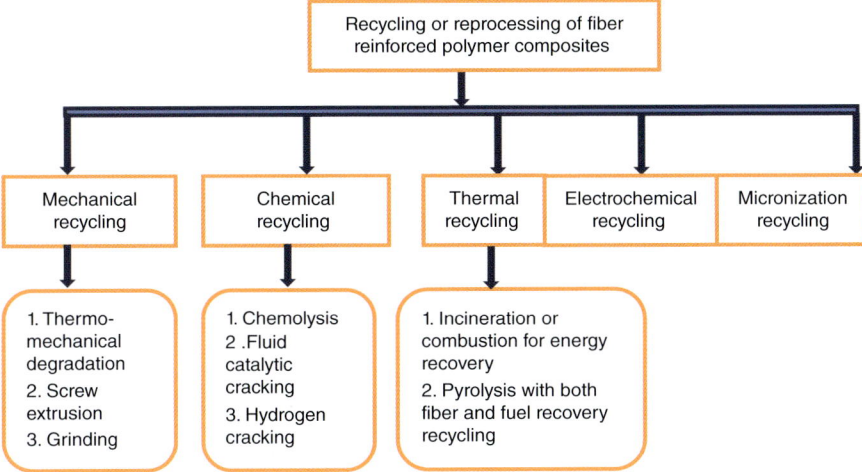

Figure 13.1 Different recycling techniques of fiber reinforced polymer composites.

modulus value at above 30 °C temperature. Figure 13.1 represents the various types of recycling techniques used for reprocessing of fiber reinforced polymer composites.

The objective of the present chapter is to summarize the existing methods of recycling or reprocessing of fiber reinforced polymer composites. Different recycling methods of fiber reinforced polymer composites are discussed in detail as also the effect of recycling methods on the properties of recycled composite materials, which is shown by the author's previous research.

13.2 Reprocessing or Recycling Methods of Fiber Reinforced Polymer Composites

Increasing plastic wastage day by day in every country and increasing pollution due to fiber reinforced polymer composites have forced the researchers and material scientists toward the field of material recycling or reprocessing, which is increasing extensively in polymer composites. Government laws related to plastic pollution also have expressed concern and encourage researchers and scientists to carry out their research work in the field of recycling of fiber reinforced polymer composites.

A number of recycling or reprocessing methods were suggested by researchers and complete experimental procedures developed. Some important reprocessing methods of fiber-based polymer composites include mechanical recycling, chemical recycling, and various degradation methods such as chemical degradation, hydrolytic degradation, and biodegradation.

Various authors carried out their research work on the recycling methods of fiber reinforced polymer composites as well as recycling of fiber separately and recycling of polymer resin as shown in Table 13.1.

Table 13.1 Recycling of natural fiber reinforced polymer composite.

Fiber/matrix	Recycling methods	Remarks	References
Sugarcane/ polypropylene composite	Mechanical recycling	Recycled composite showed decrement in tensile strength as compared to original composite	[10]
Carbon/epoxy composite	Electrochemical recycling (recycling of carbon from carbon/epoxy composite)	Process parameters such as concentration of NaCl and applied current influence the mechanical properties of recycled composite as recycled carbon fiber loses 20% of its tensile strength as compared to the original carbon fiber	[11]
Multilayered polymeric material	Micronization recycling	Recycling reduced the ductility of recycled composite material but tensile strength was enhanced by 18.43% as compared to virgin composite material	[12]
Carbon fiber reinforced polymer composite	Mechanical recycling	Mechanical recycling is suitable to reduce GHS, emission, primary energy use, and landfill waste generation	[13]
Hemp and sisal fiber reinforced polypropylene composites	Mechanical recycling	Decrement in fiber length decreased the mechanical strength of developed composites	[14]
Flax fiber reinforced polylactic acid	Mechanical recycling	Molecular weight of recycled composite enhanced due to a greater number of injection cycles	[15]
Wood fiber reinforced polyvinyl chloride (PVC) composite	Mechanical recycling	Addition of the wood fiber to PVC improves its recyclability and also enhances the mechanical strength of the developed composite	[16]

13.3 Mechanical Recycling

Mechanical recycling is also known as physical recycling. Fiber reinforced polymer composite is ground and then reprocessed using processing techniques such as extrusion and injection molding to produce a new recycled fiber reinforced polymer composite [17]. From the industrial point of view, mechanical recycling is the most suitable due to low cost and reliability.

Mechanical recycling largely depends on the type of reinforcement and polymer matrix material, the type of processing methods, and the presence of additives, which influences the mechanical properties of recycled composites. Mechanical recycling of natural fiber reinforced polymer composites includes a number of operations such as the separation of fiber reinforcement by polymer matrix, washing to remove dirt and contaminants, grinding and crushing to reduce the plastics particle size, extrusion by heat, and reprocessing into new

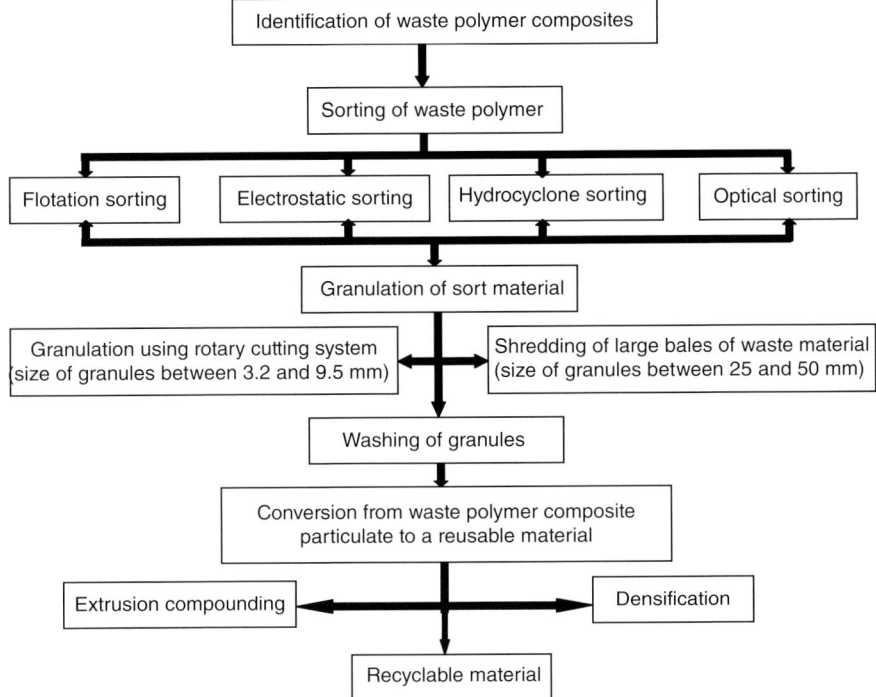

Figure 13.2 Flow diagram of mechanical recycling.

plastic goods. Figure 13.2 represents a flow diagram of the recycling route of waste fiber reinforced polymer composite during the mechanical recycling process.

Mechanical recycling of fiber reinforced polymer composite includes some important steps before reprocessing of polymer waste into new recycled polymer composite. This conversion process or phase from waste polymer composite to recycled polymer material is named "End-of-Waste." For mechanical recycling of the fiber reinforced polymer composites, the process of recycling includes the following steps, which can occur anywhere between, not at all, or multiple times during the sequence:

1. *Crushing, separation, and sorting*: Crushing comprises the breaking of waste material into tiny particles with the help of grinding, milling, etc. and separation and sorting of the different constituents of polymer composite material.
2. *Baling*: After the crushing and sorting of waste material, it is baled separately and then transported for the next process.
3. *Washing and drying*: Washing and drying process comprises the elimination of any chemical content present in the raw material that will be recycled, and after washing and drying the material goes to the next step.
4. *Extrusion and processing*: After successful crushing, separation, and washing, the raw material is extruded by screw extrusion process and then injected in injection molding machine for reprocessing and converted into the final recycled fiber reinforced polymer composites.

Various authors carried out their research on the mechanical recycling of fiber reinforced polymer composites and concluded their results by a comparative study between the properties of composites before and after recycling. Duigou et al. [15] studied the effect of recycling on the mechanical properties of flax fiber reinforced polylactic acid composites. Processing of composites was done using a single screw extruder and then molded using an injection molding machine. After recycling, composites were fabricated using six injection molding cycles and the effect of injection cycles on the mechanical, rheological, and morphological properties was investigated. The authors concluded that stiffness of the recycled composites is not influenced by injection number while molecular weight of the recycled composite material reduces after the processing cycles, which increases the crystallinity of the developed composite. The same research work has been carried out by various other researchers [18–20], who also saw the effect of recycling methods on the mechanical, rheological, and thermo-mechanical performance of the recycled composite material. Augier et al. [16] studied the mechanical recycling of PVC (polyvinyl chloride) and investigated the effect of wood fiber filler with recycled PVC on the mechanical properties of the developed wood fiber reinforced recycled PVC composite and neat recycled PVC. The authors concluded that incorporation of wood fiber into PVC improves its recyclability, and the overall mechanical properties of the developed composite remain constant except that the flexural properties of the developed composite enhanced after recycling.

13.4 Chemical Recycling

Strict governmental regulations against plastic waste strongly increase the use of recycling methods. In this chain, chemical recycling is the most vital step, which was heavily developed in Germany. Today, chemical recycling is exhaustively used by each and every plastic industry for recycling of plastic waste [21]. Chemical recycling decomposes the macromolecular structure of polymers and generates low molecular weight compounds. Chemical recycling process is performed under high temperature and in the presence of various types of catalysts. Chemical recycling consumes huge amounts of energy, and in many cases results in rather low value products.

Chemical recycling is an important recycling method of polymer reinforced composite materials. Chemical recycling changes the chemical structure of polymeric material in which polymer matrix of the composite material converts into small units of monomers through chemical reactions. A converted small unit of monomer plays a vital role in the reproduction of polymer-based composite material as a matrix material [22]. Figure 13.3 represents the chemical reactions used in chemical recycling to convert the recycled polymer into small molecules such as monomers, oligomers, and other hydrocarbon compounds.

Depolymerization of matrix by chemical dissolution reagents is carried out in chemical recycling for the liberation of fiber reinforcement. Chemical recycling process regenerates both the fiber and polymer matrix in the form of monomers. Dissolution or solvolysis depends on the solvent and can be

Figure 13.3 Different chemical reactions of chemical recycling.

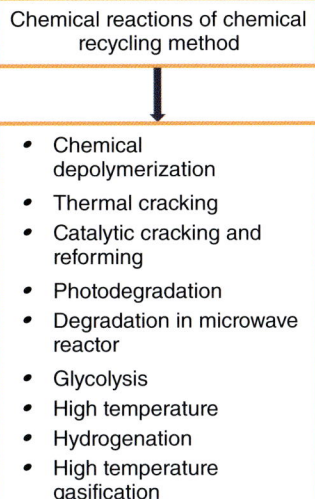

classified as glycolysis, acid digestion, and hydrolysis. Faster dissolution using water or alcohol requires high temperature and high pressure. Dissolution using acid digestion provides a very slow rate of dissolution under normal atmospheric conditions. Some supercritical fluids, water, and alcohols are potentially used as media for recycling of polymer resin and reinforced fiber. During depolymerization, water or alcohol is a good media for dissolution due to its environmentally clean property; it is easily separated from the dissolved solution by using evaporation in case of water and distillation in case of alcohol [23].

13.5 Hydrolytic Degradation of Fiber Reinforced Polymer Composite

Biopolymer composites are widely used in medical applications such as heart stents, bones, etc. due to their application their nature need to be studied. Degradation of polymers is classified as thermal, mechanical, biodegradation, and chemical degradation. Medical biopolymer composites are chemically degraded. In this section, the hydrolytic nature of degradation is studied. Hydrolytic degradation is defined as the degradation of polymer composite due to the attack of moisture. When water molecule attacks the biopolymer it starts degrading it, and formation of voids takes place [24]. These voids are the site from where the actual degradation takes place gradually as also the formation of other smaller monomer units of the polymer composite. Factors that affect the hydrolytic degradation are affinity of polymeric compound toward water, molecular weight of the polymer, various groups attached to the polymer, temperature, and the nature of the polymeric compound [25]. Benali et al. [26] developed the zinc oxide reinforced polylactide nanocomposite by the melt extrusion process. The authors discussed that polylactic acid is greatly affected by hydrolytic degradation but by the addition of zinc oxide as a filler material its degradability reduces due to the hydrophobic nature of zinc oxide. Zinc oxide prevents the attack by water molecule on polylactic acid and changes the chemical process of hydrolytic degradation. Ndazi and Karlsson [27] fabricated rice hulls reinforced polylactic acid composite by compression molding. The authors studied the effect of temperature on hydrolytic degradation through thermal and chemical analysis. The results showed that at $25\,^{\circ}C$ the properties are not much affected but at 51 and $69\,^{\circ}C$ the properties drastically change and the glass transition temperature reduces to $13\,^{\circ}C$. Gonzalez et al. [28] manufactured reduced graphene oxide reinforced polycaprolactone (rGO/PCL) composite by phase inversion method. The authors studied the hydrolytic degradation of the developed composite in order to identify whether it is suitable for scaffold. The results showed that during one year of study the molecular weight reduced from 80 to 33 kDa for pure PCL and 27 kDa for rGO/PCL composite. Martos et al. [29] investigated the four types of composites for dental application. The authors studied the effect of hydrolytic degradation on the microhardness of the developed composites. The results showed that after 30 days the microhardness of the four developed composites reduced. Bell Glass HP composite hardness reduced less as compared to Alert, Artglass, and Filtek Z250. Silva et al. [30] synthesized polyhydroxybutyrate (PHB) and polyethylene glycol (PEG) thin films and studied the effect of hydrolytic degradation by Fourier transform infrared spectroscopy. The results showed that PEG was degraded more as compared to PHB because it provides the medium for attack by water molecules.

13.6 Photodegradation of Polymer Composite

Bio-composites degrade under the action of ultraviolet (UV) light. When UV light falls on the polymer composite surface they get heated and a polymeric reaction starts, which results in breaking of bonds between polymer, formation of monomer units, and decrement in the molecular weight, all of which finally reduce the properties of the polymer composite surface [31]. Various authors carried out their study to find how the composite surface degrades and the end product behavior. Siddiquee et al. [32] fabricated glass/epoxy composites. The authors carried out the numerical simulation and experimental study to check the effect of ultraviolet (UV) light on the morphology of the developed composites. The results showed that the surface degradation of composites depends on the intensity and wavelength of UV light, and the duration for which it was exposed. Irregular surfaces are more degraded as compared to planer surface and the surface roughness was decreased up to 12.5% when exposed to UV light for 1000 hours. Abdullah et al. [33] developed jute fiber reinforced polyethylene/polypropylene reinforced polymer composite by compression molding. The authors reinforced with 5%, 10%, and 15% of jute fiber having 1 and 3 mm length. They investigated the degradation of the developed composites under soil burial, weather condition (exposed to UV radiation), and under compost. The results showed that the rate of degradation in compost was more and soil had the least. Factors that affect the phenomena of degradation are temperature, water, air, and the duration for which the study was carried out. Lopez et al. [34] investigated the effect of degradation on the mechanical characteristics of the developed composite by compression molding. They exposed kenaf/polyethylene terephthalate (PET) reinforced polyoxymethylene (POM) hybrid composite to environmental conditions (moisture, water spray, UV penetration) for 672 hours and examined the tensile, flexural, and impact strength before and after degradation. The authors showed that due to environmental conditions the tensile strength reduced by 40% for kenaf/POM and 8% for hybrid composites. Author's concluded that, due to degradation, there are significant reductions in mechanical properties. On the other hand, properties of hybrid composites are less reduced in comparison to composites before degradation. Hence hybrid composites are suitable for outdoor applications. Borsoi et al. [35] fabricated cellulose whiskers reinforced polylactic acid composites. They investigated the effect of environmental degradation on thermal and chemical properties. The results led to the conclusion that humidity facilitates hydrolytic degradation. Photodegradation provides a medium to degrade composites under soil at faster rates. Wu et al. [36] studied the effect of UV degradation on the tensile strength of curaua fiber reinforced recycled expanded polystyrene matrix. They fabricated the composite by single screw extruder by varying the percentages of fiber content and coupling agent. The results showed that degradation did not affect the tensile properties of curaua fiber composites.

13.7 Biodegradation of Fiber Reinforced Polymer Composites

Biodegradation is defined as the process in which micro-organisms attack the substrate and break it down into small constituents [37]. Biodegradation of fiber reinforced polymer composites under the soil is partially anaerobic and partially aerobic in nature. Degradation of fiber reinforced polymer composites is affected by the type of polymer used and the nature of the micro-organisms. The factors that affect polymer degradation are its molecular weight, degree of crystallinity, functional group attached, and other constituents present in it [38]. Degradation of polymer takes place in four stages. The first stage involves attack by the micro-organism on the polymer surface, second stage involves growth and successive multiplication of micro-organism by utilization of carbon, third stage involves breaking of the polymer chain into monomers, and the last stage involves degradation under aerobic and anaerobic environment. The factors related to micro-organisms that affect the degradability of fiber reinforced polymer composites are type of organisms and the environment for their growth such as the pH of soil, temperature, and humidity [39]. Petinakis et al. [40] fabricated starch/wood-flour reinforced PLA composite by twin-screw extruder. The authors studied the effect of starch and wood-flour on biodegradation and thermal decomposition by burying under compost soil and using thermal gravimetric analysis (TGA). The results showed that starch/PLA and wood-flour/PLA composites are less biodegradable as compared to cellulose but more than neat PLA. Thermal degradability of pure PLA was lowered by the addition of filler material to it. Hidayat and Tachibana [41] studied the effect of fungus (pleurotusostreatus) on the mechanical and degradation properties of kenaf fiber reinforced PLA composite. The results indicated that kenaf/PLA composite degrades 12–48% in six months, and there is an 84% decrease in flexural strength of the developed composites. Shogren et al. [24] manufactured starch/PLA/poly-hydroxyester-ether (PHEE) composites by injection molding. After one year, the authors studied the effect of starch and PHEE on the degradation rate and tensile strength of the developed composite buried under the soil. They observed that higher the starch and PHEE content higher the degradation rate. The results showed that the tensile strength of the developed composite was not much affected by burial under compost soil. Yussuf et al. [42] fabricated kenaf (20% wt) reinforced PLA and rice husk (20% wt) reinforced PLA composite by twin-screw extrusion. The authors analyzed the flexural strength, impact strength, and biodegradability of the developed composites. The results showed that kenaf fiber reinforced PLA composite had more mechanical properties than rice husk reinforced PLA composite. Degradability of neat PLA increased by 1.2% and 0.8% by the addition of kenaf and rice husk to it. Scanning electron micrographs show weak interfacial bonding between PLA and kenaf fibers. Khan et al. [43] developed jute fiber reinforced polypropylene and jute fiber reinforced natural rubber composite by compression molding. The authors analyzed the mechanical properties and degradation study of the developed composites. The results showed that jute fiber reinforced natural rubber had more mechanical properties and degradability as compared to jute fiber reinforced polypropylene composite. Table 13.2 shows the work done by various authors on the biodegradability of fiber reinforced polymer composites.

13.8 Conclusion

In the present chapter, different degradation techniques for polymer composites and how they can be reused are discussed. Various techniques such as mechanical, thermal, chemical, and biodegradation are discussed in detail and their effects on the various properties of polymer composites are studied. In mechanical methods the reinforcement can be obtained in short fiber form and can be reused in injection molding and screw molding for the fabrication of new composites with slightly lower properties than the previous one. In chemical methods, the degradation of polymer composites by the action of certain chemical reactions leads to the formation of gases, which can be used as a fuel. In certain chemical processes the reinforcement can be retracted and used again for the development of new composites. Biodegradation of natural fiber composites leads to the complete degradation of the composite while in case of

Table 13.2 Biodegradability study of fiber reinforced polymer composites.

Fiber/matrix	Study	Method	Results	References
Polylactic acid/PLA fibers	Investigated the degradability of self-reinforced composites (SRC) under different conditions	SRC was prepared by thermo-compression method: • FTIR, TGA was analyzed to study the nature of degradability	• Samples under water and compost environment degrade fast • Results indicate that SRC can be compost at their end-of-life	[44]
Date palm fiber/polyethylene	Studied the effect of basic additives Effect of treated filler on biodegradation	Neat LLDPE samples and 5 wt% of filler composite samples were prepared by two roll mills at 190 °C under compression The prepared samples were buried at a depth of 8–9 cm for 80 days under normal environment conditions	The composite samples show more degradability than neat polymer composites The neat polymer composite does not show any weight loss up to 60 days Fiber composite shows 20% weight loss after 20 days	[45]
Wood and wheat flour (WF)/polypropylene	Investigated the improvement in the biodegradability by addition of WS and WF To study biodegradability by visual inspection, tensile test, FTIR, and water absorption test	5 wt% of WS and WF PP composites were fabricated by hot press into sheets. The developed composites are exposed to four degradation conditions (i) moist soil (ii) water (iii) 10% salt solution, (iv) normal environment conditions	Tensile strength after 15 weeks showed more degradation in samples under brine solution and burial under moist soil Water absorption results indicated that neat PP composite water absorbing capacity increased by the addition of WS and WF Saw dust/PP shows more degradation under all condition then flour/PP composites WS dust is more hydrophilic in nature than WF	[46]

(Continued)

Table 13.2 (Continued)

Fiber/matrix	Study	Method	Results	References
Date palm fiber	Investigated the structural, mechanical, and degradation of untreated palm fiber and treated palm fiber	Investigated the effect of biodegradation in soil for 30, 60, and 90 days SEM, FT-IR, and tensile test were carried out	Results indicated that hemicellulose shows faster degradation as compared to lignin: • Tensile strength after 90 days of treated fiber was decreased around 43%	[47]
Jute, Sisal, and elephant grass/PLA	Development and characterization of elephant grass/PLA composites by injection molding Reinforcement variation from 5 wt% up to 25 wt%	Fiber treatment with 10% NaOH and H_2O_2 to improve the surface adhesion between fiber and matrix Biodegradability test by burial under compost soil and enzymatic environment	Mechanical properties of elephant grass/PLA composites were more than jute/PLA, Sisal/PLA, and neat PLA composite The degradation under enzymatic environment is faster than soil burial test	[48]
Sisal fiber/PLA	Fabrication of short fiber reinforced composite by injection molding Biodegradability was measured in terms of weight loss after soil burial up to 90 days	Varying the fiber % from 5 wt% up to 25 wt% Treatment of fiber by 10% NaOH and H_2O_2 Enzymatic test under 5 ml of lipase PS solution at 38 °C	The enzymatic action increased the rate of degradation Water absorption was more for untreated sisal PLA composite and least for neat PLA, that is 0.88% Untreated fiber composites are more degraded in soil than treated fiber composites	[49]

Material	Study	Process	Findings	Reference
Silk/gelatin	Investigated the effect of sisal fiber in gelatin matrix. Studied the effect of environment and soil burial on tensile strength	Gelatin films were prepared and then silk fiber was stacked between films by compression molding	By addition of silk fiber to gelatin the mechanical properties improve. The tensile strength is affected by environment conditions. Biodegradability test showed that composite loses weight by 20 wt% and silk by 52.1% within 24 hours	[50]
Sugar palm fiber and sugar palm starch/glycerol	Studied the degradation behavior under compost soil burial. Investigated the tensile strength after 24, 48, and 72 hours of soil burial	Fabricate the SPS/glycerol (70/30 wt%) and SPF/SPS composite by compression molding process	SPS shows 63.58% weight loss after 72 hours and degrades after a week: • SPF/SPS shows 56.73% weight loss after 72 hours and degrades after 10 days • SPS has more hydrophilic nature as compared to SPF/SPS; that is why it degrades at faster rate	[51]
Sisal fiber coupling agent-maleic anhydride/PLA	Correlation between molar mass and crystallinity degree during biodegradability	Sisal/PLA composites are fabricated at 10, 20, 30 wt% with or without coupling agent by compression molding. Degradability study by DSC and XRD analysis	Biodegradation in soil, hydrolytic chain scission aggravated a decrease in molar mass and small rise in crystallinity degree	[52]

synthetic fiber polymer composites the fibers are reused and their properties are not greatly affected. The final conclusion can be drawn from this chapter that polymer composites can be reprocessed or reused through the various techniques but only after a decrement in properties.

References

1 Chaudhary, V., Bajpai, P.K., and Maheshwari, S. (2017). Studies on mechanical and morphological characterization of developed jute/hemp/flax reinforced hybrid composites for structural applications. *Journal of Natural Fibers* 15: 80–97.

2 Bajpai, P.K., Ahmad, F., and Chaudhar, V. (2017). Processing and characterization of bio-composites. In: *Handbook of Ecomaterials* (ed. L.M.T. Martínez, O.V. Kharissova and B.I. Kharisov). Springer International Publishing AG https://doi.org/10.1007/978-3-319-48281-1_98-1.

3 Chaudhary, V., Bajpai, P.K., and Maheshwari, S. (2018). An investigation on wear and dynamic mechanical behavior of jute/hemp/flax reinforced composites and its hybrids for tribological applications. *Fibers and Polymers* 19: 403–415.

4 Chaudhary, V., Bajpai, P.K., and Maheshwari, S. A study on the effects of fiber-matrix interfacial adhesion on the performance of bio-composites. In: *Biocomposites Properties, Performance and Applications* (ed. A. Shahzad). New York: Nova science publishers Inc. E-ISBN – 978-1-53612-145-2.

5 Agarwal, B.D. and Broutman, L.J. (2006). *Analysis and performance of fiber composites*, 3e. Wiley ISBN-10: 0-471-26891-7.

6 Jacob, A. (2011). Composites can be recycled. *Reinforced Plastics* 55: 45–46.

7 Molnar, A. (1995). Recycling advanced composites. Final report for the Clean Washington Center (CWC).

8 Steenkamer, D.A. and Sullivan, J.L. (1998). On the recyclability of a cyclic thermoplastic composite material. *Composites Part B* 29B: 745–752.

9 Palmer, J., Ghita, O.R., Savage, L., and Evans, K.E. (2009). Successful closed-loop recycling of thermoset composites. *Composites Part A* 40: 490–498.

10 Lea, R.M., da Luz, S.M., Araujo, J.A., and Christoforo, A.L. (2015). The recycling of sugarcane fiber/polypropylene composites. *Materials Research* 18: 690–697.

11 Sun, H., Guo, G., Memon, S.A. et al. (2015). Recycling of carbon fibers from carbon fiber reinforced polymer using electrochemical method. *Composites: Part A* 78: 10–17.

12 Ashton, E.G., Kindlein, W. Jr., Demori, R. et al. (2016). Recycling polymeric multi-material products through micronization. *Journal of Cleaner Production* 116: 268–278.

13 Li, X., Bai, R., and McKechnie, J. (2016). Environmental and financial performance of mechanical recycling of carbon fibre reinforced polymers and comparison with conventional disposal routes. *Journal of Cleaner Production* 127: 451–460.

14 Bourmaud, A. and Baley, C. (2007). Investigations on the recycling of hemp and sisal fibre reinforced polypropylene composites. *Polymer Degradation and Stability* 92: 1034–1045.

15 Duigou, A., Pillin, I., A Bourmaud, A. et al. (2008). Effect of recycling on mechanical behaviour of biocompostable flax/poly(L-lactide) composites. *Composites: Part A* 39: 1471–1478.

16 Augier, L., Sperone, G., Garcia, C., and Borredon, M. (2007). Influence of the wood fibre filler on the internal recycling of poly(vinyl chloride)-based composites. *Polymer Degradation and Stability* 92: 1169–1176.

17 Hamad, K., Ko, Y.G., Kaseem, M., and Deri, F. Recycling of waste from polymer materials: an overview of the recent works. *Polymer Degradation and Stability* 98: 2801–2812.

18 Pillin, I., Montrelay, N., Bourmaud, A., and Grohens, Y. (2008). Effect of thermo-mechanical cycles on the physico-chemical properties of poly(lactic acid). *Polymer Degradation and Stability* 93: 321–328.

19 Hamad, K., Kaseem, M., and Deri, F. (2011). Effect of recycling on rheological and mechanical properties of poly(lactic acid)/polystyrene polymer blend. *Journal of Material Science* 46: 3013–3019.

20 Collins, E.A. and Metzger, A.P. (1970). Polyvinylchloride melt rheology II – the influence of the molecular weight on flow activation energy. *Polymer Engineering and Science* 10: 57–65.

21 Jiang, J., Deng, G., Chen, X. et al. (2017). On the successful chemical recycling of carbon fiber/epoxy resin composites under the mild condition. *Composites Science and Technology* 151: 243–251.

22 Ma, Y. and Nutt, S. (2018). Chemical treatment for recycling of amine/epoxy composites at atmospheric pressure. *Polymer Degradation and Stability* 153: 307–317.

23 Xu, P., Li, J., and Din, J. (2013). Chemical recycling of carbon fibre/epoxy composites in a mixed solution of peroxide hydrogen and *N,N*-dimethylformamide. *Composites Science and Technology* 82: 54–59.

24 Shogren, R.L., Doane, W.M., Garlotta, D. et al. (2003). Biodegradation of starch/polylactic acid/poly(hydroxyester-ether) composite bars in soil. *Polymer Degradation and Stability* 79: 405–411.

25 Elsawya, M.A., Kim, K., Park, J.W., and Deep, A. (2017). Hydrolytic degradation of polylactic acid (PLA) and its composites. *Renewable and Sustainable Energy Reviews* 79: 1346–1352.

26 Benali, S., Aouadi, S., Dechief, A. et al. (2015). Key factors for tuning hydrolytic degradation of polylactide/zinc oxide nanocomposites. *Nanocomposites* 1 (1): 51–61.

27 Ndazi, B.S. and Karlsson, S. (2011). Characterization of hydrolytic degradation of polylactic acid/rice hulls composites in water at different temperatures. *Express Polymer Letters* 5: 119–131.

28 Gonzalez, S.S., Diban, N., and Urtiaga, A. (2018). Hydrolytic degradation and mechanical stability of poly(ε-caprolactone)/reduced graphene oxide membranes as scaffolds for in vitro neural tissue regeneration. *Membranes* 8: 1–14.

29 Martos, J., Osinaga, P.W.R., Oliveira, E., and Castro, L.A.S. (2003). Hydrolytic degradation of composite resins: effects on the microhardness. *Materials Research* 6: 599–604.

30 Silva, R.N., Oliveira, T.A., Conceicao, I.D. et al. (2018). Evaluation of hydrolytic degradation of bionanocomposites through fourier transform infrared spectroscopy. *Polímeros* 28 (4): 348–354.

31 Lu, T., Ramos, E.S., Yi, Y., and Kumosa, M. (2018). UV degradation model for polymers and polymer matrix composites. *Polymer Degradation and Stability* 154: 203–210.

32 Siddiquee, M., Helali, M., Gafur, A., and Chakraborty, S. (2014). Investigation of an optimum method of biodegradation process for jute polymer composites. *American Journal of Engineering Research* 3: 200–206.

33 Abdullah, M.Z., Dan-mallam, Y., and Yusoff, P.S.M. (2013). Effect of environmental degradation on mechanical properties of kenaf/polyethylene terephthalate fiber reinforced polyoxymethylene hybrid composite. *Advances in Materials Science and Engineering* 1–8.

34 Lopez, G.B., Veleva, L., Gonzalez, A.V., and Owen, P.Q. (2013). Weathering and biodegradation of polylactic acid composite reinforced with cellulose whiskers. *Revista Mexicana de Ingenieria uimica* 12: 143–153.

35 Borsoi, C., Berwig, K.H., Scienza, L.C., and Zattera, A.J. (2013). The photodegradation and biodegradation of rEPS-curaua fiber composites. *Polymer Composites* 967–977.

36 Wu, C.S., Liao, H.T., and Cai, Y.X. (2017). Characterisation, biodegradability and application of palm fibre reinforced polyhydroxyalkanoate composites. *Polymer Degradation and Stability* 140: 55–63.

37 John, M.J. (2017). Environmental degradation in biocomposites. In: *Biocomposites for High-Performance Applications* (ed. D. Ray), 1–17.

38 Stepczynska, M. and Rytlewski, P. (2017). Enzymatic degradation of flax-fibers reinforced polylactide. *International Biodeterioration and Biodegradability* 1–7.

39 Oliveux, G., Dandy, L.O., and Leeke, G.A. (2015). Current status of recycling of fibre reinforced polymers: review of technologies, reuse and resulting properties. *Progress in Materials Science* 72: 61–99.

40 Petinakis, E., Liu, X., Yu, L. et al. (2010). Biodegradation and thermal decomposition of poly(lactic acid)-based materials reinforced by hydrophilic fillers. *Polymer Degradation and Stability* 95: 1704–1707.

41 Hidayat, A. and Tachibana, S. (2017). Characterization of polylactic acid (PLA)/kenaf composite degradation by immobilized mycelia of *Pleurotus ostreatus*. *International Biodeterioration and Biodegradability* 71: 50–54.

42 Yussuf, A.A., Massoumi, I., and Hassan, A. (2010). Comparison of polylactic acid/kenaf and polylactic acid/rise husk composites: the influence of the natural fibers on the mechanical, thermal and biodegradability properties. *Journal of Polymer and Environment* 18: 422–429.

43 Khan, R.A., Haque, M.E., Huq, T. et al. (2010). Studies on the relative degradation and interfacial properties between jute/polypropylene and jute/natural rubber composites. *Journal of Thermoplastic Composite Materials* 23: 665–681.

44 Gil-Castell, O., Badia, J.D., Ingles-Mascaros, S. et al. (2018). Polylactide-based self-reinforced composites biodegradation: individual and combined influence of temperature, water and compost. *Polymer Degradation and Stability* 1–25.

45 Alshabanat, M. (2018). Morphological, thermal, and biodegradation properties of LLDPE/treated date palm waste composite buried in a soil environment. *Journal of Saudi Chemical Society* 1–10.

46 Fakhural, T. and Islam, M.A. (2013). Degradation behavior of natural fiber reinforced polymer matrix composites. *Procedia Engineering* 56: 795–800.

47 Chen, C., Yin, W., Chen, G. et al. (2017). Effects of biodegradation on the structure and properties of windmill palm (*Trachycarpus fortunei*) fibers using different chemical treatments. *Materials* 10: 1–10.

48 Gunti, R., Prasad, A.V.R., and Gupta, A.V.S.S.K.S. (2016). Mechanical and degradation properties of natural fiber reinforced PLA composites: jute, sisal, and elephant grass. *Polymer Composites* 1–16.

49 Gunti, R., Prasad, A.V.R., and Gupta, A.V.S.S.K.S. (2015). Mechanical and degradation properties of successive alkali treated completely biodegradable sisal fiber reinforced poly lactic acid composites. *Journal of Reinforced Plastics and Composites* 34: 951–961.

50 Shubhra, Q.T.H., Alam, A.K.M.M., and Beg, M.D.H. (2011). Mechanical and degradation characteristics of natural silk fiber reinforced gelatin composites. *Materials Letters* 65: 333–336.

51 Sahari, J., Salit, M.S., Zainudin, E.S., and Maleque, M.A. (2014). Degradation characteristics of SPF/SPS biocomposites. *Fibres and Textiles in Eastern Europe* 5 (17): 96–98.

52 Badia, J.D., Stromberg, E., Kittikorn, T. et al. (2018). Relevant factors for the eco-design of polylactide/sisal biocomposites to control biodegradation in soil in an end-of-life scenario. *Polymer Degradation and Stability* 143: 9–19.

Index

Reinforced Polymer Composites: Processing, Characterization and Post Life Cycle Assessment,
First Edition. Edited by Pramendra K. Bajpai and Inderdeep Singh.
© 2020 Wiley-VCH Verlag GmbH & Co. KGaA. Published 2020 by Wiley-VCH Verlag GmbH & Co. KGaA.